DEVELOPMENTAL DISORDERS:
Diagnostic Criteria and Clinical Assessment

DEVELOPMENTAL DISORDERS:
Diagnostic Criteria and Clinical Assessment

Edited by

Stephen R. Hooper
University of North Carolina at Chapel Hill

George W. Hynd
University of Georgia
Medical College of Georgia

Richard E. Mattison
Washington University

LEA LAWRENCE ERLBAUM ASSOCIATES, PUBLISHERS
1992 Hillsdale, New Jersey Hove and London

Lawrence Erlbaum Associates, Inc., Publishers
365 Broadway
Hillsdale, New Jersey 07642

Library of Congress Cataloging-in-Publication Data

Developmental disorders: diagnostic criteria and clinical assessment / edited by Stephen R. Hooper, George W. Hynd, Richard E. Mattison.
p. cm.
Companion v. to: Child psychopathology.
Includes bibliographical references and indexes.
ISBN 0-8058-0329-7 (c)
1. Developmental disabilities—Diagnosis. 2. Developmentally disabled children—Psychological testing. 3. Behavioral assessment of children. I. Hooper, Stephen R. II. Hynd, George W. III. Mattison, Richard E.
[DNLM: 1. Child Development Disorders—diagnosis. WS 350.6 D489203]
RJ135.D483 1992
618.92′8—dc20
DNLM/DLC
for Library of Congress 91-6312
CIP

Printed in the United States of America
10 9 8 7 6 5 4 3 2 1

To my wonderful families—old and new. *S.R.H.*

To Ben, above all a good friend and valuable colleague who sensitized me to the many divergent issues related to childhood and adolescent psychopathology. *G.W.H.*

To my parents, my wife Christine, and my children Lara and Brian. *R.E.M.*

Contents

Foreword

It is indeed a pleasure to write the foreword to this text and its companion volume. One text focuses on child and adolescent psychopathology, and the other focuses on developmental disorders. Both volumes deal directly with evaluating the diagnostic criteria for these disorders, with a particular eye toward improving these criteria based on contemporary literature. Both volumes also focus on the clinical assessment of these disorders and related issues. Taken together, these two volumes should contribute to advancing the scientific study of child psychopathology and developmental disorders.

The scientific study of child and adolescent psychopathology does not have a very long history. Child psychiatry and the delivery of child mental health services grew up in the child guidance clinic movement of the 1920s. These clinics were designed for service delivery and operated on what became known as the "Holy Trinity Approach" to service delivery. That is, the child psychiatrist, who had no officially approved training in child and adolescent psychopathology, generally saw the child for play therapy; the psychologist saw the child for testing; and the social worker generally saw the family. These service delivery clinics did not promote research into the nature of the disorders that they were evaluating and treating and, as a result, there does not exist a serious body of scientific literature emanating from the large number of patients who were seen in those early years.

It wasn't until 1959 that the American Board of Psychiatry and Neurology accepted child and adolescent psychiatry as a legitimate board certifiable specialty. Concurrent with this, an increased number of research reports began to appear in the areas of childhood psychopathology and developmental disorders. But a perusal of the content of child psychology and child psychiatry journals of even the late 1950s and 1960s, as well as the content of programs of the annual meetings of the American Academy of Child and Adolescent Psychiatry of those years, also reveals that most of the publications and presentations were of the clinical and case report variety. Although these works were noteworthy, the scientific study of childhood psychopathology and developmental disorders was clearly in its infancy.

In addition to the few number of scientific endeavors in the field at that time, at least part of the difficulty with the early literature on child and adolescent psychopathology was the lack of a modern classification system. Relatedly, contemporary assessment approaches to the delineation of the clinical picture and possible subtypes of child and adolescent psychiatric disorders were not well developed or refined. Although the field has addressed these areas of concern over the past several decades, it is clear that many of these issues continue into the present. Some of these issues, along with a variety of classification approaches, are reviewed by Drs. Mattison and Hooper in chapter 1 of *Child Psychopathology* and by Dr. Hooper in chapter 1 of *Developmental Disorders*. Contemporary assessment approaches and related issues also are discussed more generally in each of these introductory chapters and in more detail in the subsequent chapters in each volume.

With respect to classification issues, Kraeplin's psychiatric taxonomy, published in 1883, ignored disorders of children. Nearly 70 years later, the Diagnostic and Statistical Manual of Mental Disorders,[1] the first version of the American Psychiatric Association's classification system, contained only three psychiatric diagnoses for children and adolescents: childhood schizophrenia, adjustment reaction of childhood, and childhood special symptom reactions. Subsequently, Anna Freud's Developmental Profile[2] was considered by many to be the first diagnostic classification system that dealt extensively with child psychiatric disorders. This system was based on psychoanalytic constructs and presented a variety of diagnostic categories. For example, these categories included variations of normality, transitory by-products of developmental strain, infantile neurosis, character disorders,

---

[1]American Psychiatric Association (1952). *Diagnostic and statistical manual of mental disorders*. Washington, DC: author.

[2]Freud, A. (1965). *Normality and pathology in childhood*. New York: International Universities Press.

retarded defective or nontypical personality, and disruptions of mental growth. Not surprisingly, the categories in Anna Freud's Developmental Profile or, for that matter, any of these early classification attempts, were not subjected to the type of empirical validity and reliability studies that one would demand today.

In 1966, the Group for the Advancement of Psychiatry (GAP) produced a detailed classification of child and adolescent psychiatric disorders.[3] This was revised in 1974 to reflect a system of categorical diagnoses that were primarily descriptive as opposed to the psychodynamic, etiological focus of Freud's Developmental Profile. Ten major categories comprised the GAP Classification System, but this system did not enjoy wide usage among researchers in child and adolescent psychopathology.

Two of the more widely used classification systems have been the International Classification of Diseases (ICD) and the Diagnostic and Statistical Manual of Mental Disorders (DSM). The various editions of the ICD, including the currently used ICD-9, have had greater or lesser numbers of categories dealing with child and adolescent psychopathology as the years have progressed. While the ICD-9 has a greater number of diagnoses than its predecessors, the draft of ICD-10 suggests that the section on child and adolescent psychopathology will be quite detailed and come somewhat closer to the DSM-III, DSM-III-R, and the proposed DSM-IV than its predecessors. Although it appears that the latest editions of the DSM and ICD correspond more closely to each other than in the past, there remains a number of differences between the two systems.

As noted earlier, the first version of the DSM was published in 1952 and contained only three diagnoses for children and adolescents. The second edition, DSM-II, was published by the American Psychiatric Association in 1968 and did include a section titled "Behavior Disorders of Childhood and Adolescence."[4] This section specifies six disorders: hyperkinetic reaction, overanxious reaction, withdrawing reaction, runaway reaction, unsocialized aggressive reaction, and group delinquent reaction. It was not until DSM-III was published in 1980, however, that a truly comprehensive classification of child and adolescent psychopathology was attempted in the DSM system.[5] The revised

[3]Group for the Advancement of Psychiatry. (1966). *Psychopathological disorders in childhood: Theoretical considerations and a proposed classification*. New York: Author.

[4]American Psychiatric Association (1968). *Diagnostic and statistical manual of mental disorders*. (2nd ed.). Washington, DC: Author.

[5]American Psychiatric Association (1980). *Diagnostic and statistical manual of mental disorders*. (3rd ed.). Washington, DC: Author.

version of this attempt, DSM-III-R, was published in 1987 and included some substantial changes.[6] Many of these specific changes have been reviewed elsewhere (e.g., Cantwell & Baker, 1988)[7] and are reflected in nearly all of the chapters contained within these two volumes. Some of these changes were considered positive, but some of them were not. Furthermore, many of the changes were made with a lack of empirical support. A major strength of these two volumes is that all of the contributors were encouraged to suggest specific changes in diagnostic criteria for a selected disorder, but they were encouraged to do this based on the available scientific literature for that disorder. Perhaps these efforts will contribute to improving the scientific basis for many of the childhood disorders and their associated diagnostic criteria.

It is highly likely that the DSM and ICD, and their future versions (i.e., DSM-IV and ICD-10), will be the two major systems that will set the tone for the future with regard to official classification attempts; however, both systems consist of categories of psychiatric disorders that were initially derived from clinical experience. Other approaches to the classification of child and adolescent disorders, such as the dimensional classification approach, have been developed in an attempt to address this issue. The dimensional classification approach is based on numerical scores obtained with specific behavioral assessment tools. Statistical procedures, such as factor analysis, are used to determine the tendency of specific items of behavior to co-occur. As dimensions of behavior, or factors, are identified by these statistical procedures, individuals can be classified into mutually exclusive groups using statistical clustering techniques. Systematic comparative studies have suggested that the clinical syndromes that arise from dimensional classification correspond somewhat to the clinically derived syndromes, but there is less than a perfect correspondence. More specifically, there are ICD and DSM syndromes that are not exactly reproduced in these dimensional systems, and, conversely, the reverse also is true.

Proponents of the dimensional approach suggest that because dimensional classification is empirically derived, the groups that result from this approach are more reliable, valid, and clinically homogeneous. Further, they assert that the diagnoses tend to be tied to the assessment process more directly than those produced in the DSM and ICD systems. However, there are criticisms of the dimensional approach as well. In particular, the dimensions of behavior created by the

---

[6]American Psychiatric Association (1987). *Diagnostic and statistical manual of mental disorders.* (3rd ed., rev.). Washington, DC: Author.

statistical procedures may not turn out to be either clinically or theoretically meaningful. In the future, it is likely that the dimensional and categorical approaches may be combined and, consequently, result in a more useful classification system than is currently in existence. In fact, some of the categorically defined disorders in the DSM and ICD systems already may have dimensional aspects to them. For example, the developmental learning disorders and mental retardation do have dimensional components to their classification (e.g., slow and/or disrupted learning), but the exact nature of this relationship requires further investigation.

The chapters in these two volumes reflect the major DSM types of childhood and adolescent psychopathology that are currently the focus of intense study. These include the disruptive behavior disorders (i.e., conduct disorder, attention deficit-hyperactivity disorder), schizophrenia, major depression, some of the anxiety disorders, some of the eating disorders, gender disorders, tic disorders, elimination disorders, elective mutism, mental retardation, autism, developmental learning disorders, developmental language and speech disorders, and motor skill disorders. Each chapter presents a thorough review of the ontogeny of each diagnosis and its scientific status (i.e., reliability and validity) according to contemporary literature. Suggestions are proposed for possible changes in diagnostic criteria based on this literature. It is hoped that such a detailed review of each of these disorders will contribute to the processes of revising and generating future versions of the ICD and DSM and, ultimately, result in improved classifications systems (e.g., ICD-10 and DSM-IV).

The issues for each of the disorders discussed in these two volumes differ slightly with regard to their place in the nosological system. For the disruptive behavior disorders, there is the question of whether this overall rubric should be retained for attention deficit-hyperactivity disorder, conduct disorder, and oppositional defiant disorder. For example, the role and place of attention deficit disorder without hyperactivity need to be explicated further, and its association with other disorders should receive more intense study as well. The relationship of oppositional defiant disorder to conduct disorder also needs to be delineated, particularly from a developmental perspective. Subtyping and severity ratings of all of the disruptive disorders need to be extensively studied and more fully described.

Related questions can be asserted for other disorders as well. For example, with schizophrenia there is the problem of defining a disorder based primarily on its manifestations in adulthood and late adolescence and then extending it downward to prepubertal children. Russell challenges this notion in his chapter in *Child Psychopathology* and indicates that the literature on such children is sparse. Whether these children present with the same phenomonologic picture as adults or whether there is a "pre-schizophrenic" picture in children which is phenomonologically different that eventually melds into the adult picture needs to be explored further.

Similar questions exist with regard to the mood disorders. The nosologic status of mood disorders in childhood and adolescence versus their adult counterparts needs further exploration. Systematic studies have suggested that children and adolescents do present with clinical pictures very similar to those seen in adults with mood disorders. But whether there are meaningful modifications of the diagnostic criteria that should be made in preschool children, grade school children, and adolescents remains a subject for scientific inquiry.

The relationship between childhood and adult onset anxiety disorders also needs to be explored further. For example, some of the anxiety disorders regularly begin in childhood and are listed under the category of anxiety disorders of childhood and adolescence. Others, such as obsessive-compulsive disorder, which often do occur for the first time in childhood and adolescence are listed with other anxiety disorders. Again, the question of whether there are meaningful modifications of the clinical picture that occur as a function of age and developmental stage remains to be studied.

For the eating disorders, some of the issues include whether anorexia nervosa should be subtyped according to the presence or absence of bulimic symptoms, and whether there should be a separate category of binge eating disorder that might be somewhat different from the current definition of bulimia. The relationship of the eating disorders of infancy and early childhood to the later developing eating disorders also remains an area for future research efforts. With the gender identity disorders, one of the major questions revolves around whether these disorders are on a continuum. Some investigators might suggest that the clinical separations, such as those found in the DSM-III and DSM-III-R, are rather arbitrary in nature and lack a scientific base. The clinical literature also suggests that there may be gender identity disorders that are secondary to documented Axis III conditions. Many of these issues are addressed in the chapters contained in *Child Psychopathology.*

A wide array of similar concerns can be raised with respect to the developmental disorders. For example, mental retardation could be primary in nature, or it could be secondary to a known Axis III condition. One of the classification systems currently in use (e.g., DSM-III-R) specifies severity to be based largely on IQ measures, although the diagnosis and its severity should be made on the basis of the presence of both intellectual and adaptive behavior deficiencies. Only the American Association of Mental Retardation (AAMR) classification system takes into account the level of adaptive behavior impairment with regard to severity. The AAMR classification also is much more detailed than the DSM-III-R system, although the correspondence between the two is closer than ever before. Reschly addresses many of these issues in his chapter on this topic in *Developmental Disorders*.

The pervasive developmental disorder category was created for DSM-III-R. At that time, two major subtypes were proposed: classical infantile autism and a residual category of pervasive developmental disorder childhood onset. However, recent literature suggests that conditions such as Aspergers Syndrome, Retts' Syndrome, and disintegrative disorder of childhood all may be possible inclusions under the rubric of pervasive developmental disorder. Further confusion was created in this general category of disorders with revision of the DSM-III. Specifically, the differences between the diagnostic criteria for the more classical infantile autism in the DSM-III and DSM-III-R have contributed to the likelihood of clinicians and researchers obtaining significantly different groups of children for examination and study. This undoubtedly has hindered the comparison of the research findings on this population. Future classification systems need to specify criteria that will correctly include and exclude those children who represent the more classical autistic picture. The similarities and differences of all of the pervasive developmental disorders also will require intensive study.

The developmental learning, language and speech, and motor skill disorders are treated rather simply in the DSM classification schemes, and it is not coincidence that most of the chapters in *Developmental Disorders* advocate for significant changes in diagnostic criteria for these disorders. The chapters in *Developmental Disorders* discuss in much more detail this available literature. Possible underlying etiologic factors that may affect some of these disorders also are presented. Whether these hypothesized underlying problems should be part of the core criteria is an issue for further discussion, but it is clear that the criteria associated with these disorders in the DSM system should be expanded and/or totally revised.

For any classification system of child and adolescent phenomenon to be useful, it must meet minimal standards of reliability and validity, show adequate coverage, observability, and age sensitivity, and maintain clinical feasibility (Cantwell & Baker, 1988).[7] None of the current

[7]Cantwell, D. P., & Baker, L. (1988). Issues in the classification of child and adolescent psychopathology. *Journal of the American Academy of Child and Adolescent Psychiatry, 27*, 521–533.

systems is perfect in any of these areas. Future classification systems, if they are going to advance the field, must reflect increasingly sophisticated knowledge about the disorders of childhood and adolescence. For this to occur, much more research is needed in a variety of areas. For example, one area that should receive intense attention is in the external validation of the various categories and possible subtypes of childhood and adolescent psychopathology. Research on the assessment process itself also is sorely needed, especially in the very sticky aspect of combining data from a variety of sources (e.g., parents, teachers, significant others, and the children themselves) into a diagnostic statement. In this regard, sophisticated techniques of assessment are emerging almost on a yearly basis, and this should contribute to refining diagnostic criteria and associated characteristics of these disorders. Although these efforts in assessment and external validation should contribute to increasing the reliability, validity, and clinical utility of the classification systems, there is a danger that these clinical assessment procedures and techniques (e.g., structured interviews, behavioral rating scales) will be so tightly bound to specific classification systems that they will result in a reification of those systems and the disorders that they delineate.

Although the study of child psychopathology and developmental disorders has shown considerable growth over the past several decades, it is clear that there are a large number of issues that remain to be addressed and/or resolved. These issues not withstanding, these two volumes contribute to advancing our understanding of child psychopathology and developmental disorders, and they produce an empirical basis upon which one can build for the future.

—Dennis P. Cantwell, MD.
—Joseph Campbell Professor of Child Psychiatry
University of California, Los Angeles

Preface

In what some would perhaps consider an historical piece of writing, Smoller (1985) wrote a satirical article about the etiology and treatment of childhood. In his treatise, Smoller noted that "Childhood is a syndrome which has only recently begun to receive serious attention from clinicians" (p. 3). With this conjecture, Smoller acknowledged the "growing acceptance of childhood as a distinct phenomenon" and its pending inclusion in the upcoming DSM–IV. Although the focus of this work was on the origins and treatments of this "syndrome," there was significant discussion regarding its evolution, core clinical components, associated features, and issues surrounding its assessment and diagnosis.

Although tongue-in-cheek, Smoller's work certainly highlighted several of the most important aspects of any childhood diagnosis (e.g., core components, associated features). Furthermore, although the diagnosis of "childhood" has not really captured the clinical spotlight, the study and classification of childhood developmental and psychiatric disorders truly has progressed in the past two decades. This is particularly illustrated by the fact that the DSM–I and DSM–II devoted precious little space and attention to the conceptualization of psychiatric disorders in children and adolescents, whereas more recent versions of the DSM (DSM–III and DSM–III–R) have devoted a significant amount of energy and space to these diagnostic considerations. A similar pattern can be seen in the evolution of the International Classification of Diseases (ICD) system.

Despite this increased interest in developmental and psychiatric disorders, with the evolution of how one conceives of childhood and adolescent disorders comes the potential for confusion among practicing clinicians and researchers. This confusion is created by issues related to: (a) how to diagnose psychopathology; (b) what correspondence exists between currently employed criteria and research findings with children and adolescents; (c) how the criteria were developed, with particular concerns directed toward how decisions were made to retain or eliminate specific criteria for inclusion; and (d) what particular assessment strategies or models would be most useful in arriving at a specific diagnosis. To address these issues, there have been a number of texts published on the general topic of childhood and adolescent psychopathology (e.g., Quay & Werry, 1986; Rutter & Hersov, 1985; Rutter, Tuma, & Lann, 1988; Solnit, Cohen, & Schowalter, 1986), and there have been single volumes devoted to selected child psychiatric and developmental disorders such as Attention Deficit Disorder (Barkley, 1981), affective disorders (Cantwell & Carlson, 1983), anxiety disorders (Gittelman, 1986), eating disorders (Garner, 1988), autism and pervasive developmental disorders (Cohen, Donnellan, & Paul, 1987), dyslexia (Hynd & Cohen, 1983), and speech and language disorders (Cantwell & Baker, 1987). Although the current volume was designed to complement these works, it also was designed to address the diagnostic criteria and clinical assessment of childhood and adolescent developmental disorders from a more critical perspective. In particular, this text was designed to review and critique the diagnostic criteria for most of the major developmental disorders that are used to describe children and adolescents in the DSM–III–R. As such, it is hoped that the data-based nature of these chapters will provide some guidance for the continued evolution of classification for children and adolescents (i.e., DSM–IV), particularly those with developmental disorders.

This text and its companion volume, *Child Psychopathology*, were designed to provide a critical evaluation of literature with respect to the diagnostic criteria set forth for selected child and adolescent diagnoses. The major goal of these volumes is to provide contemporary diagnostic criteria and assessment guidelines for the clinician and researcher that are based on the scientific literature in each clinical domain. Consequently, this volume and its companion were written for professionals and graduate students in the fields of clinical child psychology, child psychiatry, pediatrics, pediatric psychology, school psychology, social work, special education, and other child mental health specialties. Each chapter contains a discussion of the historical background of the diagnosis; definitional issues; a critical, selective review of the literature addressing the diagnosis in question; proposed changes in the diagnostic criteria based on the available literature; and proposed assessment models and methods based on the designated criteria. Given the scientific basis for much of this discussion of diagnostic criteria, many of these chapters also should serve to provide researchers in a wide variety of fields with practical guidelines.

This volume contains nine chapters and is organized largely around the Disorders Usually First Evident in Infancy, Childhood, or Adolescence identified in the current DSM–III–R classification system. Chapter 1 (Hooper) reviews issues and models in the classification of psychiatric and developmental disorders. This introductory chapter also discusses contemporary efforts in the classification of these disorders and related diagnostic assessment technology. The next two chapters deal with more severe developmental disorders, with chapter 2 discussing the diagnosis of mental retardation (Reschly) and chapter 3 addressing the diagnosis of autism (Mesibov & Van Bourgondien). Chapters 4 through 8 deal with more specific developmental disorders. Chapter 4 (Semrud-Clikeman & Hynd), chapter 5 (Gregg), and chapter 6 (Stanovich) discuss diagnostic issues related to the disorders of arithmetic, expressive writing, and reading, respectively. Chapter 7 (Paul) and chapter 8 (Deuel) evaluate DSM–III–R diagnoses related to the language and motor domains, respectively. The volume concludes with an Epilogue highlighting the major changes suggested by the contributors. Although it is not a comprehensive listing of topics, this volume attempts to provide specific suggestions for clinicians and researchers for the major developmental diagnoses typically used with children and adolescents. As such, it is hoped that these chapters will contribute to the ongoing evolution and refinement of an accurate and clinically useful classification system for children and adolescents with developmental disorders.

ACKNOWLEDGMENTS

In addition to the large number of expert contributors, without whom this project would never have been possible, we would like to extend our sincere gratitude to several important individuals. We would like to thank Teresa Buckner and Cheryl Hunter at the Clinical Center for the Study of Developmental and Learning for their expert and timely assistance on a number of endeavors related to this project. Our sincere appreciation also goes out to Carolyn Anderson, Barb Kaylor, and Ann Moore at the Hershey Medical Center and to Dora Ervin at the University of Georgia for their assistance in selected aspects of manuscript preparation. Finally, although we could not have completed this project without considerable assistance and support, any shortcomings of this volume remain our own doing.

Stephen R. Hooper
George W. Hynd
Richard E. Mattison

REFERENCES

Barkley, R. A. (1981). *Hyperactive children. A handbook for diagnosis and treatment.* New York: Guilford Press.

Cantwell, D. P., & Baker, L. (1987). *Developmental speech and language disorders.* New York: Guilford.

Cantwell, D., & Carlson, G. (1983). *Affective disorders in childhood and adolescence: An update.* New York: Spectrum.

Cohen, D. J., Donnellan, A. M., & Paul, R. (Eds.). (1987). *Handbook of autism and pervasive developmental disorders.* New York: Wiley.

Garner, D. M. (1988). *Diagnostic issues in anorexia nervosa and bulimia nervosa.* New York: Brunner/Mayel.

Gittelman, R. (Ed.) (1986). *Anxiety disorders of childhood.* New York: Guilford.

Hynd, G. W., & Cohen, M. (1983). *Dyslexia: Neuropsychological theory, research, and clinical differentiation.* New York: Grune & Stratton.

Quay, H. C., & Werry, J. S. (Eds.). (1986). *Psychopathological disorders of childhood* (3rd ed.). Somerset, NJ: Wiley.

Rutter, M., & Hersov, L. (Eds.), (1985). *Child and adolescent psychiatry. Modern approaches* (2nd ed.). Boston: Blackwell Scientific.

Rutter, M., Tuma, A. H., & Lann, I. S. (Eds.). (1988). *Assessment and diagnosis in child psychopathology.* New York: Guilford Press.

Smoller, J. W. (1985). The etiology and treatment of childhood. *Journal of Polymorphous Perversity, 2,* 3–7.

Solnit, A. J., Cohen, D. J., & Schowalter, J. E. (Eds.). (1986). *Psychiatry, Volume 6: Child psychiatry.* New York: Basic Books.

CHAPTER 1

The Classification of Developmental Disorders: An Overview

Stephen R. Hooper
University of North Carolina School of Medicine

Although the classification of childhood and adolescent psychiatric disorders is relatively recent, the classification of developmental disorders is an even newer enterprise. As will be seen, however, whereas there have been several efforts to place psychiatric disorders into various kinds of nosological frameworks, as illustrated by the various versions of the Diagnostic and Statistical Manual of Mental Disorders (e.g., American Psychiatric Association, 1987) and the International Classification of Diseases (e.g., World Health Organization, 1978), efforts to classify developmental disorders have been loosely organized and highly varied. The various psychiatric nomenclatures have attempted to include developmental disorders in their systems, particularly with the advent of the multiaxial system, where most developmental disorders are separated from the primary psychiatric disorders (e.g., in the DSM-III-R, developmental disorders are placed on Axis II), but historically it is clear that little attention has been devoted to the classification of developmental disorders as a whole.

In contrast to the recent increased interest in childhood and adolescent psychiatric disorders over the past two decades, there appears to be a long history of active interest in general (e.g., autism, mental retardation) as well as more specific (e.g., specific reading disorder) developmental disorders of childhood and adolescence. For example, although the actual term *learning disability* was not coined until the middle 1960s, case descriptions of children with unusual but fairly specific learning profiles were presented in the 1800s. Similarly, issues with respect to the discrimination between idiocy and insanity were debated as early as the 1700s, with the diagnostic construct of mental retardation being one of the current results of those early debates. Unfortunately, however, outside of efforts by groups primarily interested in psychiatric classification, there have been few formalized efforts to organize the spectrum of developmental disorders.

As noted in this volume's companion work (*Child Psychopathology*), the classification of childhood and adolescent psychopathology has evolved as a result of attention devoted to issues surrounding the diagnostic criteria pertinent to specific disorders. In particular, the scientific parameters of classification systems have been emphasized, and the development of objective instrumentation has improved the diagnostic capabilities of clinicians and researchers. For the spectrum of developmental disorders, this scope and sequence of activities have not occurred in any organized fashion, and consequently there has yet to be a functional nosology for developmental disorders. This has not necessarily been the result of lack of interest, poor research, or general oversight, but rather it appears to be due more to the significant dynamic complexities inherent in studying behavior, particularly the behavior of children. Further, it seems that definitional issues have plagued many of the developmental disorders.

This volume and its companion, *Child Psychopathology*, are devoted to presenting expert appraisals of the scientific merits of the child and adolescent diagnostic categories described in the DSM-III (American Psychiatric Association, 1980) and in the DSM-III-R (American Psychiatric Association, 1987), with a particular eye toward proposing refinements in the current diagnostic criteria based on the available empirical literature. For the developmental disorders, in particular, this information should be helpful in refining their diagnostic criteria and, consequently, increasing their reliability, validity, and clinical utility.

This chapter provides an overview of the classification issues relevant to increasing our understanding of the wide array of developmental disorders that have been proposed. General classification concerns are presented, and specific classification rules are discussed. Classification efforts for selected developmental disorders relevant to this volume, and issues related to their inclusion within a psychiatric classification framework also are presented.

CLASSIFICATION ISSUES AND MODELS

Classification Issues

Achenbach (1985) noted that "*classification* refers to any systematic ordering of phenomena into groups or types" (p. 151). He made a distinction between a general classification and a taxonomy, suggesting that the latter term should be reserved for classifications that "reflect intrinsic differences between cases assigned to different classes" (p. 151). It should be clear, however, that these "intrinsic differences" can be defined by any number and/or combination of parameters, such as etiologies (e.g., neurological), specific behaviors (e.g., overactivity), physical characteristics (e.g., blue eyes), and research or clinical goals, all having the single purpose of differentiating between the specific cases in a regulated fashion. Given the elusive nature of etiological agents in many forms of psychopathology, most of the modern attempts at classifying childhood and adolescent psychiatric and developmental disorders have depended largely upon a descriptive approach for determining a taxonomy for these disorders.

The Pros and Cons of Classification. Despite these concerns, there are a number of benefits that can be derived from an adequate classification system. For example, Blashfield (1984) noted that an adequate classification system can provide the vernacular necessary for professionals in the field to communicate with each other efficiently and effectively. Further, the *taxonomic* aspects of the system should provide relevant information pertaining to number and kind of symptoms, prognosis, treatment selection, response to treatment, comorbid conditions, and perhaps etiological possibilities. Finally, Blashfield described the importance of the interrelationship between adequate classification efforts and theory formulation with respect to selected disorders (Hempel, 1965). Similarly, other investigators have called for any classification system to have the following characteristics: (a) be simple, (b) be based on widely used variables and operational definitions, (c) reflect the prevailing clinical, political, and theoretical views within the field, and (d) be easy to use (Goodall, 1966; Kavale & Forness, 1987).

In contrast, some investigators have asserted potential concerns for even using a classification system. In the 1940s, Huschka (1941) advocated for an idiographic approach as he believed that individual differences would be masked by classification efforts and, subsequently, would hinder the development of our understanding of disorders. More recently, Szasz (1961, 1978) described mental illness and psychotherapy as "myths" and stated that these terms lacked meaning and that their use would contribute to significant social stigma in the patient for whom a specific diagnosis and treatment were applied. Although these concerns should be heeded, it would seem that the "pros" of an *adequate* classification system far surpass the "cons" (Kendall, 1975; Weiner, 1982) and that any related negative effects (e.g., social stigma) likely arise secondary to ignorance, misapplication, and/or abuse. Taken together, these pros and cons suggest the importance of pursuing classification efforts in an active, organized, and scientific fashion.

Reliability, Validity, and Clinical Utility. As Cantwell (1988) and others (e.g., Last & Hersen, 1987; Skinner, 1986) have noted, issues of reliability, validity, and clinical utility are paramount to the development and ultimate utility of any classification scheme. In part, the integrity of these parameters are based on the following: (a) the preciseness of the operational definitions used, (b) the nature of samples studied, (c) how well the samples are marked (i.e., age, gender, socioeconomic status, etc.), and (d) developmental considerations (Hooper & Willis, 1989). Needless to say, these descriptive issues have not been addressed thoroughly for many of the developmental disorders.

Further, even if the preceding considerations have been carefully described, the reliability of any diagnosis must be determined. If the reliability is low, or if the reliability has not yet been determined, then the usefulness of the diagnosis in terms of etiology, prognosis, and treatment will be meaningless. Quay (1986) noted that the assignment of an individual to a specific diagnostic category must be relatively consistent. This consistency should be noted within and between clinicians as well as over time. The importance of obtaining good reliability for specific diagnoses becomes even more crucial when one recognizes that reliability will set the upper limits on validity.

Relatedly, the determination of validity contributes to the usefulness of any classification system. For example, specific diagnostic categories should be distinguishable from one another and inherently related to the behaviors or constructs used to define them (e.g., specific reading disorders). As Quay (1986) aptly noted, the validity of any single diagnosis within a classification system should determine the utility of the vernacular necessary for professional communication; its accompanying list of symptoms; relevant information pertinent to etiology, prognosis, and treatment; comoroid conditions; and theory formulation. Blashfield (1984) echoed similar contingencies.

Another important feature of any classification system is its clinical utility. This feature rests in part upon its reliability and validity, but it also is determined by its comprehensiveness. Although the comprehensiveness of a classification system always should be balanced with issues of parsimony, how well any system covers the range of possible problems is important, particularly from a clinical perspective. Further, simply because a classification system is comprehensive does not guarantee that the issues of reliability and validity have been addressed in a satisfactory manner. For example, despite the fact that the DSM-III-R (American Psychiatric Association, 1987) includes more developmental disorders than its predecessor (i.e., DSM-III), field trials obtained reliability estimates only for the developmental disorder of Autism. Similarly, for the DSM-III (American Psychiatric Association, 1980), field trials were conducted for the developmental disorders of mental retardation, pervasive developmental disorders, and specific developmental disorders, although kappa coefficients were somewhat variable. Further, Rutter and Tuma (1988) and Blashfield, Sprock, and Fuller (1990) noted that the DSM-III-R contains more diagnoses than can be justified either statistically or clinically; therefore, comprehensiveness in and of itself should never be the sole criterion for determining the clinical utility of a classification system.

Classification Models

To date there have been two general approaches aimed at classifying children with developmental disorders and other forms of psychopathology into groups based on their intrinsic characteristics. One approach, the *clinical-inferential* approach, has depended largely on the clinical observations of professionals in the field. Although not systematic or scientific by design, the clinical-inferential approach creates a classification scheme that is hypothesis-driven, but typically these hypotheses are not tested directly. Quay (1986) noted that "usually it is authority, not proof, that is the benchmark" of these systems. The DSM and ICD systems were based on the clinical-inferential approach, although more recent efforts have attempted to address the need for these systems to attain more satisfactory scientific standards (Blashfield et al., 1990; Cantwell & Baker, 1988; Dingemans, 1990).

Although classification models derived from these efforts tend to make intuitive sense, there are inherent difficulties with the clinical-inferential approach. Generally, this approach suffers from methodological weaknesses, limited data-reduction strategies, and questionable validity. Further, the clinical utility of classification schemes based on this approach appears limited, particularly given the numerous interacting variables that can be associated with developmental psychopathology.

Prior to the emergence of high-speed technological assistance, the clinical-inferential approach dominated classification efforts in childhood and adolescent psychopathology; however, with the ready availability of advanced computer technologies, many of the problems manifested by the clinical-inferential approach can be addressed effectively by empirical classification techniques. Generally, these techniques are based on a quantitative view of behavior that attempts to derive "prototypic" profiles of psychopathology as opposed to all-or-none categories (Achenbach, 1985). This approach has been illustrated effectively by Achenbach (1985) in the development of a dimensional classification system for childhood and adolescent psychopathology, and it has been used to establish the heterogeneity of specific developmental disorders (Hooper & Willis, 1989; Rourke, 1985).

Although this approach may appear superior to the clinical-inferential approach, the empirical approach to classification is not without its own drawbacks. The ease with which multivariate classification techniques manage data is seductive in that it leaves the door open for investigators to fall prey to "naive empiricism." Clearly, the adequacy and strength of models derived by empirical classification methods are influenced by many a priori clinical decisions, including those regarding theoretical orientation, sample selection, and variable selection. For example, if the sample is not well marked, and the dependent variables are not selected on the basis on theoretical underpinnings, then the dimensions or groupings that emerge from the empirical analyses may not have much meaning when applied to other populations or settings. Further, the empirical derivation of dimensions stipulates the existence of these dimensions, but it does not stipulate their significance. Whether these dimensions relate significantly to etiological factors or treatment issues is what will determine their ultimate utility. Lastly, the development of any empirically based classification model also should strictly adhere to standards of reliability and validity.

The clinical-inferential and empirical approaches represent the major strategies for classifying childhood and adolescent psychopathology. In addition, specific classification models and guidelines have been proposed in an effort to control the number of concerns expressed previously for both of these approaches. Several of these models and guidelines have been proposed by Cantwell (1975), Skinner (1986), and Blashfield et al. (1990) in order to define parameters for the development of a classification scheme. Although other general models have been proposed for specific developmental disorders (e.g., Adelman & Taylor, 1986), these models and guidelines are believed to be representative of the major issues important to general classification efforts in childhood and adolescent psychopathology.

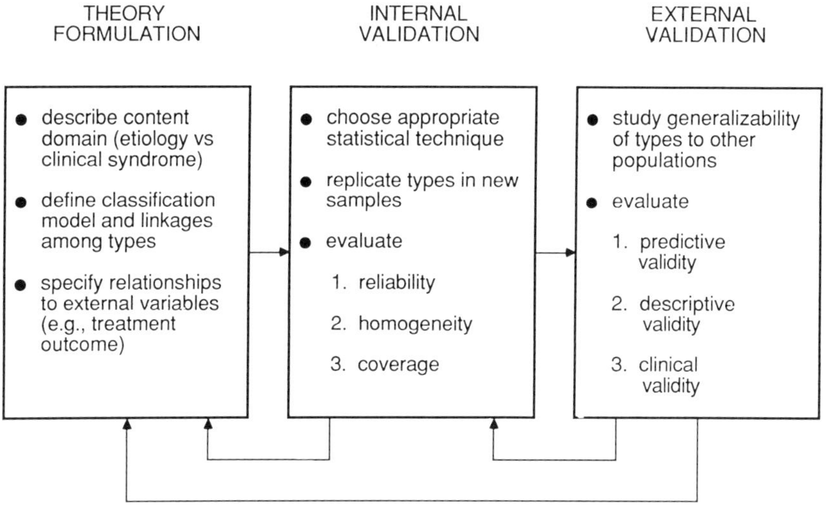

FIG. 1.1 Conceptual framework for classification research. From "Toward the integration of classification theory and methods" by H. A. Skinner, 1981, *Journal of Abnormal Psychology, 90*, 68–87. Copyright 1981 by the American Psychological Association. Reprinted by permission.

Skinner's Conceptual Framework. Skinner (1981, 1986) advanced one such conceptual framework to formulate and evaluate proposed classification models. This framework, shown in Fig. 1.1, comprises three fundamental components that, according to Skinner, should characterize any classification model. These include the following: (a) theory formulation, (b) internal validation, and (c) external validation.

The first component, theory formulation, involves decisions regarding the theory on which the classification model is based. These decisions include selecting a set of measures that coincide with the theoretical orientation of the model. Additionally, a priori hypotheses are proposed to assess the clinical and theoretical validities of identified diagnoses (e.g., response to treatment, differences on variables not included in the derivation of the diagnosis). Other hypotheses address the relationships among the many subgroups of the disorder and the relationship of the overall model to other parameters of psychopathology.

The second component, internal validation, refers to the "goodness of fit" of the theoretical model to data obtained on the identified diagnosis, preferably from a sample of subjects that differs from the original grouping from which the diagnosis was conceived. This involves selecting an appropriate classification technique, such as one of the multivariate clustering techniques (e.g., hierarchical cluster analysis), and examining its reliability, homogeneity, and robustness for the theoretical model. Currently, given the status of many classification techniques, there are few explicit rules for determining the reliability of a classification model (Blashfield, 1980; Morris, Blashfield, & Satz, 1981). Fletcher and Morris (1986), however, suggested that internally valid typologies should demonstrate the following: (a) result in the classification of the majority of the sample into the identified group, (b) result in homogeneous groupings or subgroupings, (c) be replicable across samples and techniques (e.g., clinical groupings versus Q-factor analysis) and, (d) be based on reliable variables that are appropriate to a variety of samples. In particular, most classification models proffered to date fail to address the latter two points in a substantive manner.

Finally, the third component, external validation, is important independent of internal reliability and validity. Here, the obtained model is evaluated against external criteria (e.g., family history, cognitive findings, achievement) not used in the original derivation of the diagnosis. Fletcher and Morris (1986) suggested that simple comparisons on external variables only partially addressed this component; moreover, results should be hypothesized a priori so that the utility of the classification model can be properly assessed in terms of specific parameters.

Skinner's framework for developing and evaluating classification models for psychopathology is useful because it emphasizes the conceptual or theoretical basis absent from many current models. In fact, Adams (1985) suggested that internal validation should be evaluated during theory formulation, prior to any actual statistical analyses. He argued that this would tend to maximize benefits from subsequent empirical analyses. Skinner's framework provides a promising heuristic for guiding classification research for childhood and adolescent psychopathology. Further, its application to children with specific learning disorders (i.e., learning disability subtyping) has been discussed thoroughly (Hooper & Willis, 1989).

Cantwell's Six-Stage Model. Based on the conceptualizations of Robins and Guze (1970) and Spitzer, Endicott, and Robins (1978) in their work with adult psychiatric disorders, Cantwell (1975) proposed a six-stage model for examining the validity of childhood and adolescent psychiatric disorders. As illustrated in Fig. 1.2, the first stage requires a careful description of the symptoms involved and related demographic data. This information is important for determining the diagnostic criteria for the disorder under investigation. The second stage typically involves a physical examination. This should include

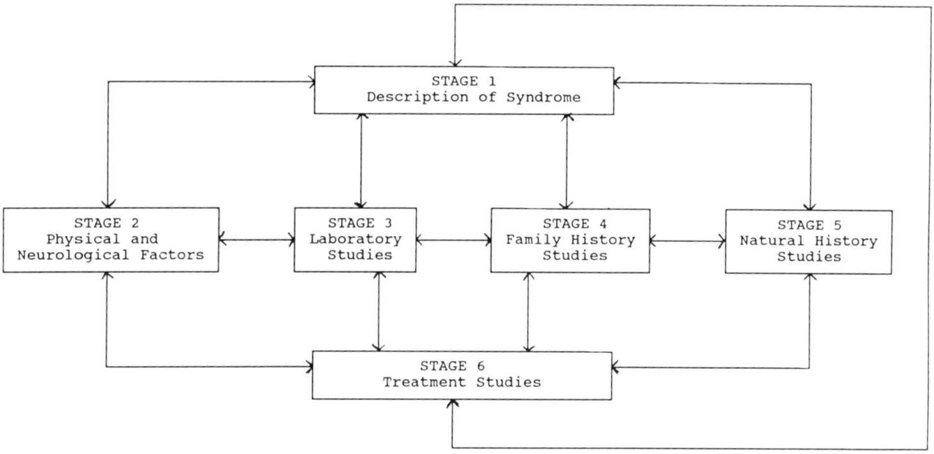

FIG. 1.2. Cantwell's six-stage model for validation of childhood psychiatric disorders. Adapted from "A model for the investigation of psychiatric disorders of childhood: Its application in genetic studies of the hyperkinetic syndrome" by D. P. Cantwell in E. J. Anthony (Ed.), *Explorations in child psychiatry* (pp. 57–79), New York: Plenum Press.

investigation of not only routine physical attributes but also exploration for minor physical and neurological (i.e., soft signs) abnormalities. Similarly, the third stage of validation requires laboratory studies in order to explore the possible biological basis of the disorder. These studies can range from routine blood tests to rule out anemia and thyroid dysfunction to electroencephalographic and neuroimaging procedures.

Stages 4 and 5 involve family studies and natural history studies of the disorder, respectively. Family studies are important in order to begin to trace the possibility of a genetic basis for the disorder. This information is particularly important given the elusive etiological aspects of many psychiatric and developmental disorders. The natural history studies are important from a prognostic perspective and serve to increase our understanding of the continuities and discontinuities of a disorder (i.e., the relationship between childhood disorders and adult disorders). Finally, Stage 6 is concerned with the response of a subject with a specific disorder to particular intervention strategies. This is crucial to the ultimate usefulness of any diagnosis. If making a specific diagnosis does not lend itself to specific treatment alternatives, then the validity of that diagnosis should be in question.

It is important to note that the stages in the Cantwell model are not necessarily hierarchical, but they are intimately related, with each stage attempting to add unique information toward the validation of a specific psychiatric disorder. The model by Cantwell has been used in whole or in part for several of the diagnoses listed in the DSM system (e.g., attention deficit disorder, affective disorders) and, as with the Skinner paradigm, it represents another model designed to improve the reliability, validity, and clinical utility of any childhood and adolescent diagnosis.

Blashfield et al.'s Guidelines. Based on previous work that was designed to address issues of psychiatric classification (Blashfield & Draguns, 1976; Millon, 1987; Quay, 1987; Spitzer & Williams, 1988), Blashfield et al. (1990) suggested several guidelines for whether a specific diagnosis should be included or excluded from the upcoming DSM-IV. Table 1.1 outlines these suggested guidelines. Specifically, Blashfield suggested that any diagnosis should satisfy five criteria prior to being considered for inclusion in the DSM-IV. First, for the last 10 years, there should be at least 50 published journal articles pertaining to the diagnosis. Further, it would be important for at least one-half of them to be empirical in nature. These investigators noted that this criterion would insure that a diagnostic category is recognized in the clinical literature and should be considered "to become part of the official language of psychopathology." The second guideline requires that a set of diagnostic criteria exist for defining the disorder. Assessment-to-treatment linkages also should be available to provide strategies for assessing and treating the various diagnostic criteria. Issues of reliability are addressed by the third suggested guideline for inclusion. Blashfield et al. stated that at least two empirical studies, conducted by independent research groups, show Kappa Coefficients $\geq .70$. This will insure that minimal standards of reliability are being met by the diagnosis. The fourth guideline requires that the proposed diagnostic category "represent a syndrome of frequently co-occurring symptoms." There should be two or more independent empirical studies demonstrating that if a patient has one of the symptoms, then there is at least a .50 probability that another symptom will be present. The fifth guideline addresses issues of clinical differentiation and requires that a diagnostic category have at least two independent studies demonstrating that it is separate and distinct from other diagnoses.

In contrast to the inclusion criteria, Blashfield et al. (1990) stated that if "any one of the exclusion guidelines are met, then an existing category in the DSM system should be considered for re-evaluation (p. 18)." However, these investigators noted that no single exclusion guideline should be the sole basis for deleting any diagnostic category. The first exclusion guideline requires that less than 20 journal articles have

TABLE 1.1
Blashfield, Sprock, & Fuller (1990) Guidelines for Including or Excluding Categories in the DSM–IV.

Inclusion Guidelines	*Exclusion Guidelines*
1. Adequate literature ($\geq$ 50 articles)	1. Inadequate literature
2. Specified diagnostic criteria	2. Extremely low coverage
3. Acceptable reliability ($K \geq .70$)	3. No diagnostic bias
4. Evidence for a syndrome	
5. Clinical differentiation	

Note. None of the exclusionary guidelines would be instituted if the category has a significant relationship with pathophysiological change, and/or the category has a clear etiology. Adapted from "Suggested guidelines for including or excluding categories in the DSM-IV" by R. K. Blashfield, J. Sprock, and A. K. Fuller, 1990, *Comprehensive Psychiatry, 31*, 15–19.

been published on the specific diagnosis within the past 10 years. As with the first guideline for inclusion, this will insure that there has been a significant scientific and clinical literature pertaining to the diagnosis. The second guideline for exclusion relates to the general category in which a specific diagnosis has been placed (e.g., developmental arithmetic disorder is one of the diagnoses under the general category of specific developmental disorders). Blashfield et al. suggest that if a specific diagnosis refers to less than five percent of the patients within the general category in at least five independent studies, then it should be considered for exclusion. The third guideline for exclusion should be invoked when at least five empirical studies have shown an undue diagnostic bias with respect to race or gender. Finally, a fourth guideline represents a "safety valve" for many diagnoses. Specifically it states that none of the previous exclusion guidelines should lead to possible deletion of diagnosis, if the diagnosis is related to pathophysiological change in the nervous system and/or has a clear etiology.

These guidelines do not represent a formal model for evaluating a psychiatric classification system, such as Skinner (1986) and Cantwell (1975) proposed, but they do address many of the criteria that are important for evaluating specific diagnoses within a classification system. Although these guidelines were derived from a subjective vantage point, they do serve to provide useful guidance for considering what diagnoses should be included or excluded in the DSM system. Further, these guidelines also insure that the basic tenets of any classification system (i.e., reliability, validity, clinical utility) would be addressed.

CLASSIFICATION EFFORTS FOR DEVELOPMENTAL DISORDERS

Despite the ongoing development of the DSM, ICD, and dimensional classification systems for childhood and adolescent psychopathology in general, the classification work involving general and specific developmental disorders has been compromised by individual efforts that have not been coordinated to any degree. Several attempts have been initiated to address this concern, such as the proposed Nosology on Disorders of Cortical Function in Children (Child Neurology Society, 1981), and the more recent efforts of the Task Force on Nosology (David, 1988), which was designed to create a classification model for preschool children with disorders of higher-order cognitive functioning, but these classification efforts remain preliminary. In addition, given the heterogeneity of many of the developmental disorders, several attempts have been made to classify children and adolescents within a selected diagnosis. Although loosely organized, these efforts have been directed at the diagnoses of mental retardation (Money, 1973), pervasive developmental disorders (Fein, Allen, & Waterhouse, 1988; Fein, Waterhouse, Lucci, & Snyder, 1985), the specific developmental disorders (Hooper & Willis, 1989; Rourke, 1985), language disorders (Aram & Nation, 1975; Wilson, Aram, Rapin, & Allen, 1988), and motor disorders (Child Neurology Society, 1981). Many of these efforts are discussed in the ensuing chapters in this volume.

Although there was a significant improvement in the number and quality of the diagnostic categories for developmental disorders in the DSM-III and DSM-III-R when compared to the DSM-I and DSM-II, it is clear from a simple scanning of these sections that little attention has been devoted to including the massive amount of information available for many of these disorders. In fact, one could argue that their inclusion within a psychiatric nosology might be inappropriate, although a variety of developmental disorders can masquerade as psychiatric problems and/or create secondary social and emotional distress that require children to be evaluated within psychiatric settings. The DSM-III-R (American Psychiatric Association, 1987, p. 40) even recognizes the controversial nature of including the specific developmental disorders within a psychiatric nosology. Although many of these children show no other signs of psychopathology, and given that they tend to be treated within educational as opposed to mental health systems (Sattler, 1988), the developers of the DSM-III-R insist that the inclusion of these disorders is consistent with their definition and conceptualization of a mental disorder. Nonetheless, it would seem that if developmental disorders, especially the specific developmental disorders, are to remain within a psychiatric nosology, then it will be important for the developers of the DSM to take advantage of the scores of volumes that have been written for each of the developmental disorders included within the DSM-III-R.

Despite this obvious oversight, however, there has been an ongoing evolution of these disorders in the DSM system. As can be seen in Table 1.2, the DSM-I contained several developmental disorders, including mental retardation, learning disturbance, and speech disturbance. Although the DSM-I did not have a formal listing for pervasive developmental disorder, it did list *schizophrenic reaction, childhood type*, which was the forerunner to autistic disorder. There was no category for children manifesting motor difficulties. Although the DSM-II reflected few changes in these disorders, there were major changes evidenced in the DSM-III and the DSM-III-R, with most of these changes coming in the form of refinements and further differentiation of disorders. For example, perhaps the most obvious changes were seen in the specific developmental disorders where the single diagnostic category of learning disturbance in the DSM-I was transformed into eight different categories in the DSM-III-R. Further, the diagnosis of developmental coordination disorder first emerged in the DSM-III-R. Another subtle change in the DSM-III-R, when compared to all previous versions of the DSM, was to place all of the developmental disorders, including mental retardation and autistic disorder, on Axis II. Perhaps further refinement and differentiation will be seen as more available evidence begins to be included into the DSM system.

ASSESSMENT TECHNOLOGY

When compared to the assessment technology that has recently evolved for children and adolescents with psychiatric disorders (e.g., structured and semistructured interviews, rating scales), the diagnostic instrumentation for determining the presence of a developmental disorder seems to have a lengthier history, dating back to before the 1800s (Sattler, 1988). In addition, it appears that there are more of these kinds of tests, and one only needs to peruse any test publisher catalog to recognize that many of the tests have been developed to assist those who work with children with developmental disorders. Despite this lengthier history, however, many of these assessment techniques have not necessarily been linked to diagnostic classification schemes in any systematic manner. Perhaps it has been a classic case of "the tail wagging the dog," with many of the tests not always being helpful in terms

TABLE 1.2
A Chronology and Evolution of Selected Child and Adolescent Developmental Disorders.

DSM-I (1952)	*DSM-II (1968)*
Mental Retardation	
Mild mental deficiency (IQ=70–85) →	Borderline mental retardation (IQ=70–85) →
Moderate mental deficiency (IQ=50–69) →	Borderline mental retardation (IQ=68–69); Mild mental retardation (IQ=52–67) →
	Moderate mental retardation (IQ=50–51) →
Severe mental retardation (IQ=<50) →	Moderate mental retardation (IQ=36–49) →
	Severe Mental Retardation (IQ=20–35) →
	Profound mental retardation (IQ<20) →
Severity not specified →	Unspecified mental retardation →
Specific Developmental Disorders	
Learning disturbance →	Specific learning disturbance →
—— →	—— →
Speech Disorders	
Speech disturbance →	Speech disturbance →
Pervasive Developmental Disorders	
Schizophrenic reaction, childhood type →	Schizophrenia, childhood type →

Note. This information was abstracted from DSM-I (APA, 1952), DSM-II (APA, 1968), DSM-III (APA, 1980), and DSM-III-R (APA, 1987).

DSM-III (1980)	*DSM-III-R (1987)*
Borderline intellectual functioning →	—
Mild-mental-retardation (IQ=50–55 to 70) →	Mild mental retardation (IQ=50–55 to 70)
Moderate mental retardation (IQ=35–49) →	Moderate mental retardation (IQ=35–40 to 50–55)
Severe mental retardation (IQ=20–34) →	Severe mental retardation (IQ=20–25–35–40)
Profound mental retardation (IQ<20) →	Profound mental retardation (IQ<20–25)
Unspecified mental retardation →	Unspecified mental retardation
Developmental reading disorder Developmental arithmetic disorder →	Developmental reading disorder Developmental arithmetic disorder Developmental expressive writing disorder
Mixed specific developmental disorder →	—
Developmental language disorder →	Developmental expressive language disorder Developmental receptive language disorder
Developmental articulation disorder →	Developmental articulation disorder
Atypical specific developmental disorder	Specific developmental NOS Developmental disorder NOS
— →	Developmental coordination disorder
Stuttering →	Stuttering Cluttering
Infantile autism Childhood onset pervasive developmental disorder →	Autistic disorder
Atypical pervasive developmental disorder	'Pervasive developmental disorder NOS

of arriving at a specific diagnosis for an individual case. This should not imply that these assessment techniques are not helpful in contributing to the diagnostic process but, rather, that the tests are not generally tied to any specific classification system such as the DSM. Although this could be viewed as a possible "glitch" in the diagnostic process, there appears to be a wide range of domains that must be assessed with respect to children and adolescents with general and specific kinds of developmental disorders. Although specific procedures associated with these domains have been reviewed extensively elsewhere (e.g., Reynolds & Kamphaus, 1990; Sattler, 1988), and a comprehensive listing of them is beyond the scope of this text, several of these general domains are mentioned here, particularly given their relevance to developmental disorders.

Cognitive Abilities

It is clear that some form of cognitive assessment is necessary for nearly all children and adolescents with a developmental disorder. This type of testing is helpful in determining a child's level of functioning and, to some extent, a profile of strengths and weaknesses depending on the test selected. Cognitive testing also is one part of the classic but controversial ability–achievement discrepancy formulas used by many State Departments of Education in their identification of children with specific learning disabilities. These tests are required by the DSM system in the diagnosis of many of the specific developmental disorders (e.g., developmental reading disorder), and they also are necessary for assisting in the diagnosis of mental retardation. Further, although the limits of their application are challenged by lower functioning individuals with autistic disorder and mental retardation, they can be quite useful in identifying a profile of strengths and weaknesses in these children as well. Several examples of the more extensively used general cognitive testing procedures for children ages 2 years and older include the Wechsler Intelligence Scale for Children-Revised, Wechsler Adult Intelligence Scale-Revised, the Wechsler Preschool and Primary Scale of Intelligence-Revised, the Stanford-Binet Intelligence Scale (fourth edition), the Kaufman Assessment Battery for Children, and the McCarthy Scales of Children's Abilities. For infants, ages birth to 2 years, the Bayley Scales of Infant Development and the Cattell Infant Intelligence Scale have dominated the field for the past several decades. A plethora of other types of tests are available to assess cognitive functioning (e.g., the Psychoeducational Profile for children with autistic disorder), and many of them are discussed in the chapters of this volume.

Achievement

Generally, this area is extremely important for all children of preschool age and older given that the major portion of their waking hours is spent in some kind of classroom setting. This information is particularly useful in determining the levels of specific academic domains in order to chart learning progress and seems particularly important in establishing learning activities and goals for achievement for all handicapped children. From a diagnostic standpoint, these data are helpful in determining the presence of many of the specific developmental disorders, although the DSM-III-R does not operationally define how impaired skill levels need to be in order to receive a diagnosis of one of these disorders. Examples of several of the more commonly used academic achievement instruments include the Woodcock-Johnson Psycho-Educational Battery-Revised, the Diagnostic Achievement Battery, the Key Math Diagnostic Arithmetic Test-Revised, the Peabody Individual Achievement Test-Revised, the Kaufman Test of Educational Achievement, Wide Range Achievement Test-Revised, and the Basic Achievement Skills Individual Screener.

Adaptive Behavior

This domain can be defined as the behavior that is necessary for satisfying the daily demands of one's environment (Grossman, 1983). It is a developmental concept that encompasses a wide range of skills (e.g., physical development, communication development, social competencies, etc.) that should be assessed in many children and adolescents with developmental disorders. Generally, the measurement of adaptive behaviors addresses two major points. The first point addresses the degree of independence that an individual is capable of having in his or her daily functioning, whereas the second assesses the level of responsibility that the individual has with respect to personal and social activities (Sattler, 1988). Although adaptive behavior typically is not assessed in children with speech disorders or specific developmental disorders, it is crucial to the diagnosis of mental retardation and likely is used frequently in children suspected of having autistic disorder. In fact, the DSM-III and DSM-III-R, in conjunction with the American Association of Mental Retardation (Grossman, 1983), require that a measure of adaptive behavior be completed prior to the diagnosis of mental retardation being applied to any individual. Although it appears that no one instrument can adequately assess all of the domains covered by adaptive behavior, several more typical examples include the American Association on Mental Deficiency Adaptive Behavior Scales, the Adaptive Behavior Inventory for Children, the Vineland Adaptive Behavior Scales, the Scales of Independent Behavior, and the Battelle Developmental Inventory.

SUMMARY

This introductory chapter provided an overview of the many classification issues pertinent to childhood and adolescent psychopathology, with a particular emphasis on developmental disorders. A variety of classification issues were addressed, including possible reasons for attempting to classify children with developmental disabilities and the ever-present and increasingly important issues of reliability, validity, and clinical utility of diagnostic conceptualizations. Several selected models for addressing these latter issues were presented as well, with Skinner's model perhaps being the most comprehensive conceptualization to date. However, the pragmatic nature of Blashfield et al.'s suggestions also warrants serious consideration for determining the inclusion or exclusion of a particular diagnosis in the upcoming DSM-IV.

A variety of classification efforts were mentioned with respect to developmental disorders. In comparison to the classification of psychiatric disorders, the efforts to classify developmental disorders have been outside of the traditional psychiatric nomenclatures (e.g., the DSM and ICD systems); however, these efforts appear noteworthy and the psychiatric taxonomies would be wise to review this information for possible inclusion prior to revisions of the taxonomies. The inclusion of this information becomes even more important given the stance that the DSM system believes that developmental disorders, particularly the specific developmental disorders, fit their definition of a "mental disorder."

Finally, this chapter touched on several broad assessment domains that have been important to the evolution of the developmental disorders. Although these domains are reviewed more extensively elsewhere, the mentioning of selected areas was viewed as important for the conceptualization of the developmental disorders. As will be seen, these domains are elaborated upon more extensively in the following chapters. Clearly, the assessment of children with developmental disorders appears somewhat advanced when compared to those with primary psychiatric disorders, but the wide variety of assessment strategies do not appear to be adequately linked to specific criteria for many of the developmental disorder diagnoses listed in the DSM or ICD systems.

REFERENCES

Achenbach, T. M. (1985). *Assessment and taxonomy of child and adolescent psychopathology.* Newbury Park, CA: Sage.

Adams, K. M. (1985). Theoretical, methodological, and statistical issues. In B. P. Rourke (Ed.), *Neuropsychology of learning disabilities: Essentials of subtype analysis* (pp. 17–39). New York: Guilford.

Adelman, H. S. & Taylor, L. (1986). The problems of definition and differentiation and the need for a classification schema. *Journal of Learning Disabilities, 19,* 514–520.

American Psychiatric Association. (1952). *Diagnostic and statistical manual of mental disorders.* Washington, DC: Author.

American Psychiatric Association. (1968). *Diagnostic and statistical manual of mental disorders* (2nd ed.). Washington, DC: Author.

American Psychiatric Association. (1980). *Diagnostic and statistical manual of mental disorders* (3rd ed.). Washington, DC: Author.

American Psychiatric Association. (1987). *Diagnostic and statistical manual of mental disorders* (3rd ed., revised). Washington, DC: Author.

Aram, D. M., & Nation, J. (1975). Patterns of language behavior in children with developmental language disorders. *Journal of Speech and Hearing Research, 18,* 229–241.

Blashfield, R. K. (1980). Propositions regarding the use of cluster analysis in clinical research. *Journal of Consulting and Clinical Psychology, 48,* 456–459.

Blashfield, R. K. (1984). *Classification of psychopathology.* New York: Plenum.

Blashfield, R. K., & Draguns, J. G. (1976). Evaluative criteria for psychiatric classification. *Journal of Abnormal Psychology, 85,* 140–150.

Blashfield, R. K., Sprock, J., & Fuller, A. K. (1990). Suggested guidelines for including or excluding categories in the DSM-IV. *Comprehensive Psychiatry, 31,* 15–19.

Cantwell, D. P. (1975). A model for the investigation of psychiatric disorders of childhood: Its application in genetic studies of the hyperkinetic syndrome. In E. J. Anthony (Ed.), *Explorations in child psychiatry* (pp. 57–79). New York: Plenum.

Cantwell, D. P. (1988). DSM-III studies. In M. Rutter, A. H. Tuma, & I. S. Lann (Eds.), *Assessment and diagnosis in child psychopathology* (pp. 3–36). New York: Guilford.

Cantwell, D. P., & Baker, L. (1988). Issues in the classification of child and adolescent psychopathology. *Journal of the American Academy of Child and Adolescent Psychiatry, 27,* 521–533.

Child Neurology Society, Task Force on Nosology of Disorders of Higher Cerebral Function in Children. (1981). *Proposed nosology of disorders of higher cerebral function in children.* Washington, DC: Author.

David, R. (1988). History and overview of the project [Abstract]. *Journal of Clinical and Experimental Neuropsychology, 10,* 80.

Dingemans, P. M. (1990). ICD-9-CM classification coding in psychiatry. *Journal of Clinical Psychology, 46,* 161–168.

Fein, D., Allen, D., & Waterhouse, L. (1988). Subtypes within pervasive developmental disorders [Abstract]. *Journal of Clinical and Experimental Neuropsychology, 10,* 80.

Fein, D., Waterhouse, L., Lucci, D., & Snyder, D. (1985). Cognitive subtypes in developmentally disabled children. A pilot study. *Journal of Autism and Developmental Disorders, 15,* 77–95.

Fletcher, J. M., & Morris, R. (1986). Classification of disabled learners: Beyond exclusionary definitions. In S. J. Ceci (Ed.), *Handbook of cognitive, social, and neuropsychological aspects of learning disabilities* (Vol. 1, pp. 55–80). Hillsdale, NJ: Lawrence Erlbaum Associates.

Goodall, D. W. (1966). Hypothesis-testing in classification. *Nature, 211,* 329–330.

Grossman, H. J. (Ed.). (1983). *Classification in mental retardation.* Washington, DC: American Association on Mental Deficiency.

Hempel, C. G. (1965). *Aspects of scientific explanation.* New York: Free Press.

Hooper, S. R., & Willis, W. G. (1989). *Learning disability subtyping. Neuropsychological foundations, conceptual models, and issues in clinical differentiation.* New York: Springer-Verlag.

Huschka, M. (1941). Psychopathological disorders in the mother. *Journal of Nervous and Mental Disorders, 94,* 76–83.

Kavale, K. A., & Forness, S. R. (1987). The far side of heterogeneity: A critical analysis of empirical subtyping research in learning disabilities. *Journal of Learning Disabilities, 20,* 374–382.

Kendall, R. E. (1975). *The role of diagnosis in psychiatry.* Oxford: Blackwell.

Last, C. G., & Hersen, M. (Eds.). (1987). *Issues in diagnostic research.* New York: Plenum.

Millon, T. M. (1987). On the nature of taxonomy in psychopathology. In C. Last & M. Hersen (Eds.), *Issues in diagnostic research* (pp. 3–85). New York: Plenum.

Money, J. (1973). Turner's syndrome and parietal lobe functions. *Cortex, 9,* 387–393.

Morris, R., Blashfield, R. K., & Satz, P. (1981). Neuropsychology and cluster analysis: Potential and problems. *Journal of Clinical Neuropsychology, 3,* 77–99.

Quay, H. C. (1986). Classification. In H. C. Quay & J. S. Werry (Eds.), *Psychopathological disorders of childhood* (3rd ed., pp. 1–34). New York: Wiley.

Quay, H. C. (1987). A critical analysis of DSM-III as a taxonomy of psychopathology in childhood and adolescence. In T. Millon & G. Klerman (Eds.), *Contemporary directions in psychopathology: Towards DSM-IV* (pp. 151–166). New York: Guilford.

Reynolds, C. R., & Kamphaus, R. W. (Eds.). (1990). *Handbook of psychological and educational assessment of children: Intelligence and achievement.* New York: Guilford.

Robins, E., & Guze, S. B. (1970). Establishment of diagnostic validity in psychiatric illness: Its application to schizophrenia. *American Journal of Psychiatry, 126,* 983–987.

Rourke, B. P. (Ed.). (1985). *Neuropsychology of learning disabilities: Essentials of subtype analysis.* New York: Guilford.

Rutter, M., & Tuma, A. H. (1988). Diagnosis and classification: Some outstanding issues. In M. Rutter, A. H. Tuma, & I. S. Lann (Eds.), *Assessment and diagnosis in child psychopathology* (pp. 437–452). New York: Guilford.

Sattler, J. M. (1988). *Assessment of children* (3rd ed.). San Diego, CA: Author.

Skinner, H. A. (1981). Toward the integration of classification theory and methods. *Journal of Abnormal Psychology, 90,* 68–87.

Skinner, H. A. (1986). Construct validation approach to psychiatric classification. In T. Millon & G. L. Klerman (Eds.), *Contemporary directions in psychopathology* (pp. 307–330). New York: Guilford.

Spitzer, R. L., Endicott, J., & Robins, E. (1978). Research diagnostic criteria: Rationale and reliability. *Archives of General Psychiatry, 35,* 773–782.

Spitzer, R. L., & Williams, J. B. W. (1988). Having a dream: A research strategy for DSM-IV. *Archives of General Psychiatry, 45,* 371–874.

Szasz, T. S. (1961). *The myth of mental illness.* New York: Harper & Row.

Szasz, T. S. (1978). *The myth of psychotherapy: Mental healing as religion, rhetoric, and repression.* New York: Doubleday.

Weiner, I. B. (1982). Classification in developmental psychopathology. In I. B. Weiner (Ed.), *Child and adolescent pathology* (pp. 42–79). New York: Wiley-Interscience.
Wilson, B. C., Aram, D. M., Rapin, I., & Allen, D. (1988). Subtypes within developmental language disorders [Abstract]. *Journal of Clinical and Experimental Neuropsychology, 10,* 80.
World Health Organization. (1978). *The International classification of diseases* (9th revision). Geneva: Author.

CHAPTER 2

Mental Retardation: Conceptual Foundations, Definitional Criteria, and Diagnostic Operations

Daniel J. Reschly
Iowa State University

Mental retardation (MR) was recognized in the Diagnostic and Statistical Manual of Mental Disorders (DSM-III-R), all prior diagnostic and statistical manuals of the American Psychiatric Association (APA), and all other comprehensive schemes for classification of human disorders and exceptional patterns of development. The various diagnostic schemes have addressed fundamental issues in conception, definitional criteria, and diagnostic operations in slightly different ways. This chapter reviews the following: (a) historical patterns in mental retardation, (b) previous mental retardation diagnostic schemes, (c) current APA and other MR diagnostic systems, (d) current research, and (e) changes in conceptual bases, definitional criteria, and diagnostic operations.

CONCEPTUAL BASES

The conceptual bases of the MR diagnostic construct are addressed in different ways. The most fundamental issue is(are) the *dimension(s)* of behavior on which all other aspects of concept, definitional criteria, and diagnostic operations depend. There appears to be widespread agreement that the fundamental dimensions of behavior in mental retardation are intellectual functioning and adaptive behavior, although one recent formulation suggested elimination of the adaptive behavior component (Zigler, Balla, & Hodapp, 1984). A second fundamental issue is *etiology,* or, more specifically, whether mental retardation is caused by biological differences, environmental factors, or some combination of both. Etiology has been emphasized less in modern conceptions of mental retardation (Grossman, 1983). A third fundamental issues is *prognosis.* Prognosis relates to the following: (a) whether or not mental retardation is viewed as a permanent condition causing lifelong substandard performance, and (b) the degree to which the kind and level of mental retardation may be related to differential outcomes. Prognosis may also be related to the likelihood of change in status depending on the availability and relative success of different treatments (Cromwell, Blashfield, & Strauss, 1975). *Developmental* considerations constitute the fourth conceptual issue of the mental retardation diagnosis. The basic issue here is whether the substandard performance first has to appear within a range of ages. The fifth conceptual issue is *level,* involving a quantitative determination of degree of substandard performance. The sixth conceptual issue is recognition of *subtypes,* related to whether different kinds of mental retardation are recognized in the diagnostic schema. The subtypes may be further related to either etiology, prognosis, or some other element of the diagnostic construct. The seventh issue is *sociocultural* status, specifically whether the diagnosis is regarded as crosssituational or, conversely, the degree to which cultural context needs to be considered when the individual is diagnosed. The final issue is *overlap* with other disorders, or comorbidity, an increasingly important consideration in recent discussions of dual diagnosis with MR and other disorders, particularly severe behavioral disorders.

A complete MR diagnostic construct would address all seven issues in some fashion. As will be seen, none of the current schemes considers all of these issues in the conceptual formulation of the mental retardation diagnosis. Conceptual bases for mental retardation must further be operationalized through definitional criteria and diagnostic operations so that professionals can make reliable and valid judgments about the performance of individuals. Ideally, each of the seven issues would be addressed through specific definitional criteria and diagnostic operations.

DEFINITIONAL CRITERIA AND DIAGNOSTIC OPERATIONS

There are four elements that must be specified in order to carry out the diagnostic construct of mental retardation. The first is a clear *definition* of each of the characteristics or issues identified in the conceptual bases of the diagnostic construct. This would require definitions of the key dimensions, the etiological considerations with the various categories of etiology, the different prognoses, and so on. The next factor is specification of the *relationship* between the key dimensions. One or more dimensions may have greater priority (e.g., intellectual vs. adaptive behavior). Critical versus subsidiary dimensions must be specified in order for the diagnostic construct to be reliable and valid. *Assessment* of the various dimensions, with appropriate guidance provided concerning the use of formal and informal (including clinical judgment) processes must be included. Finally, *decision-making* rules must be specified for each of the critical dimensions. The decision-making rules must reflect some degree of flexibility to account for imperfections in current instruments. Conceptual treatment of key dimensions, development of definitional criteria, and delineation of diagnostic operations in mental retardation have developed gradually over the past 200 years (Doll, 1962; Kanner, 1964).

HISTORICAL BACKGROUND

The history of the conception of MR and treatment of individuals with MR have been described for the western world (Doll, 1962; Kanner, 1964). Similar accounts focusing on the United States have been provided by Bruininks, Warfield, and Stealey (1978) and Mercer and Richardson (1975). From antiquity to about 1700, there was little understanding of MR or recognition of MR separate from other disorders. Harsh treatment was common, varying from suspicion, neglect, cruelty, and abuse to exploitation and extermination. Instances of formal recognition of MR as a general disorder were rare and humane treatment even less frequent.

The philosophical bases for scientific understanding and humane treatment were established in the western world by Locke and Rousseau (Doll, 1962). Locke is credited as the first to distinguish between idiocy and insanity, and both Rousseau and Locke championed the vast influence of environment in shaping human competencies (Doll, 1962, 1967; Kanner, 1964). The early beginnings of the scientific study of MR and the development of humane treatment programs rested heavily on the cultural enlightenment in 18th-century Europe.

Early Roots 1800 to 1850

The formal emergence of a diagnostic construct that we now call MR appeared in the early 19th century, principally in western Europe with French physicians and educators providing leadership roles. The

French pioneers Itard and Esquirol clearly identified a human disorder involving deficient mental operations and serious handicaps in coping with everyday demands. These formulations were clear forerunners of our current emphasis on general intellectual functioning and adaptive behavior in the MR diagnostic construct. Although Itard's efforts to cure what would now be regarded as moderate to severe MR were unsuccessful, the careful descriptions of mental processes and the specification of educational procedures that did produce considerable gains fostered the development of scientific study and humane treatment of MR (Doll, 1962, 1967). Esquirol provided a major breakthrough in classification and terminology in the 1830s with the distinction between levels of mental retardation. He identified two levels, idiot and imbecile, which were distinguished by degree of deficiency in intellectual and coping skills as well as responsiveness to intervention procedures (Doll, 1962). Esquirol's two levels of MR imbecile and idiot were remarkably similar to the current nomenclature of moderate and severe/profound MR, respectively.

Protective Environments 1850 to 1910

By the 1850s, mental retardation was clearly distinguished from other mental disorders and disease entities. Mental retardation was regarded as a condition of arrested development rather than a disease entity. The time period of 1850 to about 1900 was marked by the pioneering work of Edward Seguin in the expansion of treatment and educational programs. The principal form of treatment was institutional care: asylum care for idiots providing total support and care, and colony placement for the moderately retarded where a protective environment was established in which persons received some training and contributed to their sustenance through performing simple tasks. These protective institutional environments, although now widely regarded as inappropriate for the vast majority of even the profoundly retarded, represented major breakthroughs in public policy and medical, psychological, and educational treatment.

During the late 19th century, there was clear recognition of levels of mental retardation related to the severity of the deficits in mental operations and coping skills, but a formal definition of the condition and operational criteria to distinguish the levels were not yet developed. During the latter portion of this period, from about 1895 to 1905, public school classes for a higher level of mental retardation, now referred to as the mild level of MR, were established in nearly all major cities of the United States (MacMillan, 1982). Placement in these classes was apparently based on the persistent inability to cope with the regular school curriculum. It should be noted that these classes were established before the development of IQ tests and the introduction of the Binet method to the United States.

The mild level of MR was recognized rather early as a substantially different phenomenon than the previously identified imbecile and idiot (i.e., moderate and severe) levels. The mildly retarded were noted to be biologically normal; that is, no gross physical stigmata or other biological deficiencies were typically identifiable, a circumstance that continues today.

During this period, the direct forerunner of the current American Association on Mental Retardation was established in 1876 with the formal title of "Association of Medical Officers of American Institutions for Idiotic and Feebleminded Persons." This title captured the dominant theme during the mid- to late-19th-century (i.e., the development of protective environments that were largely separated from neighborhoods and communities). Proceedings of the annual meetings of this association have been published continuously since 1876 (Sloan & Stevens, 1976). The debates at the annual meetings of this association over the past 120 years have reflected the major problems in mental retardation terminology and classification described in the introductory section of this chapter.

An early forerunner of modern classification of mental retardation was presented by Barr at the 1905 Annual Meeting of the Association of Medical Officers (Barr, 1905; Sloan & Stevens, 1976). Barr's scheme recognized four levels of mental retardation based on severity: backward or mentally feeble, moral imbecile, idio-imbecile, and idiot. The lowest level of classification, idiot, was further subdivided into profound and superficial, with the latter being described as "improvable in self-help only." Both levels of idiot were expected to require asylum care throughout their lives. The idio-imbecile category was seen as an intermediate group between idiot and moral imbecile. Moral imbeciles were regarded as "trainable in industrial occupations; temperament bestial." This moral imbecile category was further divided into low-grade, middle-grade, and high-grade. Moral imbeciles were regarded as capable of acquiring considerable self-help skills and some vocational skills but as needing lifelong protective care in the colony form of institutionalization. Individuals in the highest grade in Barr's system, backward or mentally feeble, were described as having "mental processes normal but slow, and requiring special training and environment" (p. 35) but regarded as being capable of being "trained for a place in the world," clearly implying the potential of what we now call the mildly MR for relatively adequate functioning in normal adult environments. Although not officially adopted by the association (the first official classification scheme was developed in 1921 by this association), Barr's scheme did reflect considerable sophistication in description of MR levels, subtypes, treatment, and likely prognosis. The annual proceedings of the Association of Medical Officers, at least from about 1890 to 1910, indicated considerable interest with the various issues implicit in Barr's scheme.

Protection of Society 1905 to 1930

The era from 1905 to about 1930 was increasingly dominated by two trends, the eugenics movement and the development, dissemination, and restandardization of the Binet method of assessing intellectual functioning. The eugenics movement, which originated in England in the 1890s, involved grave concerns over the alleged prolific reproductive propensities of persons with MR, particularly the mildly MR. IQ tests were developed, widely disseminated, and used nearly universally in the United States by 1920 to assess MR, leading some to draw a causal link between IQ testing and the eugenics movement (e.g., Kamin, 1974). Actually, the eugenics movement began prior to the IQ test developments, and concerns with the reproductive propensities of the mildly MR and the alleged attendant social problems that could result were apparent at the annual meetings of the Association of Medical Officers prior to the introduction of the Binet method to the United States.

The chief proponent of the eugenics movement as applied to MR in the United States was Goddard, who published his now classic description of the descendants of one Martin Kallikak (a pseudonym), a Revolutionary War soldier (Goddard, 1912). Goddard interpreted his studies of the offspring of Martin Kallikak as demonstrating both the prolific fertility of the mildly MR as well as the clear influence of good or bad genetic stock. He further attributed the societal problems of slums, poverty, criminal behavior, and immorality to the effects of bad genetic stock and mild MR. Sarason and Doris (1969) provided a critique of Goddard's biases toward a hereditarian rather than environmental interpretation of the rather unfortunate fate of the descendants from Martin Kallikak's illicit liaison with an unschooled, impoverished tavern maid. Recently, Smith (1985) convincingly demonstrated serious flaws in Goddard's methodology, including the use of unsubstantiated reports and occasional distortion of evidence. Nevertheless, the eugenics movement led to the development of sterilization laws in nearly every state, and the widespread sterilization of MR persons was relatively common at least to the 1950s (Edgerton, 1967).

In the early 1920s, Goddard and some other leading American psychologists of the time extended the poor genetic stock and feeblemindedness arguments to persons of southern or eastern European ancestry, a link that had some influence on the development of more restrictive immigration policies in the United States in the mid-1920s. From about 1905 through the 1920s, a primary theme in the mental retardation literature was protection of society from the MR, particularly the mildly MR. This theme is in sharp contrast to the predominant theme in the late 1800s in which the principal concern was protection of mentally retarded individuals, usually the moderately, severely, and profoundly retarded, from normal environments with which they could not cope without extensive support.

The other major trend during this era was the development and widespread dissemination of the Binet method of assessing intellectual functioning. The story of the Binet method and its dissemination throughout the western world within 10 years is well known and needs not be recounted here. The influence of this method on conceptions of MR as well as on definitional criteria and diagnostic operations was nothing short of dramatic. For at least the 30-year period from 1910 to 1940, the intellectual functioning domain dominated the conceptualization of mental retardation, and the use of IQ tests was central to criteria and diagnostic operations.

In 1921, the first official classification scheme was published by what is now the American Association on Mental Retardation. The levels of moron, imbecile, and idiot were defined by the IQ ranges of 50 to 75, 25 to 50, and less than 25, respectively. Conception, definitional criteria, and diagnostic operations were based heavily on IQ. Furthermore, the underlying construct assessed by the IQ test, most frequently operationalized through application of the Stanford revision of the Binet method (Terman, 1916), was believed to be unitary, fixed, and predetermined by genetic influences (Hunt, 1961). Through the remainder of the 20th century, interpretation of what IQ tests measure gradually moved away from the notion of a unitary entity of intelligence, viewed as a fixed or unchanging characteristic and predetermined by genetic influences. Nevertheless, the notion that mental retardation should be operationalized through the IQ test, with little emphasis on other dimensions, such as social competence, persists to the present (Zigler et al., 1984).

IQ and Immutability 1930 to 1960

A wide variety of terms was used to refer to MR throughout the period 1930 to 1960. Changes in the name of the National Mental Retardation Association reflects the diversity in terminology. The association was established originally in 1876 as the Association of Medical Officers of American Institutions for Idiotic and Feebleminded Persons. From 1906 to 1933, the association was officially named the American Association for the Study of the Feebleminded. After 1934, the title American Association on Mental Deficiency was used, and then it changed to the American Association on Mental Retardation in 1988.

The term *feebleminded* was used as the superordinate category during the first third of the century. Over the same period and continuing through the 1940s and 1950s, feebleminded was sometimes used to refer to the level of MR now referred to as mild MR. Zigler et al. (1984) summarized the diverse terminology used to refer to the mild and more severe levels of MR. Terminology used to refer to the mild level included *moron, feebleminded, endogenous, garden variety, mild, educable,* and *cultural familial.* According to Zigler et al., terminology used to refer to the more severely retarded included *moderate, severe, profound, exogenous, imbecile, idiot, trainable, and subtrainable.* The superordinate terms used to refer to all levels of MR were *feebleminded* in the early part of the century, *mentally deficient* during the middle third of the century, and *mentally retarded,* with the latter being increasingly more dominant over the past 30 years.

The relative effects of environment and a broader conception of MR, incorporating a major emphasis on social competence, was apparent in the late 1930s and 1940s. First, a number of studies were published in the late 1930s, collectively referred to as the Iowa studies, indicating vastly improved outcomes for infants and toddlers earlier judged to be MR through placement in nonorphanage settings and adoptive homes. The results of the Iowa studies were initially regarded as spurious, because the IQ was believed to be fixed, and it was not until the 1960s that the potential of extremely poor environmental circumstances as an etiology of mild mental retardation was widely accepted in the American Association on Mental Deficiency (Sloan & Stevens, 1976). Despite the Iowa studies, the prevailing view in the 1930s and 1940s was that the prognosis for persons properly diagnosed as mentally retarded was permanent, lifelong disability.

Edgar A. Doll's (1934, 1941) conceptions of social competence, definition of mental retardation, and criteria for determining mental retardation were in sharp contrast to the prevailing trends in the United States emphasizing IQ. First, Doll described social competence as a complex construct similar in virtually all ways to the modern conception of adaptive behavior. In later formulations, he used the terms *personal independence* and *social responsibility,* terms used now by both the APA and the AAMR. In 1941, Doll defined mental deficiency as "a state of social incompetence obtaining at maturity, or likely to obtain at maturity, resulting from developmental arrest of constitutional origin (heredity or acquired); the condition is essentially incurable through treatment and unremediable through training" (p. 215). Doll suggested six criteria for determining mental retardation: (a) social incompetence, (b) due to mental subnormality, (c) which has been developmentally arrested, (d) which obtains at maturity, (e) is of constitutional origin, and (f) is essentially incurable. The latter two criteria, constitutional origin and incurability, became critical issues that led to a change in modern conceptions of mental retardation. Although Doll's six criteria were never accorded an official status by any organization, these criteria were highly influential throughout the 1940s and 1950s. Throughout the period, the social incompetence was generally regarded as secondary to or caused by the mental subnormality. There were well-accepted measures of intellectual functioning, (i.e., the IQ test), but measures of social competence were largely unavailable, or they were regarded as far less precise, a circumstance that still prevails in some quarters today (Zigler et al., 1984). Despite Doll's emphasis on social competence in his definition of mental retardation and the widespread acceptance of his criteria, IQ continued to dominate definitional criteria and diagnostic operations throughout the period to the present. This was due in large part to the inadequacy of measures of social competence or adaptive behavior.

The American Psychiatric Association (APA) published four editions of the *Diagnostic and Statistical Manual of Mental Disorders* (1952, 1968, 1980, and 1987). These versions are denoted by DSM-I, DSM-II, DSM-III, and DSM-III-R, respectively. The initial APA scheme published in 1952 used the generic term *mental deficiency* and specified three levels of severity: mild deficiency, moderate deficiency, and severe deficiency. The mild, moderate, and severe levels were associated with the IQ ranges of 70 to 85, 50 to 69, and 0 to 49, respectively. The IQ ranges associated with the three categories in DSM-I were particularly noteworthy in that the categories mild and moderate were each associated with considerably higher IQ ranges than typically had been used or subsequently would be used. The relationship of the APA MR scheme and the AAMD Terminology and Classification Manual during the 1950s is not at all clear. DSM-I used the generic term *mental deficiency*, as did AAMD, but the specification of levels and the use of the higher IQ cutoff score (i.e., up to 85) by APA was neither typical in AAMD schemes nor in the proceedings of the annual meetings of the American Association on Mental Deficiency (Sloan & Stevens, 1976). After 1960, the reciprocal influences of the APA and the AAMD were considerably more obvious, with the latter exerting primary influence on terminology, conceptual bases, definitional criteria, and diagnostic operations. The major reason for the greater influence of the American

Association on Mental Deficiency can be attributed to the success of the 1959–1961 manual in resolving a major conceptual problem with prior MR conceptions (i.e., incurability).

Longitudinal studies of persons classified during the school-age years as MR yielded disturbing results in the late 1940s and 1950s. By the early 1950s, the Nebraska longitudinal studies of persons classified as MR during their school years (Baller, 1936) indicated unequivocally that perhaps one half to two thirds of all mildly MR persons could achieve independent functioning and self-support during their adult years (Charles, 1953). The adult status of these individuals simply was inconsistent with the prevailing notion of the incurability of MR. For a time, this phenomenon was explained as "pseudofeeblemindedness," which implied that the original diagnosis was incorrect or that these individuals exhibited an atypical form of mental retardation not yet described adequately nor well understood. The absurdity of the concept of feeblemindedness was exposed by Benton (1956), who advocated for frank recognition that MR be regarded as a set of symptoms observed at a particular point in time rather than an incurable disease state or immutable condition. It is important to note that the issue of curability of mental retardation was restricted to the mild level, then typically defined as the IQ range of approximately 50 to 70 or 50 to 75.

1960s Conception and Definition

Since 1960, subsequent revisions of the APA DSM have closely followed changes in the AAMR *Manual on Terminology and Classification* (Grossman, 1973, 1977, 1983). These modern conceptions of MR have been virtually identical, with slightly varying degrees of emphasis on intellectual functioning and adaptive behavior. The current AAMR scheme and DSM-III-R are closely related to the AAMR revision published under the leadership of Heber (1959 to 1961). DSM-II, published in 1968, closely paralleled the AAMR scheme published in 1959 to 1961.

The AAMR 1959 to 1961 manual used the term *MR* with the following definition (Heber, 1961):

> Mental retardation refers to subaverage general intellectual functioning and is associated with impairment in one or more of the following: 1) maturation, 2) learning, and 3) social adjustment.
> In 1961 the definition was revised slightly to read that mental retardation refers to subaverage general intellectual functioning which originates during the developmental period and is associated with impairment in adaptive behavior. (p. 3)

The pervasive impact of this definition can be attributed to the development of the concept of adaptive behavior, the levels of mental retardation identified, and the evolving assumptions about developmental status, prognosis, and etiology.

The 1959–1961 AAMR classification scheme identified the behavioral dimensions of general intellectual functioning and adaptive behavior as the conceptual basis for mental retardation. Clearly, general intellectual functioning was the dominant dimension, with lesser emphasis on adaptive behavior. General intellectual functioning was operationalized primarily through the performance on an individually administered test of general ability, typically the Stanford-Binet or one of the Wechsler scales. Adaptive behavior was not clearly operationalized through this classification scheme, but the areas of maturation for preschoolers, learning for school-age children and youth, and social adjustment for adults were identified as the critical indicators of adaptive behavior. For school-age children and youth, the group to whom the diagnosis of MR is most frequently applied, low achievement as assessed by standardized measures of achievement along with referral for academic difficulties was sufficient to constitute a deficit in adaptive behavior. Professionals involved with mild MR diagnosis, typically psychologists, usually regarded referral along with teacher reports of poor academic performance as sufficient to establish the impairment of adaptive behavior. Performance on an IQ test was then studied to determine whether or not this impairment in adaptive behavior could be properly diagnosed as MR.

Five levels of MR, each defined in terms of standard deviation units, were specified in the AAMD and APA classification schemes of the 1960s. These five levels were denoted by the adjectives borderline MR, mild MR, moderate MR, severe MR, and profound MR, with the standard deviation ranges of -1 to -2, -2 to -3, -3 to -4, and greater than five standard deviations below the population mean on intellectual performance, respectively. The most controversial aspect of the 1960s classification schemes was the borderline category, defined by the IQ range of 70 to 85. Inclusion of the borderline range produced an enormously expanded proportion of persons potentially eligible for the diagnosis of MR. In comparison to the traditional cutoff scores of 70 or 75, a cutoff score of 85 renders some 16% of the population potentially eligible for the diagnosis of mental retardation rather than the traditional expectation of 2% to 3%.

The 1960s MR conceptualizations were developmentally based. Both AAMR and APA required the appearance of MR during the developmental period, defined roughly as birth to maturity. The developmental requirement excluded the diagnosis of MR with persons who later acquired behavioral patterns consistent with MR, usually secondary to other injuries or disease states (e.g., severe head injury, senile dementia).

The 1960s classification schemes focused entirely on current functioning, thereby avoiding the issues of etiology and prognosis. The 1959–1961 AAMR scheme explicitly recognized that an individual could be correctly diagnosed as MR at one point in the developmental period but function normally at a later time. The emphasis on current functioning also served as a rejection of earlier conceptions of intelligence as fixed or unitary.

The 1959–1961 AAMR scheme, as well as DSM-II, completely ignored the issue of etiology in the formulation of conceptual bases, definitional criteria, and diagnostic operation. Both noted that the etiology of MR could be either organic or environmental. Neither required that it be constitutional, either organic or acquired, a problematic provision of earlier definitions. The 1959–1961 AAMR scheme also recognized MR as a *symptom* rather than a condition or disease entity. This provision clearly permitted dual diagnosis whereby the individual's current functioning was emphasized rather than whether subaverage intellectual performance and adaptive behavior impairments were primary or secondary to other disorders such as childhood schizophrenia or neurological dysfunctions.

Finally, the 1960s classification schemes emphasized the borderline and mild levels of MR, in contrast to earlier formulations that placed primary emphasis on the moderate, severe, and profound levels of MR. Clearly, the emphasis was on primary and secondary prevention rather than control of symptomatic behavior in protected environments.

1970s Conceptualization, Definition, and Diagnosis

The 1970s classifications (Grossman, 1973, 1977; DSM-III, 1980) retained all of the salient features of the AAMD 1959–1961 classification scheme with subtle but significant changes in emphasis. First, the definition was changed (Grossman, 1973) to "mental retardation refers to *significantly* subaverage general intellectual functioning *existing concurrently* with deficits in adaptive behavior, and manifested during the developmental period" (p. 5). The word "significantly" to modify subaverage and the phrase "existing concurrently" were subtle but highly significant changes. Significantly subaverage, general intellectual functioning was operationally defined as an IQ of two or more standard deviations below the mean, thus eliminating the borderline mental retardation category from the earlier definition and reestablishing a more traditional cutoff score of IQ = 70 or below. This change reduced the population potentially eligible for the diagnosis of mental retardation from 16% to about 2% to 2.5%.

The change in emphasis regarding adaptive behavior was even more subtle but no less significant. The phrases linking adaptive behavior to general intellectual functioning were changed from "associated" to "existing concurrently." The latter clearly implies adaptive behavior as equal in importance to the general intellectual functioning dimension. Adaptive behavior in 1959–1961 was defined (Heber, 1961) as "the effectiveness of the individual in adapting to the natural and social demands of the environment" (p. 3). In 1959–1961, adaptive behavior during the school-age years was closely linked to school achievement as defined by performance on standardized measures of achievement as well as other indices. In the 1970s revisions, adaptive behavior was defined as the degree to which the individual "meets that standard independence and social responsibility expected of his age and cultural group." The criteria for adaptive behavior during the school-age years were broadened (Grossman, 1973) to include "application of basic academic skills in daily life activities and application of appropriate reasoning and judgment in mastery of the environment and social skills (participation in group activities and interpersonal relationships)" (p. 12). Grossman (1973) went on to explain that assessment of adaptive behavior during the school age years "should focus not only on the basic academic skills and their use, but also on skills essential to cope with the environment" (p. 12).

The 1970s revisions retained the same characteristics with regard to the developmental emphasis, current functioning, levels of mental retardation defined by standard deviation discrepancies (with of course the deletion of the borderline mental retardation category), etiology, and mental retardation as a disorder that can co-occur with other diagnoses. There was less emphasis in the 1970s definition on the mild level of MR.

Although neither an organic nor an environmental etiology was required for the diagnosis of mental retardation, the AAMD scheme in the 1970s (Grossman, 1973, 1977) incorporated the medical classification system developed by the International Classification of Diseases (ICD) eighth revision (World Health Organization, 1968) and the American Psychiatric Association (1968). Etiology of mental retardation was subdivided into the categories of: infection and intoxication, trauma or physical agent, metabolism or nutrition, gross brain disease, unknown diseases and conditions due to prenatal influence, chromosomal abnormality, gestational disorders, secondary to psychiatric disorder, environmental influences, and other conditions. The environmental influences category in this scheme was further delineated as psychosocial disadvantage or sensory deprivation. The criteria for the etiology of psychosocial disadvantage were evidence of retarded intellectual functioning within the immediate family, no indication of cerebral pathology or other biological basis for retarded performance, and impoverished background. The psychosocial disadvantaged etiology was regarded as applicable to the vast majority of mild MR.

The 1970s AAMR versions were influenced by the World Health Organization and the APA procedures for classification of etiology. However, the basic definition and the changes in definitional criteria for general intellectual functioning and adaptive behavior were formulated by AAMR. The 1980 revision of DSM (i.e., DSM-III) followed the AAMR changes in definitional criteria. The current (i.e., the 1980s) versions of AAMR and DSM also reflect this trend of AAMR determining the critical features of conceptualization, definitional criteria, and diagnostic operations. The current versions of AAMR and DSM are discussed in the following section.

CURRENT MR SCHEMES: DEFINITIONAL ISSUES AND CRITERIA

Highly similar MR classification systems were published in the 1980s by the AAMR and the APA. The strong influence by the AAMR classification system on the subsequent revisions of the DSM in 1980 and 1987 are clearly apparent from examination of both sets of documents. DSM-III-R and AAMR (Grossman, 1983) provide nearly identical definitions of mental retardation. The AAMR definition (Grossman, 1983) reads "mental retardation refers to significantly subaverage general intellectual functioning existing concurrently with deficits in adaptive behavior and manifested during the developmental period" (p. 1). The DSM-III-R definition is different only in the substitution of the phrases "accompanied by" for "existing concurrently with" and "onset before the age of 18" for "manifested during the developmental period" (DSM-III-R, p. 28). Both defined general intellectual functioning in terms of the results of one or more individually administered general intelligence tests. Significantly subaverage is operationalized in both definitions as an IQ of approximately 70 or below with both suggesting flexibility in interpretation of the IQ criterion such that persons with IQs up to 75 could be classified as mentally retarded, if significant deficits in adaptive functioning are attributed to intellectual performance. Both schemes define adaptive behavior in terms of personal independence and social responsibility expected for age and cultural group. Both define the developmental period as birth to the 18th birthday. An analysis of the two mental retardation classification systems in terms of conceptual bases, definitional criteria, and diagnostic operations is provided in Table 2.1 and is discussed following.

A major difference between the AAMR and DSM-III-R classification systems not represented well in Table 2.1 is the length and discussion of critical elements. The AAMR manual, *Classification in Mental Retardation* (Grossman, 1983), is approximately 150 pages, all devoted to MR. In contrast, the DSM-III-R section on MR is approximately six pages. On nearly all issues, the AAMR classification system should be viewed as the more authoritative source. The two organizations, AAMR and APA, have obviously worked together closely in formulating the MR classification system. Evidence of this close cooperation is found also in the committee membership. Two members of the AAMR committee responsible for the 1983 revision, Herb Grossman, chairman, and Dennis Cantwell, were members of the seven-person APA subcommittee on MR.

General Intellectual Functioning

Both classification systems view general intellectual functioning as determinative in MR diagnostic decisions (see Table 2.1). Both require the administration of individual tests of general intellectual functioning such as one of the Wechsler scales, the Stanford-Binet, the Kaufman Assessment Battery for Children, or the McCarthy Scale of Children's Abilities. Both express concerns about the interpretation of measures of general intellectual functioning during the preschool years, particularly during infancy. Both regard intelligence as determinative with IQ performance above 75 ruling out the diagnosis of mental retardation. The most appropriate measures of general intellectual functioning are summarized in Table 2.2.

There are two potential problems with the MR classification systems regarding general intellectual functioning. First, flexible interpretation of the IQ cutoff to allow the MR diagnosis with IQs up to 75 markedly increases the proportion of person potentially eligible. As indicated in Table 2.3, there are twice as many persons who meet the IQ cutoff score with the decision rule of 75 and below rather than the decision rule of 70 and below. Furthermore, there are more persons in the IQ range of 71 to 75 than in the entire range normally associated with mild MR (i.e., about IQ 56 to 70). However, studies of the actual numbers of persons classified as mentally retarded, even using a flexible standard such as IQ below 75, have always reported substantially fewer persons actually diagnosed as MR than the number eligible using the IQ criterion alone. Clearly, referral processes mitigate at least some of the effects of the higher IQ standards. For a variety of other reasons, discussed in the

TABLE 2.1
Analysis of APA DMS–III–R and AAMR MR Classification Systems for MR: Definitional Criteria and Diagnostic Operations

Conceptual Basis	*Definition*	*Relationship Weighting In Diagnosis*	*Assessment*	*Decision Rules*
Intelligence dimension	No — DSM Extensive discussion — AAMR	DSM & AAMR Determinative and equal to adaptive behavior	DSM & AAMR Admin. of individual test of general intellectual functioning	DSM & AAMR IQ approximately 70 up to 75
Adaptive behavior dimension	General definitions — DSM General definition and indicators by age — AAMR	DSM & AAMR Determinative and equal to intelligence	DSM & AAMR Recommend cautious use of adaptive inventories and clinical judgment for overall performance	DSM & AAMR Below age and cultural group expectations.
Developmental	DSM & AAMR Yes	DSM & AAMR Determinative	DSM & AAMR Chronological age	DSM & AAMR Birth to 18
Levels	DSM & AAMR Yes Mild Moderate Severe Profound	DSM & AAMR Implicit assumption of differences in degree—Some recognition of different prognosis and etiology by level.	IQ ranges — DSM IQ ranges and adaptive behavior descriptions for levels — AAMR	DSM & AAMR Mild=50–55 to approx. 70 Moderate=35–40 to 50–55 Severe=20–25 to 35–40 Profound ≤20 or 25 No rules for adaptive behavior.

Etiology	DSM & AAMR Biologic and psychosocial	DSM & AAMR Etiology not related to diagnosis	DSM & AAMR Both use the ICD–9 medical codes with AAMR providing more detail. Medical evaluation	DSM & AAMR Medical evaluation
Prognosis	DSM & AAMR Variable from custodial care to adequate adult functioning.	DSM & AAMR Prognosis not related to MR diagnosis	DSM & AAMR Not specified	DSM & AAMR None
Subtypes	Not specified DSM -------- ICD–9 medical codes AAMR	DSM & AAMR Subtypes not related to MR diagnosis	DSM & AAMR Medical evaluation	DSM & AAMR Medical evaluation
Sociocultural	None DSM -------- Discussion of sociocultural factors AAMR	DSM & AAMR Applied to adaptive behavior	Not discussed DSM -------- Very general guidelines provided AAMR	DSM & AAMR None
Overlap: dual diagnosis	Yes— other categories suggested and defined DSM -------- Yes—little treatment of other categories AAMR	DSM & AAMR Independent, specify each diagnosis as appropriate	Dependent on other categories DSM -------- Not specified AAMR	Dependent on other categories DSM -------- Not specified AAMR

TABLE 2.2
Comprehensive Individual Tests of General Intellectual Functioning[1]

Scale	*Age Range*	[2]*Lowest Score*	*Useful Scores for MR Diagnosis*
Bayley Scales of Mental Development (Bayley, 1969)	2 mo. to 2 yrs., 6 mo.	53	Mental development index, Psychomotor development index
McCarthy Scales of Childrens Abilities (McCarthy, 1972)	2 yrs., 6 mo. to 8 yrs., 6 mo.	53	General Cognitive Index
Kaufman Assessment Battery for Children (Kaufman & Kaufman, 1983)	2 yrs., 6 mo. to 12 yrs., 5 mo.	Varies by age; from 40–71	Mental processing composite
Stanford-Binet Fourth Edition (Thorndike, Hagen, & Sattler, 1986)	2 yrs to 23 yrs.	40	Composite score; cautious use of verbal reasoning and abstract-visual reasoning
Wechsler Preschool and Primary Scale of Intelligence (Wechsler, 1967)	4 yrs. to 6 yrs., 6 mo.	45	Full scale IQ; cautious use of verbal IQ, and performance IQ
Wechsler Intelligence Scale for Children Revised (Wechsler, 1974)	6 yrs. to 16 yrs., 11 mo.	40	Full scale IQ; cautious use of verbal IQ and performance IQ
Wechsler Adult Intelligence Scale Revised (Wechsler, 1981)	16 yrs. to 74 yrs.		Full scale IQ; cautious use of verbal IQ and performance IQ

[1]See Sattler (1988) for detailed reviews.
[2]The lowest score available using a standard score scale with the mean = 100 and the standard deviation = 15.

next section, use of the flexible IQ decision rule is critical to the identification of persons needing services in various settings. The potential impact on prevalence needs to be considered in application of flexible decision rules. Table 2.3 shows the proportion of persons who would qualify for an MR diagnosis using different IQ decision rules.

A second and more serious problem with the intellectual functioning dimension is its application to determination of levels of MR. The lowest IQ reported on virtually all recently standardized, individually administered tests of general intellectual functioning is in the range of 40 to 50. The Stanford-Binet Fourth Edition (Thorndike, Hagen, & Sat-

TABLE 2.3
Proportion of Persons Potentially Eligible Using Different IQ Decision Rules

IQ	*Normal Curve Percent*	
Below 70	2.28	
70 and Below	2.68	
Below 75	4.75	
75 and Below	5.48	
Interval	*Percent Within*	
56–60	0.30	2.51%
61–65	0.69	
66–70	1.52	
71–75	2.80	

tler, 1986) provides IQ scores down to 36 using a score scale with a mean of 100 and a standard deviation of 16. The score of 36 is equivalent to an IQ score of 40 on the Wechsler scales, where the mean is 100, and the standard deviation is 15. The norms for these tests typically are based on standardization samples of approximately 2,000 to 3,000 persons, often varying considerably in age. Thus, at a particular age level, something on the order of 200 to 400 persons may actually have been administered the test during standardization. Determination of lower scores from these tests is impossible without the use of extensive extrapolation, and that may be impossible at the lower age range for most tests. This problem was acknowledged by the authors of the Stanford-Binet. Thorndike et al. (1986) wrote: "Especially at the extremes, values had to be extrapolated beyond the available data. Even with samples of 400 or 500 cases, there were few cases that fell more than two standard deviations from the mean and almost none that went beyond three standard deviations" (p. 29). IQ scores below 55 on modern tests have a tenuous psychometric foundation. The IQs below 55 necessary to operationalize the AAMR or APA criteria for MR levels are more illusion than reality! These problems are exacerbated at the preschool age levels by the tentativeness of intellectual assessment with infants and toddlers (Sattler, 1988).

The use of IQ as one of three determinative factors in the diagnosis of MR is appropriate, with the usual cautions concerning interpretation of IQ tests and the tentativeness of IQ measurement at younger ages. The use of IQ to define the levels of MR is appropriate for the mild level but highly questionable for the more severe levels of MR. Both classification systems appear to recognize that IQ and adaptive behavior become increasingly synonymous at more severe levels of mental retardation (Meyers, Nihira, & Zetlin, 1979). Therefore, replacement of IQ with adaptive behavior information to define the lower three levels of MR would be appropriate, if adequate measures of adaptive behavior are available.

Adaptive Behavior

A deficit in adaptive behavior is the second determinative feature of the MR diagnosis in the AAMR and APA classification systems (see Table 2.1). The language linking adaptive behavior to general intellectual functioning in AAMR (i.e., "existing concurrently") implies a slightly stronger relationship to the MR diagnosis than the APA language (i.e., "accompanied by"). However, concurrent deficits in both general intellectual functioning and adaptive behavior are clearly specified in both systems. Both classifications also provide general definitions of adaptive behavior involving the concepts of personal independence and social responsibility according to expectations for age and cultural group.

The AAMR classification system provides extensive discussion of the concept of adaptive behavior, assessment procedures, and the variety of considerations needed in reaching a clinical judgment of adaptive behavior status. The AAMR also provides indicators of adaptive behavior expectations by age of the individual. During infancy and early childhood, the four indicators are sensory motor skills development, communication skills (including speech and language), self-help skills, and socialization (i.e., development of the ability to interact with others). During childhood and early adolescence, three additional indicators are suggested: application of basic academic skills in daily life activities, application of appropriate reasoning and judgment in mastery of the environment, and social skills (i.e., participation in group activities and interpersonal relationships). During late adolescence and early adult life, the prior seven indicators are applicable with one additional criterion, vocational performance and social responsibilities. The AAMR indicators provide some guidance to diagnosticians, but because adaptive behavior is a determinative component, further guidance needs to be provided in the classification systems concerning domains of adaptive behavior, measurement, and decision rules. In the absence of further specification, it is likely that the MR diagnosis will continue to be determined to a greater extent by current intellectual functioning than adaptive behavior. This outcome is clearly not consistent with the intent of either classification system.

Developmental

Both classification systems regard developmental characteristics of the individual as determinative in the diagnosis of mental retardation. The developmental status is operationally defined as the age range of birth to age 18. No particular problems appear to exist with this criterion, and further specification appears to be unnecessary.

Levels

Both the AAMR and APA classification systems recognize four levels of mental retardation, mild defined by the IQ range of 50–55 to approximately 70; moderate defined by the IQ range of 35–40 to 50–55; severe defined by the IQ range of 20–25 to 35–40; and profound involving the IQ range of less than 20 or 25 (see Table 2.1). In the DSM-III-R, the levels are defined in terms of IQ ranges with no further specification of adaptive behavior status. The AAMR system provides description of the upper levels of adaptive behavior associated with each of the levels, but little guidance is provided concerning the assessment of adaptive behavior, and no decision rules are established to guide assignment of individuals to the varying levels.

The problems with defining the levels by performance on IQ tests were discussed in a prior subsection. Most currently available IQ tests do not provide scores at the lower ranges, particularly at the severe and profound levels. IQ and adaptive behavior are virtually synonymous at these lower levels; thus, the levels would be defined more clearly through further specification of critical adaptive behavior milestones rather than IQ scores. Furthermore, the quality of adaptive behavior measures has improved considerably during the 1980s, thus, making it feasible to change the MR classification system's relative emphasis on IQ and adaptive behavior in constituting the levels of MR.

The second problem with the current specification of levels is the implicit assumption that the differences in mental retardation are primarily a matter of degree rather than kind. In fact, as noted in the following section, mild MR is qualitatively different from the more severe levels. Further effort is needed to improve communication about the fundamental nature of MR at the mild versus more severe levels.

Discussion later in the chapter will address the issues related to levels and types of MR.

Etiology

Both classification systems recognize two broad classes of etiology: biologic and psychosocial. DSM-III describes five classes of etiological influences including hereditary factors, early alterations of embryonic development, prenatal and perinatal problems, physical disorders, and environmental influences in mental disorders. Both recognize that precise etiology is impossible to determine in a substantial proportion, perhaps 30% to 40%, of cases. AAMR provides an extensive system, adapted from ICD, for statistical reporting of etiological factors. However, determination of etiology is not a determinative requirement for the diagnosis of mental retardation in either system (see Table 2.1). Establishment of the requirement that etiology be determined in all MR diagnoses is impractical given the present state of knowledge. However, etiology should be part of the basis for distinguishing mild from more severe levels of MR (see later discussion).

Prognosis

Neither the DSM-III-R or AAMR classification systems attach a particular prognostic outcome to the diagnosis of mental retardation (see Table 2.1). Both thereby avoid the problems with pseudofeeblemindedness, a major concern with MR classification systems prior to 1960. Both diagnostic systems recognize that prognosis in MR is highly variable, dependent on level of mental retardation and the effectiveness of treatment, as well as the context within which the individual must cope. Both note the relatively optimistic prognosis for most persons with mild MR, which is in sharp contrast to the typical prognosis for persons with moderate, severe, and profound levels of mental retardation. The latter virtually always require some degree of protective services throughout the lifespan. Despite the markedly different typical etiologies and prognoses for the mild versus severe levels of MR, the same diagnosis, MR, remains attached to all levels in the current classification systems. Later discussion will address separation of the mild from the other levels of MR.

Subtypes

The APA classification system is largely silent on the issue of specific subtypes. Both systems recognize levels and use the ICD medical codes to delineate further the forms of mental retardation. These medical codes are largely related to etiological conditions, typically specific syndromes such as Down's syndrome or Tay Sac's disease. The subtypes are not related to the MR diagnosis, and medical evaluations are typically used for assessment and decision rules. Some MR diagnoses are known subtypes that are related to etiology and prognosis (e.g., Tay Sac's), but many are not. Specification of subtype for all MR diagnosis is not realistic.

Sociocultural

Although APA recognizes sociocultural factors as possible etiologies of mental retardation, there is no further discussion of the nature of these factors or their use in MR diagnosis (see Table 2.1). AAMR provides an extensive discussion of the sociocultural factors that are, by definition, to be applied to the determination of deficits in adaptive behavior. However, the guidelines provided by AAMR regarding assessment of sociocultural factors are very general, and no decision rules are suggested. Sociocultural context is an important influence on expectations for behavior as well as the opportunities to acquire intellectual and adaptive behavior competencies. Consideration of sociocultural factors is particularly germane to the mild level of MR, where there is less likely to be clear biological etiologies or pronounced deficits in adaptive behavior and the high probability of performance on intellectual measures influenced by the quality of environmental influences. Both schemes need to provide further direction concerning sociocultural factors, a topic discussed in a later section.

Overlap: Dual Diagnosis

Both diagnostic systems recognize that MR can occur in conjunction with other disorders (see Table 2.1). Each suggests that MR should be diagnosed on the basis of the current functioning of the individual and that existence of the other disorders should be diagnosed concurrently using the appropriate criteria for these disorders. The presence or absence of other disorders does not affect the appropriateness of the diagnosis of MR. The assessment and decision rules for the other disorders are specified in DSM-III-R but largely ignored in AAMR.

Summary

The American Psychiatric Association's DSM-III-R (APA, 1987) and the American Association on Mental Retardation's manual, *Classification in Mental Retardation* (Grossman, 1983) provide comprehensive classification systems for the MR diagnosis. The AAMR classification system is more detailed and in some cases provides greater precision. There has been an entirely laudable effort to achieve a high degree of consistency between the two systems. Both systems resolve the problems associated with etiology and prognosis that caused struggles in the 1950s. The resolution, essentially ignoring both in the diagnosis, is not entirely satisfactory due to substantial differences between the mild and more severe levels of MR. Both systems establish three determinative standards for the MR diagnosis, general intellectual functioning, adaptive behavior, and developmental period. Both reflect consideration of other factors, none of which are determinative, such as etiology, prognosis, subtypes, and degrees of impairment. Recent research in several areas, detailed in the next section, provides the basis for suggestions regarding improvement of the mental retardation classification systems.

REVIEW OF THE LITERATURE

Research in the 1970s and 1980s has established a number of disturbing trends concerning the importance of unrecognized system factors, special problems associated with sociocultural status and the domination of the intellectual dimension in MR diagnosis. Furthermore, significant advances have occurred in the understanding of the construct of adaptive behavior and, most importantly, in the technical adequacy of available adaptive behavior inventories. These trends and developments are especially pertinent to the conceptualization, definitional criteria, and diagnostic operations for the mild level of MR. Changes in MR classification systems need to be accomplished in order to reflect the present state of knowledge.

Systems Factors

All diagnoses of mental disorders are influenced markedly by the overall context within which the individual's behavior is judged and evaluated. This general principle is especially apparent with mild MR but pertinent to all MR levels. For example, advances in treatment procedures have reduced, to some unknown extent, the number of persons performing at the profound and severe levels of MR. Moreover, some moderately retarded persons, such as persons with Down's syndrome, through effective treatment ultimately perform at the lower end of the mild MR range.

Mild MR is largely a school phenomenon (MacMillan, Meyers, & Morrison, 1980). The vast majority of persons classified as mildly MR receive that diagnosis during the school-age years of approximately 6 to 18. Prevalence data in mental retardation reflect this phenomenon. Prevalence gradually increases through the school-age years, reaching a peak at approximately 14 to 15, then slowly declines thereafter (MacMillan, 1982; Reschly & Jipson, 1976; Robinson & Robinson, 1976).

Complex factors determine the likelihood of identification of persons as mildly MR in school settings. MacMillan et al. (1980) described variations concerning teachers' tolerance for deviant behaviors as well as the degree of available support for low-achieving students in regular classroom situations. Students with identical characteristics in regard to eligibility for MR diagnosis may or may not be referred. After referral, there are variations within systems concerning screening of referrals and implementation of classification criteria through diagnostic operations. For example, use of formal measures of adaptive behavior appear to reduce the numbers of persons classified as MR compared to using IQ tests alone. Finally there are differences in placement decisions depending on the nature of State Department of Education classification/placement criteria and the availability of programs (Patrick & Reschly, 1982).

In addition to the variations MacMillan et al. described, other system variations, such as the mean IQ and achievement levels in a particular school, markedly influence likelihood of referral. For example, Reschly and Jipson (1976) reported MR prevalence data for racial-ethnic groups. According to IQ criteria and school achievement, approximately 35% of all native American Indian students in their sample would have met the criteria for mild MR. However, these students were located on an isolated reservation where average achievement levels were unfortunately low. The *actual* prevalence of about two percent (i.e., students classified and placed in programs for the MR) was not substantially different from the other settings in which this research was conducted.

Research during the 1970s and 1980s also revealed substantial variations *within* states over time. Several articles have described the changes in the population of students classified as mildly MR in California as a result of the change in the IQ cutoff score from 79 to 70 and the associated impact of the *Larry P. v. Riles* (1979) court decision banning certain uses of IQ tests (Forness, 1985; MacMillan & Meyers, 1980; Reschly, 1986). Prior to the change in the IQ criterion from 79 to 70, some 50,000 California students were in educational programs for the mildly MR. By the time of the *Larry P.* court decision, this number was reduced to under 20,000 through the change in the IQ criterion. Further reductions have occurred at least in part due to the impact of the Larry P. decision. MacMillan (1988) presented convincing arguments establishing that the underlying problems in cognitive functioning, school achievement, and adaptive behavior exhibited by children and youth formerly classified as mildly MR in California have not been resolved and, for the most part, are now being ignored due to changes in systems factors.

Mild MR is largely a school phenomenon, a fact recognized in both DSM-III-R and AAMR; therefore, the legal criteria used to define MR by State Departments of Education through special education rules exert vast influence on MR prevalence. All states provide definitions of handicapping conditions including MR. These definitions differ from the more official classification systems such as DSM-III-R and AAMR to varying degrees (Patrick & Reschly, 1982). There are enormous variations between states and prevalence of mental retardation. The variations range from a low of about .5% to a high of approximately 4.8% (MacMillan, 1988; Patrick & Reschly, 1982). Intuitively, one would expect these variations to be related to the stringency of the definitional criteria and diagnostic operations. In fact, Patrick and Reschly (1982) reported no correlation between the IQ cutoff score and prevalence. Prevalence was predicted by the state's mean education level and mean per capita income. State Department of Education MR criteria have been changed during the 1980s. Caution should be exercised in generalizing these findings to the early 1990. Nevertheless, these results clearly indicate the effects of sociocultural and system factors on the diagnosis of MR.

The prevalence of MR in school settings has been declining over the past 15 years. Each state is required by the Education of the Handicapped Act (EHA, 1975) to report the number of students served in each of 11 handicapping conditions including MR. Although there are vast differences in reported MR prevalence among the states, the overall national prevalence has declined by 33% since 1976. This decline in prevalence is especially startling in view of the fact that many students with moderate, severe, and profound levels of MR were first admitted to public school programs in the late 1970s as a result of the EHA requirements. The most recent data on national MR prevalence suggest an overall rate of approximately 1.5% and a range of .5% to 4.8% (Tenth Annual Report to Congress, 1988). Conventional estimates suggest that approximately .3% to .5% of the general population is properly diagnosed as moderately, severely, or profoundly retarded. These recent data suggest that, in the school-age population, the prevalence of mild MR nationally is about 1%, a rate substantially below traditional estimates in authoritative sources (Grossman, 1983; MacMillan, 1982; Robinson & Robinson, 1976). This astonishing decline in MR prevalence is best explained by special problems with the diagnostic construct of mild MR.

The Special Problem of Mild Mental Retardation

Momentous events have occurred in the legal arena concerning mild MR. Four federal court trials occurred between 1979 and 1986 concerning whether overrepresentation of Black students in special education programs for the mildly MR constituted discrimination in violation of the EHA and the equal protection clause in the 14th Amendment to the United States Constitution. Three of the four cases were decided in favor of the defendants, local school districts, or State Departments of Education (Reschly, Kicklighter, & McKee, 1988a, 1988b, 1988c). A fourth highly influential case, although a minority outcome in the decisions, banned the use of IQ tests with Black students (*Larry P. v. Riles*, 1979, 1984, 1986). Although three of the four federal district court decisions upheld traditional MR conceptualizations, definitional criteria, and diagnostic operations despite disproportionate impact on Black students, the litigation had a decidedly chilling effect on implementation of the MR diagnostic construct in school settings.

Analysis of the *Larry P.* court decision as well as the documents and testimony presented in court by plaintiffs in the other cases revealed a number of implicit issues and underlying assumptions (Reschly et al., 1988b, 1988c). First, the actual overrepresentation of Black students in programs for the mildly MR was often exaggerated and misundertood. In fact, the actual overrepresentation rate, although usually two to three times the prevalence for nonminority students, typically involved percentages in the range of 2% to 4% of all Black students classified and placed in school programs for the mildly MR vs. 1% to 2% for White students. Second, plaintiff's allegations typically reflected significant misconceptions about the construct of MR. Plaintiffs clearly regarded "true" MR as requiring biological anomaly, comprehensive incompetence, and permanent disability. The fact that Black students classified as mildly MR did not exhibit these characteristics was used by plaintiffs to argue that the disproportionate representation was caused by biases in IQ tests and other infirmities in definitional criteria and diagnostic operations. Judge Peckhan in *Larry P.* found these arguments persuasive, leading to his ultimate conclusion that the named plaintiffs were *not* mentally retarded despite IQ scores that met the criterion established by AAMR and DSM-III-R and severe difficulties in learning manifested over their entire school careers. Although federal court judges in Illinois, Georgia, and Florida were not persuaded by plaintiffs' arguments, the widespread misunderstanding of the meaning of mild MR continues to frustrate psychologists and educators responsible for implementing the MR diagnostic construct. Other implicit assumptions and underlying issues, although less pertinent to this discussion, were critical in these cases, including the effort to reject Jensen's interpretation of group differences in intellectual functioning (Jensen, 1969), the presumed ineffectiveness of special education for the mildly retarded (Dunn, 1968; Mercer, 1973), the alleged exclusive role of IQ tests in classification (ignoring the influence of prior educational achievement and referral) (Mercer, 1979), the limited conceptions of bias in tests, (Hilliard, 1983; Reschly, 1986) and, most importantly, the deleterious effects of labels (Jones & Wilderson, 1976; Mercer, 1973).

Research on labeling effects has declined in the 1980s. Nevertheless, assumptions about the deleterious effect of labels, such as the MR diagnosis, have continued to be widespread throughout the social sciences. MacMillan, Aloia, and Jones (1974) provided the best treatment of the research on the effects of formal classification, particularly mild MR. Most of the assumptions about the deleterious effects of labels are not substantiated by empirical research. Two generalizations about the MR label can be identified confidently. First, persons bearing the label, particularly the mildly mentally retarded, find the label demeaning and noxious (Edgerton, 1967; Mercer, 1973, 1979). Second, many people who come into daily contact with the mildly MR misunderstand the meaning of the MR diagnostic construct. The prevailing understanding of MR is that it means biological anomaly, comprehensive incompetence, and permanent disability. These two "findings" about MR raise serious questions about the meaning of this diagnostic construct as presently conceptualized. Cromwell et al. (1975) pointed to two meanings of diagnostic constructs, the formal meaning established by authoritative sources such as AAMR and APA, and the informal but perhaps more influential meanings harbored by other professionals and the general citizenry. Clearly, there is a substantial discrepancy between the two meanings of the construct of mild MR.

Table 2.4 summarizes the substantial contrast between mild versus the more severe levels of MR. The overwhelming proportion of persons classified or eligible to be classified as mildly MR using the DSM-III-R or AAMR classification systems display no identifiable biological anomaly. Moreover, their deficits in adaptive behavior are most often observed in school settings or in the context of tasks requiring cognitive competencies with written language or mathematical concepts (Koegel & Edgerton, 1984). Finally, persons classified as mildly MR using the AAMR or APA classification systems typically perform at adequate levels during the adult years. The adult adjustment may be marginal,

TABLE 2.4
Mild MR Contrasted with Moderate, Severe, and Profound MR

	Level of MR	
Qualitative Dimensions	*Mild MR*	*Moderate Severe Profound MR*
Identifiable biological anomaly Physical stigmata	No	Yes
Comprehensive incompetence (All roles and settings)	No	Yes
Permanent disability (Need for protective environment)	No	Yes

and the older notion that the mildly MR simply disappeared into the general adult population is not consistent with the longitudinal data (Baller, Charles, & Miller, 1967; Koegel & Edgerton, 1984; Zetlin, 1988). However, most meet the criteria of self-support and independent functioning despite their cognitive limitations, rendering them ineligible for the MR diagnosis as adults according to the AAMR criteria. Persons with moderate, severe, and profound levels of MR represent a stark contrast on these three critical attributes. Virtually all persons with MR at these levels display biological anomalies, exhibit comprehensive incompetence, and require lifelong protective services (MacMillan, 1982). The critical problem in the MR diagnostic construct is that the characteristics of the more severely MR, constituting a relatively small percent of the population that should be identified as retarded according to the AAMR and APA classification systems, are widely and irrevocably associated with the mildly MR. These misconceptions are an obvious disservice to the mildly MR and a fatal flaw in the MR diagnostic construct established through the authoritative systems. An obvious although complex solution is to change the classification schemes so that the mild level is clearly distinguished from the lower levels (see later discussion).

Domination of IQ

Despite admonitions to the contrary in the AAMR and APA classification systems, most diagnoses of MR are dominated by the results of IQ tests. We have to look no further than research on persons with MR published in the flagship journal, the *American Journal of Mental Re-*

tardation (AJMR), for confirmation of this trend (Smith & Polloway, 1979), a trend that has continued through the 1989 issue of *AJMR*. Research on MR diagnosis in the school context suggests that there has been increasing use of adaptive behavior as part of the basic diagnosis during the 1980s (Reschly, Robinson, Volmer, & Wilson, 1988). One could argue, as did the AAMR, (Grossman, 1983), that "adaptive behavior appraisal is informally, if unintentionally, employed when authorities or parents request a clinical assessment to determine whether the child is mentally retarded" (p. 42). This method of appraising adaptive behavior, although common, accompanied by the universal use of individually administered measures of general intellectual functioning as part of the MR diagnosis, suggests that IQ has the dominant role. This dominant role is inconsistent with the MR classification systems as well as the concerns expressed in federal courts about biases in conventional tests and IQ-dominated decision making.

IQ Scales and Levels of Mental Retardation

The use of IQ scores as the principal basis for defining the levels of MR becomes increasingly untenable at progressively lower levels of functioning. Modern IQ tests, with the exception of the now outdated 1972 norms for the Stanford-Binet, now rendered obsolete through the publication of the fourth edition of the Stanford-Binet, rarely report scores below 40–50. Yet the moderate, severe, and profound levels of MR are defined by IQ ranges. Psychometrically, there have always been questions about whether scores at the very low ranges were sound. The factor structure and content of intelligence or adaptive behavior instruments attempting to assess performance at very low levels are virtually identical. The current measurement technology for general intellectual functioning as well as research on the constructs of intelligence and adaptive behavior suggest that the latter should be used as the primary basis for differentiating levels of MR. The levels are useful; current operational criteria-based IQ are flawed.

Sociocultural Factors

Recognition of the sociocultural variables related to the MR diagnosis was apparent at the turn of the century, and research identifying those factors has become increasingly influential throughout the 20th century. Mercer's Riverside, California studies (1973, 1979) indicated unequivocally that mild MR was closely associated with poverty. Heber's

(1970) and Garber and Heber's work associated with the controversial "Milwaukee project" (Garber, 1988) implicated the effects of poverty and low maternal IQ. Zigler, in his two-group conception of mental retardation (Zigler, 1967; Zigler et al. 1984), argued that performance in the range of mild MR represented the lower tail of the normal distribution of intelligence rather than simply the effects of poverty. Although the causal sequence is impossible to unravel definitively, the statistical association between poverty and mild mental retardation is rather dramatic.

Further complexity in unraveling sociocultural factors and MR was discussed by Edgerton (1981). Although virtually all cultures recognize some members as considerably less capable, the conceptual bases and standards by which these judgments are made vary considerably. Edgerton's discussion of the residents of Duddie's Branch persuasively illustrated the notion that virtually all members of this isolated community in Appalachia might be regarded as mildly MR using the contemporary standards of advanced, technological societies.

The question raised in the literature cited in this section is whether the diagnosis of MR should take into consideration the sociocultural context of the individual. Both APA and AAMR suggest that adaptive behavior be judged in relation to age and cultural setting. The AAMR classification scheme suggests that sociocultural factors should be considered, but performance on measures of intelligence and judgment of adaptive behavior "dictate the classification to be applied" (Grossman, 1983, p. 48). Furthermore, the AAMR (Grossman, 1983) rejected the notion of adding additional IQ points to the results of intelligence tests to account for the possible effects of poverty and cultural differences, a notion that Mercer (1979) proposed. In fact, the Mercer procedure for adding IQ points had relatively little effect on eligibility for the mild MR diagnosis in studies of White and minority students primarily from urban areas of the west and southwest (Heflinger, Cook, & Thackrey, 1987; Reschly, 1981). More extreme cultural differences, such as near total isolation from mainstream American culture as experienced by some very isolated Appalachian groups, by recent immigrants from underdeveloped countries, or by persons with little or no competence in the English language, must be addressed more carefully. The critical challenges in MR diagnosis are to describe the sociocultural differences and then to evaluate the individual's status in light of expectations and opportunities for the development of various competencies. The present classification systems provide insufficient guidance regarding the evaluation and consideration of sociocultural factors (see previous section).

Adaptive Behavior Conception

Gradual improvement in the elucidation and specification of adaptive behavior is apparent in reviewing the 1959–1961, 1973, 1977, and 1983 revisions of the AAMR *Manual on Terminology and Classification*. The research base now provides further support for additional improvements in conceptualization and specification. Reschly (1987, 1990) reviewed the adaptive behavior literature including factor analysis studies, treatment programs, adult adjustment, and the content of adaptive behavior instruments. Four domains of adaptive behavior were suggested: independent functioning, social functioning, functional academic skills, and vocational/occupational competencies. McGrew and Bruininks (1989) and Bruininks, McGrew, and Maruyama (1988) reviewed the factor analysis studies of measures of adaptive behavior applied with populations varying in age and level of performance. Their review suggested five domains for adaptive behavior: personal independence, social functioning, functional academic or applied cognitive skills, vocational/community, and motor/physical development dimensions.

These different domains have varying importance depending on the age of the individual. For example, the vocational/community domain has greater salience at the adolescent and adult ages, and the motor/physical development dimension is more important for younger and/or lower performing individuals. These domains also have varying applicability to defining or distinguishing among the levels of MR. Persons with mild MR typically have the most difficulty with functional academic skills or underlying cognitive skills necessary for adequate performance in a variety of roles *in or outside* the school setting. Many mildly MR individuals also have difficulty with social functioning. In contrast, the profoundly retarded typically exhibit significant deficits over all of the dimensions of adaptive behavior with deficiencies of greatest concern in the areas of motor/physical developmental dimensions (e.g., mobility) and personal independence (e.g., self-help skill such as toileting and feeding). Although conceptualizations of adaptive behavior do vary—Mercer (1979) represents a sharp contrast to Reschly (1982, 1990) or McGrew & Bruninks (1989)—recent formulations of social competence generally include some consideration of underlying functional academic or cognitive skills as well as recognition of the importance of assessing individuals in different settings (e.g., both in school and out of school) (Reschly, 1990). Although not resolved definitively, recent conceptualizations of adaptive behavior as well as instruments developed during the 1980s suggest an emerging consensus that adaptive behavior includes a complex set of skills that can be organized into four or five domains.

Adaptive Behavior Assessment

In 1984, three technically adequate adaptive behavior measures were published (Adams, 1984; Bruninks, Woodcock, Weatherman, & Hill, 1984; Sparrow, Balla, & Cicchetti, 1984). Each of the scales yields a composite score as well as subtest scores that can be quite readily organized around the five domains of adaptive behavior suggested in the prior subsection (Reschly, 1990) (see Table 2.5). The publication of these scales provides markedly greater resources for the systematic assessment of adaptive behavior than were available when the most recent revision of the AAMR classification system was developed.

The discussion of adaptive behavior assessment would be incomplete without consideration of some of the problems with the available inventories. All depend heavily on third-party respondents, either through interview or behavioral ratings. The potential response biases of third party respondents must be considered in an interpretation of information gathered through use of these inventories. Furthermore, the scales tend to focus on limited settings, either the home or the

TABLE 2.5
Comprehensive, Well-Standardized Adaptive Behavior Inventories[1]

Scale	*Age Range*	*[2]Lowest Score*	*Useful Score for MR Diagnosis*
Comprehensive Test of Adaptive Behavior (Adams, 1984)	Birth to 21 yrs.	Percentile ranks to PR=1 for retarded and non-retarded groups.	Composite score
Scales of Independent Behavior (Bruininks, Woodcock, Weatherman, & Hill, 1984)	Birth to adult	16	Broad independence composite score
Vineland Adaptive Behavior Scales-Survey and Expanded Forms-Interview Edition (Sparrow et al., 1984)	Birth to adult	20	Adaptive behavior composite

[1]See reviews by Reschly (1990).
[2]The lowest score available using a standard score scale with the mean = 100 and the standard deviation = 15.

school. Each of these scales reflects certain limitations with regard to appropriateness for various age levels or levels of MR. Taken together, however, one or more of the scales typically provides useful information regardless of client age or level of performance. As with the assessment of general intellectual functioning, the limitations of these scales and considerations of a broad variety of other information are essential to proper interpretation.

Summary

Recent literature concerning the vast influence of system factors on MR diagnosis suggests further development of MR classification systems. The special problems of mild MR are the arena in which the systems factors are most often observed. Mild MR is qualitatively different from the moderate, severe, and profound levels of MR, a fact not adequately addressed by current classification systems. Furthermore, sociocultural factors need to be considered in determining the appropriateness of the diagnosis of MR when significantly subaverage general intellectual functioning and deficits in adaptive behavior are observed. The conceptual bases and measurement technology for adaptive behavior have improved significantly during the 1980s. It is time now for adaptive behavior to become equal to general intellectual functioning in definitional criteria and diagnostic operations.

PROPOSED CHANGES IN DIAGNOSTIC AND ASSESSMENT PROCEDURES

Further development of the MR classification systems is needed to represent systems factors and the special problems of mild MR to accommodate a greater emphasis on adaptive behavior, to redefine levels of mental retardation according to adaptive behavior, and to consider further sociocultural factors. Each will be discussed in turn. These proposed changes are summarized in Table 2.6.

Change in Terminology

The current MR classification systems attempt to represent all levels of mental retardation on a continuum with the known differences in etiology and prognosis loosely associated with the adjectives describing the levels. Most persons interpret these levels as suggesting differences in degree (or quantitative differences), not in kind (i.e., qualitative dif-

TABLE 2.6
Summary of Proposed Changes in Mental Retardation Classification Schemes[1]

Conceptual Issue	*Nomenclature*	*Definition*	*Relative Weighting*	*Assessment*	*Decision Rules*
Terminology	[1]General educational handicap	Revision of current AAMR and DSM-III-R definitions	Equal weighting of IQ or general achievement and adaptive behavior	General intellectual functioning or general achievement and adaptive behavior	IQ or general achievement below 75 or 80 and adaptive behavior composite below 80.
Terminology	Mental retardation	Current DSM-III-R and AAMR definitions.	Equal weighting of IQ and adaptive behavior.	General intellectual functioning and adaptive behavior	IQ and adaptive behavior below 55 to 60
Levels	Moderate MR Severe MR Profound MR	Current DSM-III-R and AAMR Definitions	Use adaptive behavior composite to define the levels	General intellectual functioning and adaptive behavior	*Moderate*: IQ < 55 or 60 and adaptive behavior composite of 40 to 60 *Severe*: IQ < 55 and adaptive behavior composite of 25 to 39 *Profound*: IQ < 55 and adaptive behavior composite < 25
Developmental	Continue current practices	Age	Determinative	The chronological age when deficits are diagnosed	Birth to 18

(continued)

TABLE 2.6
(Continued)

Conceptual Issue	*Nomenclature*	*Definition*	*Relative Weighting*	*Assessment*	*Decision Rules*
Etiology	Biological and development delay or psychosocial	Attach to terminology, biological for mental retardation and psychosocial or developmental delay for general educational handicap	Not determinative (i.e., does not have to be specified)	Medical evaluation and sociocultural assessment	Absence of physical symptoms indicates developmental delay or psychosocial etiology. Psychosocial indicated by poverty, low educational level of family, and presence of low functioning in immediate family.
Subtypes	ICD–9 medical codes	ICD–9	Not determinative	Medical evaluation	ICD–9 criteria
Sociocultural	Define as the effects of poverty	Related to specification of etiology, which is optional.	In rare instances, extreme cultural differences may rule out	Sociocultural characteristics based on social history and	Assessment of language and other cultural differences.

	and/or cultural differences.		diagnosis of general educational handicap but will rarely affect mental retardation.	current economic, language, and family patterns.	
Prognosis	Likely adult status with available intervention	Optional: probable relationship to level; general educational handicap associated with potential for independent functioning and self-support in adult years. Mental retardation nearly always requires lifelong protective services.	Desirable but not determinative	Current strengths and weaknesses in functional skills	Adaptive behavior measures and ecological assessment.

[1]The term *general educational handicap* is used rather than *mild mental retardation* throughout the "Summary" in Table 2.6.

ferences). Most unfortunate is the attribution of the characteristics of the more severely retarded to the mildly retarded (Goodman, 1989). Misconceptions about mild MR have been prominent influences on the dramatic decline in the use of the mild MR diagnosis in school settings and on the enormously expensive and divisive litigation concerning overrepresentation of minorities in special education programs. A change in terminology for the mild MR diagnosis is desperately needed to protect persons from pernicious misconceptions and unrealistically low expectations. A dramatic change in terminology would also contribute to reversing the declining prevalence of mild MR and allow schools and other agencies to develop the special programs needed by persons formally classifiable as mildly MR (MacMillan, 1988). These persons clearly need interventions (Edgerton, 1967; Koegel & Edgerton, 1984; Zetlin, 1988; Zetlin & Hosseini, 1989).

Proposing a change in terminology for what is now called mild MR in the current classification systems leaves the question of what term. The term should be as free as possible from perjorative connotations but summarize the symptom complex, etiology, and prognosis of mild MR to the greatest extent possible. A number of investigators suggested alternatives: Dunn (1973) suggested "general learning disabilities"; Polloway & Smith (1983) suggested "educationally retarded"; Reschly, (1979, 1988a) advocated complete separation of the mild from the more severe levels through use of a term such as "general educational handicap" or "educationally disabled."

Many other possibilities do exist, but my strong preference is to avoid use of the word "retarded." This position was strongly advocated through personal communication to Ruth Luckasson of the University of New Mexico, who currently chairs a special AAMR committee responsible for revising the 1983 classification system (Reschly, personal communication, September 8, 1988b). The decision on terminology for the mild level of MR should involve persons who have primary interests in and professional responsibilities for children and youth who might be eligible for the mild MR diagnosis (e.g., school psychologists, special educators). Persons affiliated primarily with medical settings typically have far less contact with mild MR, because their MR client populations typically involve persons with MR at the moderate, severe, and profound levels. The change in terminology from mild MR, in addition to fostering the development of services for children and youth with significant problems, would also contribute to the revitalization of research on persons with mild MR, a problem identified by other influential investigators as well (Haywood, 1979; Prehm, 1985).

Enhancing Adaptive Behavior

A second vital change is to establish greater emphasis on adaptive behavior through specification of definitional criteria, assessment procedures, and decision rules. Major advances have occurred in the conceptualization and assessment of adaptive behavior. The domains of adaptive behavior described in the prior section have varying degrees of importance depending on the age of the individual. Cutoff scores, dependent on the age of the individual, need to be specified for each of the relevant domains in adaptive behavior in order for adaptive behavior to have the intended determinative and equal influence on diagnostic decisions as information on general intellectual functioning.

Determination of dual decision rules to be used simultaneously with adaptive behavior and general intellectual functioning is obviously complex. First, adaptive behavior and general intellectual functioning in the mild range of MR typically have correlations in the range of .3 to .5. Establishment of the same cutoff scores for both could have the effect of drastically reducing the population for whom the diagnosis is intended. Several alternatives are possible. First, the same cutoff score for both adaptive behavior and general intellectual functioning could be established, but that cutoff score would need to be considerably higher than the current IQ criterion of approximately 70 (widely understood to include the range up to 75). In order to reach the target population of approximately 2% of children and youth, a higher cutoff score for both, perhaps as high as 80, would need to be established. Second, parallel criteria could be established whereby performance on one of the dimensions would have to be in the range of a standard score of 70 to 75 and performance on the other dimension would have to be below a slightly less stringent criterion such as 80 or 85. Thus, a person with an IQ of 80 and adaptive behavior below 70 might quite properly be regarded as eligible, or vice-versa. The needs of persons in these ranges of functioning on adaptive behavior and general intellectual functioning are well established (MacMillan, 1988). The critical issue is not establishing a precise cutoff score for one or both but rather that cutoff scores be established on some rational basis calculated to reestablish the population of persons for whom the classification of mild MR was intended in Grossman (1983) and DSM-III-R (1987).

Adaptive behavior will be used more widely in MR diagnosis when the authoritative classification systems establish definitional criteria, describe assessment measures, and specify decision rules. These advances will have the additional desirable outcome of fostering greater use of adaptive behavior information in the formulation of treatment plans.

Criteria for Levels of Mental Retardation

Definitional criteria, assessment procedures, and decision rules relying heavily on IQ tests are increasingly inappropriate to define the levels of MR, specifically moderate, severe, and profound. Modern IQ tests do not provide, nor can they provide, the necessary scores associated with these levels. Extrapolation far beyond observed data points to define the levels according to IQ assumes a relationship that was never observed during the standardization of the test and, with many tests, could never be observed, because the tests do not have sufficient floors to assess intellectual functioning at the lower levels. Moreover, intelligence and adaptive behavior are virtually synonymous at younger ages and/or lower levels of functioning (Meyers et al., 1979).

The obvious solution to operationalizing levels of MR is to place far greater emphasis on adaptive behavior. Two of the recently developed adaptive behavior scales—the Scales of Independent Behavior (Bruininks et al., 1984) and the Vineland Adaptive Behavior Scales (Sparrow et al., 1984)—provide data-based scores down to the standard score of 20 (see Table 2.4) (using standard score scales comparable to the common IQ scale, mean = 100, SD = 15). The descriptions of the top adaptive behavior levels associated with levels of MR that already have been provided in Grossman (1983) could be formally attached to the levels of MR as definitional criteria, thereby providing further assistance to diagnosticians in the interpretation of the results of adaptive behavior measures. Treatment needs and intervention objectives for the moderately, severely, and profoundly retarded typically involve skills in the domains of motor development or mobility, independent functioning, and social functioning. These areas are rarely assessed in widely used measures of general intellectual functioning. Definition of levels of mental retardation by adaptive behavior ranges, rather than IQ ranges, would have the effect of more precise description, more accurate diagnosis, and potentially closer linkages between assessment, diagnosis, and treatment. These changes would increase the ecological validity of the MR diagnosis.

Sociocultural Factors

Measurement of sociocultural factors has improved (Mercer, 1979), but incorporation of this information in MR diagnoses remains problematic. Adding points to IQ test results or to any other measurement of functioning suggests levels of competence not attained by the individual. Moreover, the Mercer procedure had relatively little influence on the proportions of minority students potentially eligible for the mild MR diagnosis (Heflinger et al., 1987; Reschly, 1981).

Extreme sociocultural differences, particularly involving extremely limited exposure to the mainstream culture, should constitute possible exclusion of the MR diagnosis pending the use of more appropriate measures of functioning. The most obvious examples are recent immigrants from underdeveloped countries who have experienced extreme poverty circumstances and limited educational opportunities. A MR diagnosis based on administration of conventional IQ and adaptive behavior measures seems absurd in such instances, but, in some individual cases, that has occurred. The sociocultural exclusion factor should be limited to persons with markedly different language and environmental experiences. It is not applicable to the vast majority of minority children and youth in the United States who have been exposed to the general culture through television, formal schooling, and other significant experiences.

SUMMARY AND DIRECTIONS

The MR diagnostic construct has evolved gradually over the past 200 years. There is broad consensus on the critical dimensions used as the conceptual basis for the diagnosis. The recent classification systems developed by the APA and AAMR represent the vast improvements in the scientific understanding and the provision of effective services to persons with MR during the 20th century. Further revision of these classification systems is necessary to resolve serious problems with the mild level of the current MR classification system. The role of adaptive behavior in diagnostic operations needs to be enhanced based on the significant advances in conception and measurement accomplished during the 1980s. These advances in conception and measurement of adaptive behavior also provide the basis for a more accurate and useful scheme for defining and diagnosing the levels of mental retardation.

REFERENCES

Adams, G. (1984). *Comprehensive test of adaptive behavior.* San Antonio, TX: Psychological Corporation.

American Psychiatric Association. (1952). *Diagnostic and statistical manual of mental disorders.* Washington, DC: Author.

American Psychiatric Association. (1968). *Diagnostic and statistical manual of mental disorders* (2nd ed.). Washington, DC: Author.

American Psychiatric Association. (1980). *Diagnostic and statistical manual of mental disorders* (3rd ed.). Washington, DC: Author.

American Psychiatric Association. (1987). *Diagnostic and statistical manual of mental disorders* (3rd ed.–rev.). (DSM-III-R). Washington, DC: Author.

Baller, W. R. (1936). A study of the present social status of a group of adults who, when they were in elementary schools, were classified as mentally deficient. *Genetic Psychology Monographs, 18,* 165–244.

Baller, W., Charles, D., & Miller, E. (1967). Mid-life attainment of the mentally retarded. *Genetic Psychology Monographs, 75,* 235–329.

Barr, M. W. (1905). Classification of mental defectives. *Journal of Psycho-Asthenics,* 9(2), 35.

Bayley, N. (1969). *Bayley Scales of Infant Development: Birth to two years.* San Antonio, TX: Psychological Corporation.

Benton, A. (1956). The concept of pseudofeeblemindedness. *Archives of Neurology and Psychiatry, 75,* 379–388.

Bruininks, R., McGrew, K., & Maruyama, G. (1988). Structure of adaptive behavior in samples with and without mental retardation. *American Journal of Mental Deficiency, 93,* 265–272.

Bruininks, R. H., Warfield, G., & Stealey, D. S. (1978). The mentally retarded. In E. L. Meyer (Ed.), *Exceptional children and youth* (pp. 196–261). Denver, CO: Love.

Bruininks, R. H., Woodcock, R. W., Weatherman, R. F., & Hill, B. K. (1984). *Interview manual: Scales of independent behavior.* Allen, TX: DLM Teaching Resources.

Charles, D. C. (1953). Ability and accomplishment of persons earlier judged mentally deficient. *Genetic Psychology Monographs, 47,* 3–71.

Cromwell, R., Blashfield, R., & Strauss, J. (1975). Criteria for classification systems: In N. Hobbs (Ed.), *Issues in the classification of children* (pp. 4–25). San Francisco, CA: Jossey-Bass.

Doll, E. A. (1934). Social adjustment of the mental subnormal. *Journal of Educational Research, 28,* 36–43.

Doll, E. A. (1941). The essential of an inclusive concept of mental deficiency. *American Journal of Mental Deficiency, 46,* 214–219.

Doll, E. E. (1962). A historical survey of research and management of mental retardation in the United States. In E. P. Trapp & P. Himmelstein (Eds.), *Readings on the exceptional child* (pp. 21–68). New York: Appleton-Century-Crofts.

Doll, E. E. (Ed.). (1967). Historical review of mental retardation 1800–1965: A symposium. *American Journal of Mental Deficiency, 72,* 165–189.

Dunn, L. (1968). Special education for the mildly retarded: Is much of it justifiable? *Exceptional Children, 35,* 5–22.

Dunn, L. M. (Ed.). (1973). *Exceptional children in the schools: Special education in transition* (2nd ed.). New York: Holt, Rinehart & Winston.

Edgerton, R. B. (1967). *The cloak of competence: Stigma in the lives of the mentally retarded.* Berkeley, CA: University of California Press.

Edgerton, R. B. (1981). Another look at culture and mental retardation. In M. J. Begab, H. C. Haywood, & H. L. Garber (Eds.), *Psychosocial influences in retarded performance. Volume I: Issues and theories in development* (pp. 309–323). Baltimore: University Park Press.

Education of the Handicapped Act. (1975). PL94–142, 20 U.S.C. 1400–1485, 34 CFR–300. (As amended to October 8, 1986)

Forness, S. R. (1985). Effects of public policy at the state level: California's impact on MR, LD, and ED categories. *Remedial and Special Education,* 6(3), 36–43.

Garber, H. L. (1988). *The Milwaukee Project: Preventing mental retardation in children at risk.* Washington, DC: American Association on Mental Retardation.

Goddard, H. H. (1912). *The Kallikak family.* Norwood, MA: Norwood.

Goodman, J. F. (1989). Does retardation mean dumb? Children's perceptions of the nature, cause, and course of mental retardation. *Journal of Special Education, 23,* 313–329.

Grossman, H. J. (Ed.). (1973). *Manual on terminology and classification in mental retardation.* Washington, DC: American Association on Mental Deficiency.

Grossman, H. J. (1977). *Manual on terminology and classification in mental retardation.* Washington, DC: American Association on Mental Deficiency.

Grossman, H. J. (Ed.). (1983). *Classification in mental retardation.* Washington, DC: American Association on Mental Deficiency.

Haywood, H. C. (1979). What happened to mild and moderate mental retardation? *American Journal of Mental Deficiency, 83,* 429–431.

Heber, R. (1959). A manual on terminology and classification in mental retardation. *American Journal of Mental Deficiency Monograph Supplement, 64*(2).

Heber, R. (1961). Modification of the "Manual on terminology and classification in mental retardation." *American Journal of Mental Deficiency, 65*(4), 499–500.

Heber, R. (1970). *Epidemiology of mental retardation.* Springfield, IL: Thomas.

Heflinger, C. R., Cook, V. J., & Thackrey, M. (1987). Identification of mental retardation by the System of Multicultural Pluralistic Assessment: Nondiscriminatory or nonexistent? *Journal of School Psychology, 25,* 177–183.

Hilliard, A. G. (1983). IQ and the courts: *Larry P. vs. Wilson Riles* and *PASE vs. Hannon. Journal of Block Psychology, 10,* 1–18.

Hunt, J. (1961). *Intelligence and experience.* New York: Ronald Press.

Jensen, A. R. (1969). How much can we boost IQ and scholastic achievement? *Harvard Educational Review, 39,* 1–123.

Jones, R., & Wilderson, F. (1976). Mainstreaming and the minority child: An overview of issues and a perspective. In R. Jones (Ed.), *Mainstreaming and the minority child.* Reston, VA: Council for Exceptional Children.

Kamin, L. J. (1974). *The science and politics of IQ.* New York: Halsted.

Kanner, L. (1964). *A history of the care and study of the mentally retarded.* Springfield, IL: Thomas.

Kaufman, A., & Kaufman, N. (1983). *Kaufman Assessment Battery for Children* (K–ABC). Circle Press, MN: American Guidance Service.

Koegel, P., & Edgerton, R. B. (1984). Black "six hour retarded children" as young adults. In R. B. Edgerton (Ed.), *Lives in process: Mildly retarded adults in a large city* (pp. 145–171). Washington, DC: American Association on Mental Deficiency.

Larry P. v. Riles (1979, 1984, 1986). 343 F. Supp. 1306 (N. D. Cal. 1972) (preliminary injunction). aff'd 502 F. 2d 963 (9th cir. 1974); 495 F. Supp. 926 (N. D. Cal. 1979) (decision on merits) aff'd (9th cir. no. 80-427 Jan. 23, 1984). Order modifying judgment, C–71–2270 RFP, September 25, 1986.

MacMillan, D. (1982). *Mental retardation in school and society* (2nd ed.). Boston, MA: Little, Brown.

MacMillan, D. L. (1988). Issues in mild mental retardation. *Education and Training of the Mentally Retarded, 23,* 273–284.

MacMillan, D., Jones, R., & Aloia, G. (1974). The mentally retarded label: A theoretical analysis and review of research. *American Journal of Mental Deficiency, 79,* 241–261.

MacMillan, D., & Meyers, C. E. (1980). Larry P: An educational interpretation. *School Psychology Review, 9,* 136–148.

MacMillan, D., Meyers, C. E., & Morrison, G. (1980). System-identification of mildly mentally retarded children: Implications for interpreting and conducting research. *American Journal of Mental Deficiency, 85,* 108–115.

McCarthy, D. (1972). *Manual for the McCarthy Scales of Children's Abilities*. New York: Psychological Corporation.

McGrew, K., & Bruininks, R. (1989). The factor structure of adaptive behavior. *School Psychology Review, 18*, 64–81.

Mercer, J. R. (1973). *Labeling the mentally retarded*. Berkeley, CA: University of California Press.

Mercer, J. (1979). *System of Multicultural Pluralistic Assessment technical manual*. New York: Psychological Corporation.

Mercer, J. R., & Richardson, J. G. (1975). Mental retardation as a social problem. In N. Hobbs (Ed.), *Issues in the classification of children* (Vol. 2, pp. 463–496). San Francisco, CA: Jossey-Bass.

Meyers, C. E., Nihira, K., & Zetlin, A. (1979). The measurement of adaptive behavior. In N. R. Ellis (Ed.), *Handbook of mental deficiency* (pp. 431–481). Psychological theory and research. Hillsdale, NJ: Lawrence Erlbaum Associates.

Patrick, J., & Reschly, D. (1982). Relationship of state educational criteria and demographic variables to school-system prevalence of mental retardation. *American Journal of Mental Deficiency, 86*, 351–360.

Polloway, E. A., & Smith, J. D. (1983). Changes in mild rental retardation: Population, programs, and perspectives. *Exceptional Children, 50*, 149–159.

Prehm, H. J. (1985, January). *Educational and training of the mentally retarded*. Midyear report to the board of directors, CEC–MR.

Reschly, D. (1979). Nonbiased assessment. In G. Phye & D. Reschly (Eds.), *School psychology: Perspectives and issues* (pp. 215–253). New York: Academic Press.

Reschly, D. (1981). Evaluation of the effects of SOMPA measures on classification of students as mildly mentally retarded. *American Journal of Mental Deficiency, 86*, 16–20.

Reschly, D. (1982). Assessing mild mental retardation: The influence of adaptive behavior, sociocultural status and prospects for nonbiased assessment. In C. Reynolds & T. Gutkin (Eds.), *The handbook of school psychology* (pp. 209–242). New York: Wiley-Interscience.

Reschly, D. J. (1986). Economic and cultural factors in childhood exceptionality. In R. T. Brown & C. R. Reynolds (Eds.), *Psychological perspectives on childhood exceptionality: A handbook* (pp. 423–466). New York: Wiley-Interscience.

Reschly, D. J. (1987). *Adaptive behavior*. Tallahassee, FL: Bureau of Education for Exceptional Students, Florida Department of Education.

Reschly, D. J. (1988a). Assessment issues, placement litigation, and the future of mild mental retardation classification and programming. *Education and Training of the Mentally Retarded, 23*, 285–301.

Reschly, D. J. (1990). Adaptive behavior. In A. Thomas & J. Grimes (Eds.), *Best practices in school psychology* (2nd ed., pp. 29–42). Washington, DC: National Association of School Psychologists.

Reschly, D. J., & Jipson, F. J. (1976). Ethnicity, geographic locale, age, sex, and urban–rural residence as variables in the prevalence of mild retardation. *American Journal of Mental Deficiency, 81*, 154–161.

Reschly, D. J., Kicklighter, R. H., & McKee, P. (1988a). Recent placement litigation, Part I: Regular education grouping: Comparison of *Marshall* (1984, 1985) and *Hobson* (1967, 1969). *School Psychology Review, 17*, 7–19.

Reschly, D. J., Kicklighter, R. H., & McKee, P. (1988b). Recent placement litigation, Part II: Minority EMR overrepresentation: Comparison of *Larry P.* (1979, 1984, 1986) with *Marshall* (1984, 1985) and *S–1* (1986). *School Psychology Review, 17*, 20–36.

Reschly, D. J., Kicklighter, R. H., & McKee, P. (1988c). Recent placement litigation, Part

III: Analysis of differences in *Larry P., Marshall,* and S–1 and implications for future practices. *School Psychology Review, 17,* 37–48.

Reschly, D. J., Robinson, G. A., Volmer, L. M., & Wilson, L. R. (1988). *Iowa Mental Disabilities Research Project Final Report and Executive Summary.* Des Moines, IA: Iowa Department of Education, Bureau of Special Education.

Robinson, N., & Robinson, H. (1976). *The mentally retarded child* (2nd ed.). New York: McGraw-Hill.

Sarason, S. B., & Doris, J. (1969). *Psychological problems in mental deficiency* (4th ed.). New York: Harper & Row.

Sattler, J. M. (1988). *Assessment of children* (3rd ed.). San Diego, CA: Sattler.

Sloan, W., & Stevens, H. A. (1976). *A century of concern: A history of the American Association on Mental Deficiency.* Washington, DC: American Association on Mental Deficiency.

Smith, J. D. (1985). *Minds made feeble: The myth and legacy of Kallikaks.* Rockville, MD: Aspen Systems.

Smith, J. D., & Polloway, E. A. (1979). The dimension of adaptive behavior in mental retardation research: An analysis of recent practices. *American Journal of Mental Deficiency, 84,* 203–206.

Sparrow, S., Balla, D. A., & Cicchetti, D. V. (1984). *Vineland Adaptive Behavior Scales.* Circle Pines, MN: American Guidance Service.

Tenth Annual Report to Congress on the Implementation of the Education of the Handicapped Children Act (1988). Washington, DC: United States Department of Education, Office of Special Education and Rehabilitation Services.

Terman, L. M. (1916). *The measurement of intelligence.* Boston: Houghton-Mifflin.

Thorndike, R. L., Hagen, E. P., & Sattler, J. M. (1986). *Technical manual: Stanford–Binet Intelligence Scale–Fourth Edition.* Chicago: Riverside.

Wechsler, D. (1967). *Manual for the Wechsler Preschool and Primary Scale of Intelligence.* San Antonio, TX: Psychological Corporation.

Wechsler, D. (1974). *Manual for the Wechsler Intelligence Scale for Children–Revised.* San Antonio, TX: Psychological Corporation.

Wechsler, D. (1981). *Manual for the Wechsler Adult Intelligence Scale–Revised.* San Antonio, TX: Psychological Corporation.

World Health Organization. (1978). *International classification of diseases* (9th rev.) (ICD–9). Geneva: Author.

Zetlin, A. G. (1988). Adult development of the mildly retarded. In M. C. Wang, M. C. Reynolds, and H. J. Walberg (Eds.). *The handbook of special education: Research and practice* (Vol. 2, pp. 77–90). Oxford, UK: Pergamon.

Zetlin, A. G., & Hosseini, A. (1989). Six post school case studies of mildly learning handicapped young adults. *Exceptional Children, 55,* 405–411.

Zigler, E. (1967). Familial mental retardation: A continuing dilemma. *Science, 155,* 292–298.

Zigler, E., Balla, D., & Hodapp, R. (1984). On the definition and classification of mental retardation. *American Journal of Mental Deficiency, 89,* 215–230.

CHAPTER 3

Autism

Gary B. Mesibov
Mary E. Van Bourgondien
University of North Carolina
School of Medicine

A relatively new diagnostic category described within the past half century (Kanner, 1943), autism has been regaining lost time by generating considerable interest and confusion (Schopler, 1983). Early conceptualizations of autism as a psychogenic disturbance mislead investigators for many years (Bettelheim, 1967). Recent advances in our understanding and treatment of this disorder, however, have been substantial. In its short history, autism has gone from one of the most poorly understood to one of the most systematically investigated of the childhood disorders (Rutter & Schopler, 1988).

The history and evolution of the autism syndrome is the subject of this chapter. The chapter begins by tracing the historical background of autism. Following this brief review, we examine important issues in the evolution of the definition, emphasizing current changes contained in DSM-III-R. We also examine the definition of autism in light of current research. Finally, we review existing diagnostic instruments and the supporting evidence for each of them.

HISTORICAL BACKGROUND

Leo Kanner (1943) first used the term *infantile autism* to describe a group of 11 children who were thought to have a rare psychiatric disorder representing the earliest form of schizophrenia. He singled out this group because they seemed more similar to one another than to the schizophrenic children they were generally associated with. These children exhibited a variety of social, communicative, and behavioral peculiarities that Kanner was able to describe with precision. The social difficulties included an inability to relate to their parents from the beginning of life and a preference for objects over people. Communication problems ranged from muteness to echolalia to pronoun reversals to impaired comprehension. The behavioral difficulties Kanner noted included stereotypic movements, such as finger flicking; and repetitive, ritualized behaviors, such as lining up objects, routinized unimaginative play, peculiar attachments and fascinations with objects, and extreme resistance to environmental changes.

Kanner made several interesting observations and inferences about these children and their families. He noted that their disorder began in infancy, which was unlike previously described forms of childhood psychosis, which usually developed later in life. He observed that many of the children had unusual peak skills in rote memory, musical ability, and numerical manipulation (Kanner, 1973). The peak skills suggested normal intellectual potential to Kanner. Of his original 11 children, 8 were boys, which led to his belief that autism was more common in males than in females.

Along with his behavioral observations of the children, Kanner noted that their families usually excelled both educationally and professionally. A high percentage of the parents were physicians, and several were renouned in their field. A disproportionate number of this group were in the upper middle and upper classes socioeconomically. Kanner thought that these family characteristics were important contributing factors to the autism syndrome.

In addition to recognizing autism as a separate disability and identifying its salient features, Kanner's most important contribution was in his precise descriptions of these children and their behaviors; some of his descriptions are among the most accurate and comprehensive in the literature. His small sample size, however, led to several inaccurate formulations concerning their family patterns and IQ levels. For example, Kanner thought autistic children were of average or above average intelligence. His psychogenic interpretations of autism have also been refuted by subsequent investigators (Rutter & Schopler, 1978).

Immediately following Kanner, much of the autism work was psychoanalytically oriented. Investigators examined unconscious mental conflicts (Boatman & Szurek, 1960), emotional detachment (Despert, 1951), parental depression (Ruttenburg, 1971), and stresses involving separation issues for the children (Rank, 1949). Of the psychoanalytic theorists, the one with the greatest impact has been Bettelheim (1967), who stressed the causative role of parents and the need for "parent-ectomies," defined as removal of the children from the environments causing their autism and placements in residential programs away from their parents. During the years when psychoanalytic theory dominated the field of autism, little progress was made, and many parents suffered anguish and guilt.

The first major book asserting that autism was organic and not an emotional disorder was Rimland's classic work in 1964. Following its publication, the professional climate changed rapidly with more emphasis on the organic problems of autism and less attention to parental pathology. Concurrently the literature began to reflect the notion that parental pathology and autism were unrelated. An early study by Pitfield and Oppenheim (1964) found no differences in parental rejection between parents of autistic children, normal children, and children with Down's syndrome. Schopler and Loftin (1969) reported that the guilt-inducing psychoanalytic therapy was causing the maladaptive behavior of parents of autistic children rather than the parents causing the maladaptive behavior of their children. These studies concluded that parents of psychotic children had been inappropriately scapegoated by the professionals working with them (Schopler, 1971). Lennox, Callias, and Rutter (1977) demonstrated that parents of autistic children did not differ from parents of normal children on scores measuring thought disorders, and Cox, Rutter, Newman, and Bartak (1975) were also unable to find any differences between these parents. DeMyer et al. (1972) interviewed parents of autistic, normal, and developmentally delayed children and found no differences in infantile acceptance, nurturing warmth, feeding, or tactile stimulation, whereas Byassee and Murrell (1975) found no differences in family interactions in families with autistic children compared with nonhandicapped controls. Other studies confirmed that parents of autistic children were no different from parents of normal children or children with related disabilities (Creak & Ini, 1960; Kolvin, Garside, & Kidd, 1971; Netley, Lockyer, & Greenbaum, 1975). During the 1970s, the unequivocal conclusion became that pathological family factors are not a sufficient cause for the development of autism (Cantwell, Baker, & Rutter, 1977).

As psychoanalytic theory was losing its appeal because of accumulating evidence refuting it, autism became widely recognized as an organic disorder. Consequently, efforts intensified to more clearly and succinctly define the autism syndrome. In 1956, Eisenberg narrowed the symptoms of autism to only two: extreme aloneness and preoccupation with sameness from the beginning of life. Although his effort to simplify the diagnostic criteria was appropriate and desirable, he unfortunately omitted the central problems of language and communication. Ornitz and Ritvo (1968) later added disturbances of perception to the list of essential characteristics. This new emphasis on sensory behaviors—something not included in Kanner's original definition—became an important component of the autism definition that has been studied by other investigators but has never gained the recognition that the social, communicative, and repetitive characteristics have maintained.

Of the attempts to clarify the autism diagnosis in the 1960s and early 1970s (Rendle-Short, 1969; Schain & Yannet, 1960; Tinbergen & Tinbergen, 1972; Wing & Ricks, 1976), the most influential was Creak's (1963), which emphasized nine major points: gross impairment of emotional relationships with people, apparent unawareness of one's own identity, preoccupation with particular objects, sustained resistance to change, abnormal perceptual experiences, acute and seemingly illogical anxiety, lost or never-acquired speech, distortion in motility patterns, and a background of serious retardation in which islets of normal or even exceptional skills may appear. The importance of the Creak criteria is that they were the first ones to be empirically tested. Although the results of the empirical investigations were disappointing, this represented a major advance over earlier systems that were totally descriptive. Creak's system was also important because it accelerated the emergence of organically oriented theories by placing the burden on the psychodynamic formulations to verify themselves, which they have not been able to do.

The work of Rutter (1966, 1978) has been the most influential in leading the field to its current definition of autism. He systematically examined differences between autistic and other psychiatrically impaired children of the same sex, age, and general intelligence level and found three characteristics that were seen more frequently in the autistic children: a failure to develop social relationships; language delays and impaired comprehension, echolalia, and pronoun reversals; and ritualistic or compulsive behaviors, designed to maintain consistency in their environments. Four additional symptoms were more likely in the autistic group but not always present: stereotypic repetitive movements (particularly hand and finger mannerisms), short attention spans, self-injurious behaviors, and delayed bowel control.

Concurrent with Rutter, the National Society for Autistic Children approved a similar definition, with small modifications (Ritvo & Freeman, 1977, 1978). Similar to Rutter, the Society highlighted onset prior to 30 months of age, impaired communication, and impaired social relationships. The Society did not, however, include insistence on sameness, instead substituting disturbances of developmental rates or sequences and disturbances in responses to sensory stimuli. These modifications were designed to highlight the peak skills frequently observed in these children and also to emphasize the sensory processes Ornitz and Ritvo (1968) identified.

CURRENT DIAGNOSTIC SYSTEMS

The definitions of Rutter and the National Autism Society were essentially those that were incorporated into DSM-III, classifying autism as a Pervasive Developmental Disorder (PDD). Infantile autism was included among the PDDs, because multiple aspects of psychological functioning are impaired, and the disability is quite severe. As with Rutter's and the National Society's definitions, DSM-III identified the essential features as a lack of responsiveness to others, gross impairment in communicative skills, and bizarre responses to aspects of the environment. DSM-III added a Residual State for older adolescents or adults who no longer met the full criteria for Infantile Autism but showed symptoms of the condition including oddities of communication and social awkwardness.

DSM-III did not list infantile autism as the only PDD; Childhood Onset Pervasive Developmental Disorder (COPDD) (Full Syndrome and Residual State) and Atypical Pervasive Developmental Disorder were also included. COPDD was added for individuals with social deficits similar to those in Infantile Autism, but whose onset was after 30 months of age. Atypical Pervasive Developmental Disorder was designed for children with distortions in the development of social skills and language.

DSM-III-R (APA, 1987) has remained consistent with DSM-III (1980) while introducing modifications in the description of autism. The consistency is in still viewing autism as a Pervasive Developmental Disorder involving multiple psychological functions and characterized by deviant development in social relationships, communication, and behavior. Autistic Disorder is the only PDD that DSM-III-R describes in detail, however, with the diagnosis of Pervasive Developmental Disorder Not Otherwise Specified (PDDNOS) reserved for those who do not meet the full criteria for autism. Eliminated from DSM-III-R are Childhood Onset Pervasive Developmental Disorder (Full Syndrome and Residual), Atypical Pervasive Developmental Disorder and Infantile Autism, Residual State. The term *Infantile Autism* was changed to *Autistic Disorder*, recognizing autism as a lifelong disability.

In DSM-III-R, there are specific reasons for placing diagnoses on Axis I or Axis II: Axis I diagnoses are mental illnesses, and Axis II diagnoses are developmental, having an onset in childhood or adolescence and usualy persisting in a stable form into adult life. Although Axis I and

Axis II diagnoses are mental disorders, the distinction is that disorders of Axis I are more florid and unstable than the long-standing developmental problems of Axis II. Given these current distinctions, the shift of PDD from Axis I to Axis II is a positive change. Autism is more accurately categorized as Axis II, and this placement might highlight autism as a developmental disability.

The DSM-III-R diagnostic criteria are shown in Table 3.1. In order to meet the criteria for autism, at least 8 of the 16 items must be present, including at least 2 items from Part A, 1 from Part B, and 1 from Part C. A criterion is only met if the behavior is abnormal for the person's developmental level. The examples under social interaction and communication are ordered so that the first ones listed are more likely to apply to younger or more handicapped individuals, and the later ones are more common in older or less handicapped people with autism.

Differences between DSM-III-R and DSM-III are subtle but significant. In the descriptions of social behavior, for example, the emphasis is on a lack of reciprocity in these relationships rather than the total absence of social behavior. Communication problems emphasize verbal and nonverbal difficulties as well as lack of imaginative capabilities. Age of onset has been eliminated as a defining characteristic.

Although the specific changes from DSM-III to DSM-III-R are few, the implications are substantial. DSM-III-R has further clarified the boundaries of autism, helping clinicians to determine if behaviors previously described as autistic-like truly constitute part of the autism syndrome. In discussing other implications, we examine each aspect of the definition separately.

Social abnormalities characterizing autism led Kanner to coin the term, and they remain the most interesting and unique aspects of this puzzling disorder. In DSM-III, the failure to develop interpersonal relationships and lack of responsiveness to other people were highlighted and emphasized. DSM-III-R also describes the autistic lack of interest in others but expands the concept to include impairments in reciprocal social interactions. This extends the category of deviant social behavior to interactive relationships that lack reciprocity or a normal give and take. DSM-III-R also recognizes lack of empathy as an important aspect of autistic people's social difficulties.

Although DSM-III and DSM-III-R describe both verbal and nonverbal communication difficulties as characteristic of autism, DSM-III-R incorporates nonverbal communication difficulties more explicitly into the definition. In DSM-III-R, the emphasis shifts toward problems of communication rather than simply speech or language. Communication problems can either be verbal or nonverbal and encompass a wider range of individuals than in DSM-III. DSM-III-R also includes difficul-

TABLE 3.1
Diagnostic Criteria for 299.00 Autistic Disorder

	E–2	*BRIAAC*	*BOS*	*ASIEP (ABC)*	*CARS*
A. Qualitative impairment in reciprocal social interaction as manifested by the following:					
1. Marked lack of awareness of the existence or feelings of others (e.g., treats a person as if he or she were a piece of furniture; does not notice another person's distress; apparently has no concept of the need of others for privacy).	+	+	+	+	+
2. No or abnormal seeking of comfort at times of distress (e.g., does not come for comfort even when ill, hurt, or tired; seeks comfort in a stereotyped way, e.g., says "cheese, cheese, cheese" whenever hurt).	–	–	–	–	+
3. No or impaired imitation (e.g., does not wave bye-bye; does not copy mother's domestic activities; mechanical imitation of others' actions out of context).	–	–	–	–	+
4. No or abnormal social play (e.g., does not actively participte in simple games; prefers solitary play activities; involves other children in play only as "mechanical aids").	–	–	–	+	+
5. Gross impairment in ability to make peer friendships (e.g., no interest in making peer friendships; despite interest in making friends, demonstrates lack of understanding of conventions of social interaction, for example, reads phone book to uninterested peer).	–	–	–	+	–
B. Qualitative impairment in verbal and nonverbal communication, and in imaginative activity, as manifested by the following:					
1. No mode of communication, such as communicative babbling, facial expression, gesture, mime or spoken language.	+	+	–	–	+
2. Markedly abnormal nonverbal communication, as in the use of eye-to-eye gaze, facial expression, body posture, or gestures to initiate or modulate social interaction (e.g., does not anticipate being held, stiffens when held, does not look at the person or smile when mak-	+	–	+	–	+

(continued)

TABLE 3.1
(Continued)

	E–2	*BRIAAC*	*BOS*	*ASIEP (ABC)*	*CARS*
ing a social approach, does not greet parents or visitors, has a fixed stare in social situations).					
3. Absence of imaginative activity, such as playacting of adult roles, fantasy characters, or animals; lack of interest in stories about imaginary events.	–	–	–	+	–
4. Marked abnormalities in the production of speech, including volume, pitch, stress, rate, rhythm, and intonation (e.g., monotonous tone, question-like melody, or high pitch).	+	–	–	+	+
5. Marked abnormalities in the form or content of speech, including stereotyped and repetitive use of speech (e.g., immediate echolalia or mechanical repetition of television commercial); use of "you" when "I" is meant (e.g., using "You want cookie?" to mean "I want a cookie"); idiosyncratic use of words or phrases (e.g., "Go on green riding" to mean "I want to go on the swing"); or frequent irrelevant remarks (e.g., starts talking about train schedules during a conversation about sports).	+	–	+	+	+
6. Marked impairment in the ability to initiate or sustain a conversation with others, despite adequate speech (e.g., indulging in lengthy monologues on one subject regardless of interjections from others).	+	+	–	–	+
C. Markedly restricted repertoire of activities and interest, as manifested by the following:					
1. Stereotyped body movements (e.g., hand-flicking or twisting, spinning, head-banging, complex whole-body movements).	+	+	+	+	+
2. Persistent preoccupation with parts of objects (e.g., sniffling or smelling objects, repetitive feeling of texture of materials, spinning wheels of toy cars) or attachment to unusual objects (e.g., insists on carrying around a piece of string).	+	–	+	+	+

(continued)

TABLE 3.1
(Continued)

	E–2	BRIAAC	BOS	ASIEP (ABC)	CARS
3. Marked distress over changes in trivial aspects of environment (e.g., when a vase is moved from usual position).	+	–	+	+	+
4. Unreasonable insistence on following routines in precise detail (e.g., insisting that exactly the same route always be followed when shopping).	+	–	–	+	+
5. Markedly restricted range of interests and a preoccupation with one narrow interest (e.g., interested only in lining up objects, in amassing facts about meteorology, or in pretending to be a fantasy character).	+	+	–	+	+

ties in imaginative ability as part of the communication impairment: absence of symbolic or fantasy play with toys, absence of playacting of adult roles, and imaginative activity that is generally restricted in content and repetitive in form. Although these difficulties have frequently been noted by other investigators (Wing, Yeates, & Brierley, 1977), this is the first time that they have been included in a major definition of autism. The addition of these limitations in imaginative ability is welcome, although it is not clear why they were included under communication problems rather than restricted behavioral patterns.

Bizarre and restrictive behavioral patterns have been consistent features of the autism definition over time and are retained by DSM-III-R. Manifestations of these difficulties in higher functioning autistic people are more evident in the new diagnostic system. For example, the description (APA, 1987) of these behavioral peculiarities gives the following example:

> In older children, tasks involving long-term memory, for example, recall of the exact words of songs heard years before, train timetables, historical dates, or chemical formulae, may be excellent, but the information tends to be repeated over and over again, regardless of the social context and the inappropriateness of the information. (p. 35)

There are other implications of these changes from DSM-III to DSM-III-R as well. The new definition more clearly includes those clients—especially high-functioning ones—previously on the periphery and often described as "autistic-like." By emphasizing poor reciprocity in social relationships, problems with communicative reactions, and lack of imaginative conceptualizations, more of the verbal high-functioning autistic people who communicate and relate—although neither normally nor easily—are now included in the definition of autism.

DSM-III-R also increases the likelihood that higher-functioning individuals will be diagnosed as autistic by eliminating the category from DSM-III called "schizoid disorder of childhood." In DSM-III, this category described children having no same-aged friends, no apparent interest in making friends, no apparent pleasure from peer interactions, and generally avoiding peer contacts. Several of these characteristics describe high-level autistic people; therefore, they are more likely to be diagnosed as autistic now that this diagnostic category has been eliminated.

The removal of Autism, Residual State as a diagnostic category and the elimination of age of onset as a diagnostic criterion are also important. By having a Residual State category for those whose autistic characteristics diminish in adulthood, DSM-III implied that autistic people could be cured. Current experience, however, suggests that the problems of adjustment and the need for ongoing support and assistance continue in adulthood (Schopler & Mesibov, 1983), although specific symptoms might decrease. This change is especially important now, because the issue of autistic people being cured has been a recent controversy (Lovaas, 1987; Schopler, Short, & Mesibov, 1989). The elimination of the Residual State category in DSM-III-R will help to avoid the unrealistic hopes and expectations that these claims generate.

The elimination of the age of onset criterion is important for establishing the reliability of the autism category; this change had been proposed by investigators several years ago (Volkmar, Stier, & Cohen, 1985). Although it does not change the popular belief that autism has an early onset—almost always before age 3—it should minimize troublesome conceptual, methodological, and reliability problems.

The major signs of autism are not always present until age 2 or even 3; thus, children are usually not referred until this age or later. Age of onset, therefore, is generally determined by parental report. Although parents can be and often are perceptive in describing their children's behavior and in understanding their strengths and problems, parents cannot always accurately identify the precise age when a specific behavior or difficulty began.

Another problem with the age of onset criteria has been determining precisely those behaviors that are being examined. The question has been whether early indicators of autism have to be autistic characteristics per se or could as easily be early problems of slow development and learning. In other words, does early onset have to mean autistic characteristics before age 3 or simply any developmental difficulties? Sometimes these are very difficult to distinguish in young children, and the new definition—no longer requiring these distinctions—has eliminated a chronic source of confusion.

A difficulty with eliminating age of onset as a criterion is those children who develop autistic behaviors at older ages (after 5) as a result of an accident or other trauma. Although most investigators acknowledge significant differences between this group and those with the more traditional autism syndrome, DSM-III-R now includes them in the definition of autism. One suggested solution is to include this later onset group as a Pervasive Developmental Disorder in the Not Otherwise Specified category (Volkmar & Cohen, 1986), but this issue remains unresolved.

Presently there is a consensus on the three general characteristics of autism: qualitative impairment in reciprocal social interactions, qualitative impairment in verbal and nonverbal communication and imaginative activity, and markedly restricted repertoire of activities and interests. Some believe that a fourth general category called cognitive difficulties (e.g., organization, sequencing, attention, concreteness) should be added, but that was not a major concern during the DSM-III-R deliberations.

The inclusion of specific criteria within each category, however, did raise questions. These criteria, included by the DSM-III-R Task Force, were designed as examples of the general categories at a variety of developmental levels. Once the 16 criteria were selected, a national field trial (Spitzer & Siegel, 1989) was conducted to determine the optimal number of items required to maximize the sensitivity and specificity of the criteria. The field trial included 506 children from twelve different sites (DSM-III-R, 1987). At each site, instructions to the clinicians were as follows: (a) make a clinical diagnosis based on your own best judgment, (b) make a diagnosis of Infantile Autism or Childhood Onset Pervasive Developmental Disorder based on DSM-III criteria, and (c) complete a symptom checklist that includes the 16 criteria from DSM-III-R. In these trials, a cutoff of at least 8 of the 16 DSM-III-R items maximized sensitivity and specificity. The Task Force added that at least 2 of the 16 characteristics must come from the social area and 1 each from the other 2 areas to insure the presence of all 3 of the central features in everyone diagnosed as having autism.

The field trials also indicated that clinicians using the DSM-III-R criteria diagnosed more children as autistic compared with their own clinical diagnoses or DSM-III. These results demonstrated that DSM-III-R reflects the broader definition of autism that many clinicians desire.

The field trials were not without problems, however. Reliability of individual items and the overall diagnosis were not assessed. Comparison samples varied among the different field sites. Some comparison groups included other types of developmental disabilities, whereas others included more conduct disordered children. The criteria for the clinical diagnoses was not stipulated making comparisons to this standard difficult to interpret.

Currently there are only two published studies of the DSM-III-R criteria, one based on a reanalysis of a subgroup of the field trial (Siegel, Vukicevic, Elliott, & Kraemer, 1989) and the other adding an independent sample (Volkmar, Bregman, Cohen, & Cicchetti, 1988). In a reanalysis of 60 cases from the field trial, signal detection theory methodology was applied to evaluate the sensitivity and specificity of the criteria (Siegel et al., 1989). The results indicated that two of the criteria ("marked lack of awareness of the existence of feelings of others" and "persistent preoccupation with parts of objects") are better predictors of the autism diagnosis than the cutoff of 8 of 16 items. This reanalysis suggests that several of the criteria are redundant because of high correlations with other criteria. A number of items were also significantly related to the subjects' mental ages.

Volkmar and his colleagues (1988) examined the reliability of the DSM-III-R criteria and compared the sensitivity and specificity of DSM-III and DSM-III-R criteria for autism. The interobserver agreement using Kappa-related statistics indicated excellent (Kappa $\geq$ .75) agreement for 12 of the items, good agreement (Kappa between .60 and .74) for two of the items, and fair agreement (Kappa between .40 and .59) for the remaining two items. Consistent with the field trial and previous research (Volkmar, Cohen, & Paul, 1986), the DSM-III criteria identified a smaller proportion of individuals as autistic compared to clinicians' diagnoses. In contrast, the DSM-III-R criteria identified significantly more subjects as autistic than did the clinical diagnoses.

Based on these limited empirical data, it is difficult to reach any clear conclusions regarding the utility of the DSM-III-R criteria for autism. More documentation regarding the interrater and test–retest reliability of these items is needed. Although the interrater agreement in the one published report (Volkmar et al., 1988) is high, the authors stated that the clinical raters were very experienced in diagnosing autism. How these criteria are applied by less sophisticated clinicians is yet to be determined.

The empirical data provide preliminary support for the clinical impression that the DSM-III-R changes have broadened the group of individuals identified as autistic. As others have noted (Volkmar et al., 1988), this broadening of the diagnostic category has different implications for clinicians and researchers. From a clinical perspective, children who were previously described as "autistic-like" or as not fitting into any well-defined PDD category are now more likely to receive a diagnosis of autism and access to appropriate services. The efficacy of specific interventions is linked more to the individualized assessment of the child than to the diagnostic label (Schopler, Reichler, & Lansing, 1980); therefore, diagnosing a child as autistic when he or she may not be is likely to have minimal negative effects on the individual. The clinical concern will be the availability of appropriate assessment and treatment options for a population that is already underserved.

The impact of broadening the diagnosis on research is more problematic. The limited data suggest that more higher functioning individuals may be identified because of the changes. The high correlations between many of the criteria and mental retardation (Siegel et al., 1989) suggest that more individuals who are mentally retarded will also be identified as autistic. Adding more variability into an already heterogeneous group is problematic from a research perspective.

In summary, there are some advantages and some drawbacks to the changes from DSM-III to DSM-III-R. On the positive side, the autism category has been expanded, adding odd social interactors and communicators to those who refuse to initiate. The nature of the social, communicative, and behavioral peculiarities accompanying autism have been more clearly described, whereas unreliable and misleading criteria like age of onset and Autism, Residual State have been eliminated. In these ways, DSM-III-R is clearer and more progressive.

On the other hand, significant problems remain. The 16 criteria selected for inclusion have not been adequately validated and might not represent the best choices. More careful study of these and other possibilities is strongly indicated. Differences between autism and other pervasive developmental disabilities remain elusive and in need of further investigation. Although progress is being made, there is clearly more work to be done.

DIAGNOSTIC INSTRUMENTS

Paralleling the evolution of the current diagnostic system have been attempts to identify the most relevant characteristics of autism through diagnostic and assessment instruments. Although many of these instruments have been irrelevant or cumbersome, some have been useful (Morgan, 1988; Parks, 1988). In this section, we summarize the most important of these instruments and discuss the extent to which each addresses questions of reliability and validity.

Diagnostic Checklist for Behavior–Disturbed Children–Form E–2.

Rimland was among the first to develop an objective checklist to diagnose autism. The Diagnostic Checklist for Behavior–Disturbed Children Form E–2 is a revision of an earlier scale (E–1). It is an 80-item, multiple-choice measure that is completed by parents based on their children's behavior prior to age 5. The E–2's emphasis on early development comes from Rimland's interest in having a measure of classical autistic symptoms as defined by Kanner (1943) and on research suggesting that these symptoms are less clearly distinguishable after age 5 (Lotter, 1978; Rimland, 1968). For each item that is representative of autism on the E–2, the child receives +1 point, whereas each endorsed nonautistic characteristic is scored as a −1 point. The total score is the sum of all 80 items.

In a study of 2,218 children, the scores were reported to range from −42 to +45, with a cutoff point of +20 for diagnosing a child as autistic (Rimland, 1971). In an initial study, this cutoff identified only about 10% of the sample as autistic, a percentage that corresponds to Rimland's and Kanner's notion that only 10% of the children labeled as autistic actually have the syndrome (Parks, 1988; Rimland, 1971).

Rimland's scale has several methodological shortcomings (Morgan, 1988; Parks, 1988). There have not been any studies of the interrater reliability, test–retest reliability, or the internal consistency of the E–2. The measure depends on parents' memories of early development, which makes the lack of demonstrated reliability a significant concern. The closest approximation to a reliability study was done by Prior and Bence (1975) comparing parent and staff ratings on the E–2 for nine children with autism in a day program. Although clearly not a reliability study (Parks, 1988), the results indicated that staff members report more abnormal behaviors than do parents.

As Morgan (1988) stated, the content validity of the E–2 is based primarily on the inclusion of items related to Kanner's (1943) original criteria for autism and Rimland's (1964) review of the literature regarding the developmental, medical, and behavioral characteristics of these children. Most of the efforts to establish the validity of this instrument have focused on its discriminant validity. In one of the earliest studies, DeMyer, Churchill, Pontius, and Gilkey (1971) compared the E–1 and E–2 with four other diagnostic approaches that were in use at that time. Although they found that the E–1 scale was able to distinguish between psychotic (early schizophrenic and autistic) and nonpsychotic children, the E–2 scale was not able to make distinctions between these groups or to discriminate within the group of psychotic children.

In a more recent study (Teal & Wiebe, 1986), the E–2, the Autism Screening Instrument for Educational Planning (ASIEP; Krug, Arick, & Almond, 1979) and the Childhood Autism Rating Scale (CARS; Schopler, Reichler, DeVellis, & Daly, 1980) were used to distinguish between autistic and trainable mentally retarded students. The results indicated that the E–2 form accurately identified 85% of the autistic students and 95% of the mentally retarded children for a pooled group accuracy of 90%. Although the E–2 significantly discriminated between the groups, the other two scales were better predictors of group membership.

The construct validity of the E–2 has also been investigated by examining biological differences between those children with high E–2 form scores and those with lower scores. The results of these studies have been inconclusive. A preliminary study showed that children with high E–2 scores also exhibited high serotonin outflow (Boullin, Coleman, O'Brian, & Rimland, 1971). Efforts to replicate these results (Boullin et al., 1982; Yuwiler et al., 1975), however, have been unsuccessful. Based on the hypothesis that there is a genetic error in metabolic functioning, Rimland (1973) investigated the relationship between E–2 scores and responses to megavitamin therapy. Responders were distinguishable from nonresponders based on E–2 scores, although this study has not yet been replicated.

The reliability and validity data suggest that the E–2 form may have some research value in identifying a small subset of children who meet Kanner's original criteria for autism (Parks, 1988). It is less useful in clinical settings, however, where the objective is to establish a diagnosis that is useful for the development of appropriate treatment programs.

Behavior Rating Instrument for Autistic and Other Atypical Children (BRIAAC)

The Behavior Rating Instrument for Autistic and Other Atypical Children (BRIAAC; Ruttenberg, Dratman, Fraknoi, & Wenar, 1966; Ruttenberg, Kalish, Wenar, & Wolf, 1977) is based on observations of autistic children in a psychoanalytically oriented treatment program. The measure consists of eight scales: relationship, communication, drive for mastery, vocalization, speech and sound reception, social functioning, body movement, and psychobiologic development. Each scale is divided into 10 levels, which range from behaviors that are typical in a normal 4-year-old to behaviors common in autistic children. The scale is completed after extensive direct observation of the child.

In a study examining the interrater reliability of the scales, Wenar and Ruttenberg (1976) had seven pairs of raters observe 113 autistic children and obtained correlation coefficients of .85 to .93. In the same study, they found intercorrelations of the scales ranging from .54 to .86, indicating that the scales were tapping a common entity yet had sufficient variability to justify the different scales. Factor analyses revealed a high loading on a single factor, which Wenar and Ruttenberg (1976) termed resistance to realistic participation in activities. In a related study, Cohen and his colleagues (1978) found a single factor that reportedly accounted for 69% of the variance.

Concurrent validity has been assessed by comparing the ranked BRIAAC scores on 26 autistic and atypical children with the rankings of an expert clinician. There were significant correlations for the total score and three of the scales: relationship, vocalization, and sound and speech reception. The remaining scales were not significantly related to clinicians' rankings. As Parks (1988) pointed out, the fact that the correlations were based on rankings and not actual scores makes the lack of correlation on the other five scales more problematic.

Discriminative validity studies have demonstrated that the BRIAAC scores can distinguish between autistic, mentally retarded, and normal children (Wolf, Wenar, & Ruttenberg, 1972). Unfortunately there is no evidence that this measure can discriminate between different types of behaviorally disturbed children. Cohen et al. (1978) found that the scores on the BRIAAC did not distinguish between groups they labeled as primary and secondly autism, early childhood psychosis, developmental aphasia, and mental retardation.

Although the BRIAAC seems to use some of the same criteria as the E–2 form, Cohen et al. (1978) found no correlation between these two measures. The possibility that this measure may be less helpful in diagnosing autism is consistent with the development of this tool as a measure of progress in a treatment program (Wenar & Rutterberg, 1976).

Behavior Observation System

The Behavior Observation System was first developed by Freeman, Ritvo, Guthrie, Schroth, and Ball (1978) as an objective behavioral assessment tool for diagnosing autism and for documenting changes over time. Based on their initial studies, Freeman and her colleagues (Freeman, Ritvo, & Schroth, 1984) developed a revised version of the BOS, which includes 24 objectively defined behaviors divided into four groups: solitary, relation to objects, relation to people, and language. For the observation period, the child is placed in a room with age-appropriate toys, a child's table and chair, and a chair for the adult observer. The session is videotaped and later rated by trained observers, who assess the occurrence or nonoccurrence of the 24 behavior categories in 10-second intervals.

Freeman et al. (1984) reported 80% interrater agreement during observer training, and interobserver agreement correlation coefficients from .23 to 1.00 with coefficients at .70 for 16 of the 24 scales during the study. The low agreement on six of the eight scales was related to difficulties in accurately measuring low-frequency behaviors.

The revised BOS discriminates both between the behaviors of low-functioning autistic children and mentally retarded children, and also between the behaviors of high-functioning autistic children and normal children (Freeman et al., 1984). There is considerable overlap in the behaviors between the groups, however, such that no one behavior or group of behaviors predicts diagnosis with complete accuracy. Freeman et al. (1984) concluded that behavioral observation systems such as the BOS must capture the quality of social interactions—and not just the quantity—if they are going to accurately diagnose autism.

Adrien, Ornitz, Barthelemy, Sauvage, and Lelord (1987) used a French modification of the BOS to compare the behaviors of normal, severely retarded, and autistic children who were also severely retarded. Controlling for mental age, they found substantial overlap between the groups but also several behaviors that distinguished the autistic children from the others. These behaviors included items identified in DSM-III-R: eye contact, social smile, and inappropriate use of objects. Several sensory behaviors that are not included in DSM-III-R also distinguished the group with autism.

The Autism Screening Instrument for Educational Planning

The Autism Screening Instrument for Educational Planning (ASIEP) was developed by Krug, Arick, and Almond (1979) as an easy-to-use assessment tool to facilitate educational placement. The ASIEP has five subcomponents: the Autism Behavior Checklist, a Sample of Vocal Behavior, Interaction Assessment, Educational Assessment of Functional Skills, and a Prognosis of Learning Rate. To date the only subcomponent that has been examined empirically to any extent is the Autism Behavior Checklist (ABC) (Morgan, 1988; Parks, 1988).

The ABC provides a behavioral profile that can be used to distinguish individuals with autism from other handicapping conditions (Krug et al., 1980). The checklist is completed by teachers and includes 57 behaviors that are grouped into five categories: sensory, relating, body and object use, language, and social self-help. The behaviors were selected from Rimland's E–2 form, the British Working Party Checklist, Creak's (1964) nine points, the BRIAAC, Rendle-Short and Clancy's checklist (1968), Lotter's (1974) checklist, and Kanner's (1943) original list of symptoms. Each item on the checklist is weighted from 1 to 4 with higher weights indicating an item is more indicative of autism. The weightings were based on empirical data from 1,041 completed checklists (Krug et al., 1980). Limited reliability data is available on the scale. Krug et al. (1980) reported 95% agreement by 42 raters of 14 children. The small number of cases limits the generalizability of this study. In a recent study of interrater reliability utilizing the ABC, Volkmar and colleagues (1988) reported acceptable levels of overall probability of agreement (>70%) and the ratio of kappa to maximum kappa (>.40) for 17 of the 57 items.

The discriminate validity of the ABC has been examined in three studies (Krug et al., 1980; Teal & Wiebe, 1986; Volkmar et al., 1988). In the first study (Krug et al., 1980), the ABC was reported to distinguish between autistic, deaf-blind, emotionally disturbed, mentally retarded, and normal children. Teal and Wiebe (1986) found that the ABC and the interaction skills and educational skills subcomponents of the ASIEP were able to correctly identify 100% of the autistic group and 95% of the mentally retarded group with a pooled group accuracy of 97.5%. This measure was more accurate than the E–2 form but not quite as accurate as the CARS in predicting group membership. Volkmar et al. (1988) compared diagnoses based on scores with clinical diagnoses on a sample of 157 adolescents and adults. The ABC correctly identified 80.9% of the autistic sample and 61.9% of the non-autistic developmentally disabled subjects. The authors suggested that false negatives were more common in autistic subjects who were relatively high-functioning. Based on this data and the generally poor reliability of most of its scales, Volkmar et al. (1988) suggested that the ABC may be useful as a screening tool but that the ABC scores alone are not sufficient for diagnosing autism.

Childhood Autism Rating Scale

The Childhood Autism Rating Scale (CARS; Schopler et al., 1980) was developed based on Kanner's (1943) primary features of autism, Creak's nine points (Creak, 1964), and information about symptoms of young children with autism. Although the scales and primary scoring criteria have remained the same, recent revisions have increased its flexibility by extending the conditions under which the CARS can be used (Schopler, Reichler, & Renner, 1988).

The CARS is comprised of 15 scales: relating to people, imitation, emotional response, body use, object use, adaptation to change, visual response, listening response, taste, smell, and touch response and use, fear or nervousness, verbal communication, nonverbal communication, activity level, level and consistency of intellectual response, and general impressions. Each scale is scored from 1 to 4, with 1 indicating normal or age-appropriate behavior and 4 indicating severely abnormal behavior. Ratings from the individual scales are totaled, with the range of total scores being from 15 to 60 points. Based on data from a study of 537 children (Schopler et al., 1980), the total scores are divided into three categories: 15–29.5 is considered nonautistic; 30–36.5 is mildly to moderately autistic, and 37–60 is severely autistic.

The scales are usually completed while observing a child during a formal assessment session that includes both structured and unstructured time. The latest edition of the scale is reported to be appropriate for use by professionals from a variety of backgrounds and can be based on data obtained from parent interviews, school observations, medical records, (Schopler et al., 1988), or the direct observation of a formal assessment.

The CARS has been demonstrated to be very reliable. Based on the ratings of two independent observers of 280 cases, the average interrater reliability was .71, with the range being from .55 for intellectual consistency to .93 for human relatedness (Schopler et al., 1980). In a more recent study involving the assessment of 44 children and adolescents with autism (Garfin, McCallon, & Cox, 1988), the interrater reliability on the CARS was reported to be .80 or better.

The assessment of the internal consistency of the scale has yielded a reliability coefficient of .94 (Schopler et al., 1980). Garfin and her colleagues (Garfin et al., 1988) reported alpha coefficients of .79 for the sample of children and .73 for the group of adolescents. They found that the Inconsistency in Intelligence scale has a negative impact on the internal consistency and recommended that this scale be modified or dropped.

Test–retest reliability was calculated for 91 cases that were assessed during two rediagnostic evaluations that were 1 year apart (Schopler et al., 1988). The correlation between the total CARS scores for these two evaluations was .88, and the means were not significantly different. The diagnoses based on these scores were in agreement 82% of the time, and the kappa value was .64.

Concurrent validity has been assessed by comparing the CARS scores with clinicians' ratings of autism and with independent assessments by a child psychiatrist and a child psychologist. The correlation coefficients were .84 and .80, respectively. The sample included children with autism as well as other communication handicaps and developmental disabilities.

The CARS also successfully discriminates between children who are autistic and those who are trainable mentally retarded (Teal & Wiebe, 1986). The CARS rating successfully predicted group membership with 100% accuracy, which was more accurate than either of the other measures (the E–2 form and the ASIEP) used in the study. Garfin et al. (1988) found that CARS total scores significantly discriminated between autistic and mentally retarded adolescents. Only 1 out of 40 subjects was incorrectly identified.

Additional studies have documented that the completion of the CARS can be based on parent interviews, classroom observations, and chart reviews as well as observations during standardized assessments (Schopler et al., 1988). In one study, CARS ratings for 41 children were completed independently by one rater who interviewed the parents and a second rater who observed a psychoeducational evaluation of the child. The correlation coefficient was .82. Diagnostic agreement (autistic versus nonautistic) among raters was 90% for a coefficient kappa of .75.

In a subsequent study, CARS ratings based on classroom observations of 20 children were compared to CARS ratings of the same children by a second rater who observed them during a psychoeducational evaluation in the clinic. The correlation coefficient was .73, and the raters had a diagnostic agreement rate of 86% for a kappa value of .86.

In the final study, CARS scores were based on chart reviews of 61 cases, which had already been rated on the CARS during the administration of the Psychoeducational Profile. Again the correlation coefficient of .82 indicated good agreement and the diagnosis was the same in 82% of the cases yielding a kappa value of .63.

In order to assess the appropriateness of the CARS for use by professionals with different backgrounds, the ratings of 18 visitors to TEACCH clinics who were medical students, pediatric residents and interns, special educators, school psychologists, speech pathologists, and audiologists were compared to the ratings of the clinical directors (PhD level clinical or school psychologists). The correlation coefficient was .83, and the two groups agreed on a diagnosis 92% of the time for a kappa value of .81 (Schopler et al., 1988).

Recent studies with the CARS have examined its applicability as a diagnostic tool with adolescents and adults with autism. In a study of 59 subjects who had been assessed with the CARS prior to age 10 and then again after age 13 (Mesibov, Schopler, Schaffer, & Michal, 1989; Van Bourgondien & Mesibov, 1989), the average CARS total score decreased by 2.2 as the subjects got older. Based on this information, the authors suggested modifying the cut-off scores for adolescents and adults, making the cut-off score for mild to moderate 28 instead of 30.

With this modification, 92% of the sample initially diagnosed were still identified as autistic when they were adolescents and adults. The Garfin et al. (1988) study found no significant differences in CARS scores between children and adolescents with autism, although the average total score for adolescents was 1.7 points lower than the average score for children. The Garfin study was cross-sectional, however, whereas Mesibov and his colleagues followed the same subjects longitudinally.

COMPARISON OF MEASURES TO DSM-III-R CRITERIA

Table 3.1 compares the content of each of the measures with the 16 criteria for autism in DSM-III-R. As indicated by the table, none of the measures assesses all of the items in the current diagnosis. The CARS, ABC, and E–2 scales tap the most items, whereas the BOS and the BRIAAC address the fewest number of items. Although the E–2 scale includes items related to a number of criteria, this measure assesses the presence and absence of discrete symptoms as opposed to general concepts (e.g., reciprocal social interactions).

Although all of the measures assess the general issue of a child's awareness of the existence of feelings in others, they generally do not address the other items listed under the reciprocal social interactions category; none of the measures assesses the child's comfort-seeking behavior, and only the CARS and the ABC address some of the social play and imitation items. The inclusion of more social items in these two measures may be because they are the more recent ones and reflect the shift in thinking that has occurred regarding the nature of social difficulties. At the time that Rimland first developed the E–2 form, the emphasis was on defining the bizarre behaviors that characterized infantile psychoses. As the field has evolved, a greater understanding of the cognitive deficits has shifted the emphasis toward impairments in social and communication development (Rutter & Schopler, 1987).

As noted in DSM-III-R, in making the diagnosis of autism, the clinician must determine whether the behavior is abnormal for the individual's developmental level. With the possible exception of the E–2 scale, all of the measures to some extent take into account the importance of distinguishing between developmental delays and developmental deviance. The BRIAAC compares the child's functioning—regardless of age—to the behaviors of typical 3- to 4-year-old children. During the BOS, the situation is structured so that the child is given developmentally appropriate materials. The scales on both the ABC and CARS were constructed to make higher scores indicative of deviance in development as opposed to delays in development. As mentioned earlier, discriminant validity studies have been done on all five measures to demonstrate that they can distinguish between mentally retarded children and children with autism.

Based on the reliability and validity studies as well as the overlap with DSM-III-R criteria for autism, the CARS has the most empirical support for its utility in diagnosing autism (Morgan, 1988). The CARS also parallels DSM-III-R most closely with seven of its scales relating to primary characteristics of autism and four relating to secondary characteristics. As Rutter and Schopler (1987) recommended, however, a comprehensive approach to diagnosis requires more than a simple rating scale and should include an in-depth parent interview that can elicit key diagnostic features. If the goal is to generate treatment and programming suggestions, other assessment strategies and techniques should be added as well (Parks, 1988).

SUMMARY AND FUTURE DIRECTIONS

Since its identification in 1943, our understanding of the autism syndrome has evolved rapidly and made it one of the most carefully researched and thoroughly understood of the diagnostic categories in DSM-III-R. Given the enormous range of abilities that are included under autism and the subtleties of the social and communication problems that these people have, this is a major accomplishment. In addition, several rating scales have been developed to aid in the diagnostic process, some of which have demonstrated reliability and validity and also adhere closely to the characteristics outlined in DSM-III-R.

The changes in the categorization and description of autism in DSM-III-R generally correspond with both the empirical data and current clinical notions about the disorder. Greater emphasis has been placed on autism as a lifelong, developmental disability that manifests itself in a variety of ways. The definition of the major category of social difficulties has been expanded to include impairments in reciprocal interactions. Likewise, the communication domain now includes nonverbal as well as verbal deficits. Eliminating the age of onset and a number of poorly defined diagnoses under PDD should add to the reliability of diagnoses in this area.

In spite of the impressive progress that has been made, however, there is still work to be done and some difficult issues to be resolved as we prepare for future diagnostic systems. The goal of this work will be further clarification of this diagnostic category and its relationship to other pervasive developmental disorders. More empirical studies to document the interrater and test–retest reliability of the criteria are needed. Further study should also address the issue of redundancy within the criteria. Studies that compare DSM-III-R criteria with established diagnostic instruments may provide a more reliable standard with which to evaluate the sensitivity and specificity of the diagnostic criteria. Perennial questions about how well very high- and/or low-functioning clients and those with related problems fit into the autism syndrome continue.

Comparisons with other developmentally disabled groups matched for mental age will help make distinctions at the lower IQ levels. There is also a need for further clarification and refinement of the specific social, communication, and behavioral problems that define autism as a separate disability. Areas not addressed in the current definition include problems with receptive communication processes and specific cognitive difficulties such as organization, sequencing, and attention.

Clarifying the diagnosis of PDDNOS would also be beneficial. This diagnosis is currently used to describe individuals who are not described by any other DSM-III-R diagnoses and are similar to those who meet the criteria for autistic disorder. Documenting which of the criteria for autism these individuals do meet will help in the identification of the group as a whole as well as possible subgroups.

Although the remaining questions are difficult, and the issues are both subtle and complex, we can expect that the past 45 years of progress and current interest in diagnosis and autism will be most helpful in addressing these and other important concerns.

REFERENCES

Adrien, J. L., Ornitz, E., Barthelemy, C., Sauvage, D., & Lelord, G. (1987). The presence or absence of certain behaviors associated with infantile autism in severely retarded autistic and nonautistic retarded children and very young normal children. *Journal of Autism and Developmental Disorders, 17,* 407–416.

American Psychiatric Association. (1980). *Diagnostic and statistical manual* (3rd ed.). Washington, DC: Author.

American Psychiatric Association. (1987). *Diagnostic and statistical manual* (3rd ed., rev.). Washington, DC: Author.

Bettelheim, B. (1967). *The empty fortress.* New York: Free Press.

Boatman, M. J., & Szurek, S. (1960). A clinical study of childhood schizophrenia. In D. Jackson (Ed.), *The etiology of schizophrenia* (pp. 389–440). New York: Basic.

Boullin, D. J., Coleman, M., O'Brian, R. A., & Rimland, B. (1971). Laboratory predictions of infantile autism, based on 5–hydroxytryptamine efflux from blood platelets and their correlation with the Rimland E–2 scores. *Journal of Autism and Childhood Schizophrenia, 1,* 63–71.

Boullin, D. J., Freeman, B. J., Geller, E., Ritvo, E., Rutter, M., & Yuwiler, A. (1982). Toward the resolution of conflicting findings (letter to the editor). *Journal of Autism and Developmental Disorders, 12,* 97–98.

Byassee, J., & Murrell, S. (1975). Interaction patterns in families of autistic, disturbed, and normal children. *American Journal of Orthopsychiatry, 45,* 473–478.

Cantwell, D. P., Baker, L., & Rutter, M. (1977). Families of autistic and dysphasic children: Vol. 2: Mothers' speech to the children. *Journal of Autism and Childhood Schizophrenia, 7,* 313–327.

Cohen, D. J., Caparulo, B. K., Gold, J. R., Waldo, M. C., Shaywitz, B. A., Ruttenberg, B. A., & Rimland, B. (1978). Agreement in diagnosis: Clinical assessment and behavior rating scales for pervasively disturbed children. *Journal of the American Academy of Child Psychiatry, 17,* 589–603.

Cox, A., Rutter, M., Newman, S., & Bartak, L. (1975). A comparative study of infantile autism and specific developmental receptive language disorder: II. Parental characteristics. *British Journal of Psychiatry, 126,* 146–159.

Creak, E. M. (1963). Childhood psychosis: A review of 100 cases. *British Journal of Psychiatry, 109,* 84–89.

Creak, M. (1964). Schizophrenia syndrome in childhood: Further progress report of a working party (April, 1964). *Developmental Medicine and Child Neurology, 6,* 530–535.

Creak, M., & Ini, S. (1960). Families of psychotic children. *Journal of Child Psychology and Psychiatry, 1,* 156–175.

DeMyer, M. K., Churchill, D. W., Pontius, W., & Gilkey, K. M. (1971). A comparison of five diagnostic systems for childhood schizophrenia and infantile autism. *Journal of Autism and Childhood Schizophrenia, 1,* 175–189.

DeMyer, M. K., Pontius, W., Norton, J. A., Barton, S., Allen, J., & Steele, R. (1972). Parental practices and innate activity in normal, autistic, and brain-damaged infants. *Journal of Autism and Childhood Schizophrenia, 2,* 49–66.

Despert, J. (1951). Some considerations relating to the genesis of autistic behavior in children. *American Journal of Orthopsychiatry, 21,* 335–350.

Eisenberg, L. (1956). The autistic child in adolescence. *American Journal of Psychiatry, 112,* 607–612.

Freeman, B. J., Ritvo, E. R., Guthrie, D., Schroth, P., & Ball, J. (1978). The Behavioral Observation Scale for Autism: Initial methodology, data analysis, and preliminary findings on 89 children. *Journal of the American Academy of Child Psychiatry, 17,* 576–588.

Freeman, B. J., Ritvo, E. R., & Schroth, P. C. (1984). Behavior assessment of the syndrome of autism: Behavior Observation System. *Journal of the American Academy of Child Psychiatry, 23,* 588–594.

Garfin, D. G., McCallon, D., & Cox, R. (1988). Validity and reliability of the Childhood Autism Rating Scale with autistic adolescents. *Journal of Autism and Developmental Disorders, 18,* 367–378.

Kanner, L. (1943). Autistic disturbances of affective contact. *Nervous Child, 2,* 217–250.

Kanner, L. (1973). *Childhood psychosis: Initial studies and new insights.* Washington, DC: Winston.

Kolvin, I., Garside, R., & Kidd, J. (1971). Studies in the childhood psychoses IV. Parental personality and attitude and childhood psychoses. *British Journal of Psychiatry, 118,* 403–406.

Krug, D. A., Arick, J. R., & Almond, P. J. (1979). Autism Screening Instrument for Educational Planning: Background and development. In J. Gilliam (Ed.), *Autism: Diagnosis, instruction, management, and research* (pp. 64–78). Austin: University of Texas at Austin Press.

Krug, D. A., Arick, J., & Almond, P. (1980). Behavior checklist for identifying severely handicapped individuals with high levels of autistic behavior. *Journal of Child Psychology and Psychiatry, 21,* 221–229.
Lennox, C., Callias, M., & Rutter, M. (1977). Cognitive characteristics of parents of autistic children. *Journal of Autism and Childhood Schizophrenia, 7,* 243–261.
Lotter, V. (1974). Factors related to outcome in autistic children. *Journal of Autism and Childhood Schizophrenia, 4,* 263–276.
Lotter, V. (1978). Follow-up studies. In M. Rutter & E. Schopler (Eds.), *Autism: A reappraisal of concepts and treatment* (pp. 475–496). New York: Plenum.
Lovaas, O. I. (1987). Behavioral treatment and normal educational and intellectual functioning in young autistic children. *Journal of Consulting and Clinical Psychology, 55,* 3–9.
Mesibov, G. B., Schopler, E., Schaffer, B., & Michal, N. (1989). Use of the Childhood Autism Rating Scale with autistic adolescents and adults. *Journal of the American Academy of Child and Adolescent Psychiatry, 28,* 538–541.
Morgan, S. (1988). Diagnostic assessment of autism: A review of objective scales. *Journal of Psychoeducational Assessment, 6,* 139–151.
Netley, C., Lockyer, L., & Greenbaum, G. (1975). Parental characteristics in relation to diagnosis and neurological status in childhood psychosis. *British Journal of Psychiatry, 127,* 440–444.
Ornitz, E. M., & Ritvo, E. R. (1968). Perceptive inconsistency in early infantile autism. *Archives of General Psychiatry, 18,* 76–98.
Parks, S. L. (1988). Psychometric instruments available for the assessment of autistic children. In E. Schopler & G. Mesibov (Eds.), *Diagnosis and assessment in autism* (pp. 123–136). New York: Plenum.
Pitfield, M., & Oppenheim, A. N. (1964). Child rearing attitudes of mothers of psychotic children. *Journal of Child Psychology and Child Psychiatry, 5,* 51–57.
Prior, M., & Bence, R. (1975). A note on the validity of the Rimland Diagnostic Checklist. *Journal of Clinical Psychology, 31,* 510–513.
Rank, B. (1949). Adaptation of the psychoanalytic technique for the treatment of young children with atypical development. *American Journal of Orthopsychiatry, 19,* 130–139.
Rendle-Short, J. (1969). Infantile autism in Australia. *Medical Journal of Australia, U2,* 245–249.
Rendle-Short, J., & Clancy, H. G. (1968). Infantile autism. *Medical Journal of Australia, 1,* 921–922.
Rimland, B. (1964). *Infantile autism: The syndrome and its implication for a neural theory of behavior.* New York: Appleton-Century-Crofts.
Rimland, B. (1968). On the objective diagnosis of infantile autism. *Acta Paedopsychiatarica, 35,* 146–161.
Rimland, B. (1971). The differentiation of childhood psychoses: An analysis of checklists for 2,218 psychotic children. *Journal of Autism and Childhood Schizophrenia, 1,* 161–174.
Rimland, B. (1973). The effect of high dosage levels of certain vitamins on the behavior of children with severe mental disorders. In D. R. Hawkins & L. Pauling (Eds.), *Orthomolecular psychiatry* (pp. 513–539). San Francisco: Freeman.
Ritvo, E. R., & Freeman, B. J. (1977). National Society for Autistic Children definition of the syndrome of autism. *Journal of Pediatric Psychology, 2,* 146–148.
Ritvo, E. R., & Freeman, B. J. (1978). National Society for Autistic Children definition of autism. *Journal of Autism and Developmental Disorders, 8,* 162–167.
Ruttenberg, B. A. (1971). A psychoanalytic understanding of infantile autism and its treatment. In D. Churchill, G. Alpern, & M. DeMyer (Eds.), *Infantile autism: Proceed-*

ings of the *Indiana University Colloquium* (pp. 145–184). Springfield, IL: Thomas.

Ruttenberg, B. A., Dratman, M. L., Fraknoi, J., & Wenar, C. (1966). An instrument for evaluating autistic children. *Journal of the American Academy of Child Psychiatry, 5,* 453–478.

Ruttenberg, B. A., Kalish, B. I., Wenar, C., & Wolf, E. G. (1977). *Behavior rating instrument for autistic and other atypical children* (rev. ed.). Philadelphia: Developmental Center for Autistic Children.

Rutter, M. (1966). Prognosis: Psychotic children in adolescence and early adult life. In J. K. Wing (Ed.), *Early childhood autism: Clinical, educational, and social aspects* (pp. 83–99). Oxford: Pergamon.

Rutter, M. (1978). Diagnosis and definition of childhood autism. *Journal of Autism and Childhood Schizophrenia, 8,* 139–161.

Rutter, M., & Schopler, E. (Eds.). (1978). *Autism: A reappraisal of concepts and treatment.* New York: Plenum.

Rutter, M., & Schopler, E. (1987). Autism and pervasive developmental disorders: Concepts and diagnostic issues. *Journal of Autism and Developmental Disorders, 17,* 159–186.

Rutter, M., & Schopler, E. (1988). Autism and pervasive developmental disorders: Concepts and diagnostic issues. In E. Schopler & G. B. Mesibov (Eds.), *Diagnosis and assessment in autism* (pp. 15–36). New York: Plenum.

Schain, R. J., & Yannet, H. (1960). Infantile autism: Analysis of 50 cases and a consideration of certain relevant neurophysiologic concepts. *Journal of Pediatrics, 57,* 560–567.

Schopler, E. (1971). Parents of psychotic children as scapegoats. *Journal of Contemporary Psychotherapy, 4,* 17–22.

Schopler, E. (1983). Introduction: Can an adolescent or adult have autism? In E. Schopler & G. B. Mesibov (Eds.), *Autism in adolescents and adults* (pp. 3–10). New York: Plenum.

Schopler, E., & Loftin, J. M. (1969). Thought disorders in parents of psychotic children: A function of test anxiety. *Archives of General Psychiatry, 20,* 174–181.

Schopler, E., & Mesibov, G. B. (Eds.). (1983). *Autism in adolescents and adults.* New York: Plenum.

Schopler, E., Reichler, R. J., DeVellis, R. F., & Daly, K. (1980). Toward objective classification of childhood autism: Childhood Autism Rating Scale (CARS). *Journal of Autism and Developmental Disorders, 10,* 91–103.

Schopler, E., Reichler, R. J., & Lansing, M. (1980). *Individualized assessment and treatment for autistic and developmentally disabled children (Vol. 2). Teaching strategies for parents and professionals.* Baltimore: University Park Press.

Schopler, E., Reichler, R. J., & Renner, B. R. (1988). *The Childhood Autism Rating Scale (CARS).* Los Angeles: Western Psychological Services.

Schopler, E., Short, A., & Mesibov, G. B. (1989). Relation of behavioral treatment to "normal functioning": Comment on Lovaas. *Journal of Consulting and Clinical Psychology, 57,* 1–3.

Siegel, B. S., Vukicevic, J., Elliott, C. R., & Kraemer, H. C. (1989). The use of signal detection theory to assess DSM-III-R criteria for autistic disorder. *Journal of American Academy of Child and Adolescent Psychiatry, 28,* 542–548.

Spitzer, R. L., & Siegel, B. (1989). The *DSM-III-R field trial of pervasive developmental disorders.* Unpublished manuscript.

Teal, M. B., & Wiebe, M. J. (1986). A validity analysis of selected instruments used to assess autism. *Journal of Autism and Developmental Disorders, 16,* 485–494.

Tinbergen, E. A., & Tinbergen, N. (1972). Early childhood autism: An ethological approach. In *Advances in Ethology, 10,* Supplement to *Journal of Comparative Ethology.* Berlin and Hamburg: Verlag Paul Barry.

Van Bourgondien, M., & Mesibov, G. B. (1989). Diagnosis and treatment of adolescents and adults with autism. In G. Dawson (Ed.), *Autism* (pp. 367–385). New York: Guilford.

Volkmar, F. R., Bregman, J., Cohen, D. J., & Cicchetti, D. V. (1988). DSM-III and DSM-III-R diagnoses of autism. *American Journal of Psychiatry, 145,* 1404–1408.

Volkmar, F. R., Cicchetti, D. V., Dykens, E., Sparrow, S., Leckman, J. F., & Cohen, D. J. (1988). *An evaluation of the Autism Behavior Checklist, 18,* 81–98.

Volkmar, F. R., & Cohen, D. J. (1986). Current concepts: Infantile autism and the pervasive developmental disorders. *Journal of Developmental and Behavioral Pediatrics, 7,* 324–329.

Volkmar, F. R., Cohen, D. J., & Paul, R. (1986). An evaluation of DSM-III criteria for infantile autism. *Journal of American Academy of Child and Adolescent Psychiatry, 25,* 190–197.

Volkmar, F. R., Stier, D., & Cohen, D. J. (1985). Age of onset of pervasive developmental disorder. *American Journal of Psychiatry, 142,* 1450–1452.

Wenar, C., & Ruttenberg, B. A. (1976). The use of BRIAC for evaluating therapeutic effectiveness. *Journal of Autism and Childhood Schizophrenia, 6,* 175–191.

Wing, L., & Ricks, D. M. (1976). The etiology of childhood autism: A criticism of the Tinbergen's ethological theory. *Psychological Medicine, 6,* 533–544.

Wing, L., Yeates, S. R., & Brierley, L. M. (1977). Symbolic play in severely mentally retarded and in autistic children. *Journal of Child Psychology and Psychiatry, 18,* 167–178.

Wolf, E. G., Wenar, C., & Ruttenberg, B. A. (1972). A comparison of personality variables in autistic and mentally retarded children. *Journal of Autism and Childhood Schizophrenia, 2,* 92–108.

Yuwiler, A., Ritvo, E., Geller, E., Glousman, R., Schneiderman, G., & Matsuno, D. (1975). Uptake and efflux of serotonin from platelets of autistic and nonautistic children. *Journal of Autism and Childhood Schizophrenia, 5,* 83–98.

CHAPTER 4

Developmental Arithmetic Disorder

Margaret Semrud-Clikeman
Massachusetts General Hospital

George W. Hynd
University of Georgia
Medical College of Georgia

ARITHMETIC DISORDERS: SERIATION, CALCULATION, AND CONCEPTS

Arithmetic disabilities have not been explored as completely as disorders in reading skill development. Possibly due to the increasing need for a mathematically literate population (Smith, 1983), only recently have empirical studies begun to elucidate mathematical abilities. It has also been noted that learning disabled children who are identified as reading disabled often have arithmetic delays as well (Reid & Hresko, 1981).

Rosner (1973) found that arithmetic readiness in preschool children involves visuospatial organization and coordination. A child who is able to separate blocks in a uniform fashion as well as point to blocks as he or she counts is able succeed more readily than a child who is not able to organize and plan his motoric movements (Johnson & Myklebust, 1967; Kaliski, 1962). Michael, Guilford, Fruchter, and Zimmerman (1957) postulated that the ability to manipulate and represent spatial relationships mentally is directly linked to elementary mathematic operations. A child who can visualize a quantity can more readily access that visualization to solve a problem. For instance, "six apples plus three apples" requires an understanding of the concept of six as well as the ability to relate six to three. The child then can more readily count three more apples from the six already visualized.

Developmental differences appear in the ability to visualize quantity. For example, Groen and Parkman (1972) studied the processes that children used during mental addition of two integers. They found that younger children utilized counting to solve the problems using the smallest number as the starting point, presumably because this was the quantity they could most easily visualize. Older children and adults utilized the largest quantity and used their past retrieval of learned facts. Several other studies support counting models for children through third grade (Svenson, 1975b; Svenson & Broquist, 1975; Svenson, Hedenborg, & Lingman, 1976). Therefore, success in arithmetic processes involves the ability to conceptualize quantity, utilize seriation abilities, and retrieve previously learned material.

Larsen and Hammill (1975) reviewed over 600 studies and found that visual discrimination, memory for visual sequences, and visual-motor coordination correlated highest with arithmetic success. Verbal abilities were also found to be highly correlated with success in arithmetic with increasing chronological age. Higher-level logical reasoning skills, sequencing abilities, and comprehension appear to be highly related with the ability to conceptualize relationships regarding money, time, measurement, and complex word problem solving (McLeod & Crump, 1978).

Thus, success in arithmetic skill acquisition differs depending on developmental stage, experiential background, and skill development. In addition, anxiety, concentration and attention difficulties, cultural attitude, or poor teaching can also impact on a student's arithmetic aptitude and skill development (Gaddes, 1980). From a neurological perspective, mathematics clearly requires the integration of visual, motor, language, and attentional processes, perhaps with each process having a different point of ascendancy. It is thought to involve parietal, temporal, and occipital regions in both cerebral hemispheres (Luria, 1980).

Language skills also appear to be important for mathematics success, as mathematics is just another type of language in number form (Kosc, 1981). Word problems in later grades directly involve reading and reading comprehension. The more complex the problem, the more difficult for a child to solve (Larsen, Parker, & Trenholme, 1978). For example, word problems in fifth grade may involve several steps, each requiring a different arithmetic operation as well as requiring the ability to decode the words directing the problem solution. Thus, if a child has deficits in reading skills, it becomes a chore to not only decode the words but to understand the required operations needed.

This chapter seeks to explore the nosological history of arithmetic disability through DSM-III (APA, 1980) and DSM-III-R (APA, 1987) as well as through the development of the learning disability field. As the preceding paragraphs indicate, several subsystems are necessary for success in mathematics. These subsystems complicate the definition of an arithmetic disability. Definitional issues are addressed in relation to the available research in this area. A review of relevant literature is presented to elucidate current empirical findings with an eye to the relevance and appropriateness of the DSM-III and DSM-III-R criteria. Finally, suggestion for diagnosis and assessment and some discussion of intervention are provided.

HISTORICAL BACKGROUND AND DEFINITIONAL ISSUES

The existence of children with primary arithmetic disability has been largely ignored by researchers until recently (Weinstein, 1980). One of the early case studies of arithmetic disability was in 1922 (Eliasberg & Feuchtwanger, 1922). Problems included confusion of arithmetic signs (+ or −), an inability to copy numbers, problems with number naming, and number repetition. These studies were with adults who had demonstrated brain lesions. Although there are descriptions of children with arithmetic learning disabilities, these studies are usually case studies (Buswell & John, 1926). More recently, a study by Homan (1970) further pinpointed specific arithmetic disabilities in children ranging from difficulties with place value to calculation deficits. However, not many researchers have solely studied developmental arithmetic disabilities.

The preceding difficulties have been compounded by definitions provided by DSM-III and DSM-III-R that are so encompassing that they are of little use. Attempts by PL 94–142 (Federal Register, 1977) to give arithmetic difficulties an equal emphasis with reading and written language disabilities has been a small but positive step in addressing the need for treatment-based definitions. Table 4.1 presents the major differences among definitions provided by PL 94–142, DSM-III, and DSM-III-R.

Arithmetic disabilities were given equal emphasis along with reading, language, and writing disorders with the passage of Public Law 94–142 in 1977. This definition stated that "Specific learning disability means a disorder in one or more of the basic psychological processes involved in understanding or in using language, spoken or written, which may manifest itself in an imperfect ability to listen, think, speak, read, write, spell, or do mathematical calculations" (Federal Register, Dec. 29, 1977, p. 65083).

TABLE 4.1
Definitional Differences Among the Three Nosologies for Arithmetic Disabilities

Criteria	*P.L. 94–142*	*DSM-III*	*DSM-III-R*
Concurrent disabilities allowed	No restriction	Reading and spelling disabilities generally not as pronounced	No restriction
Specification of type arithmetic disability	Calculation or reasoning	No	Linguistic, perceptual, attention, mathematical skills of counting, following directions, and learning multiplication tables
Language involvement addressed	Yes	No	Yes
Degree of discrepancy from IQ	Severe; specified by individual states—usually is 15–20 standard score point difference or > 1.5 standard deviation	Achievement significantly below ability—not further specified. Requires individual IQ test	Achievement significantly below ability—not further specified. Requires individual IQ test
Presumed neurological involvement	Includes examples such as perceptual handicaps, dyslexia, developmental aphasia, and minimal brain dysfunction	No	No
Exclusions	Primary cause may not be: 1) Visual, hearing, motor handicap 2) Mental retardation 3) Emotional disturbance 4) Environmental, cultural, or economic disadvantage	Inadequate schooling	1) Mental retardation 2) Visual & hearing handicaps

The law's regulations developed the following identification criteria:

(a) A team may determine that a child has a specific learning disability if:
1. The child does not achieve commensurate with his or her age and ability levels in one or more areas . . . when provided with learning experiences appropriate for the child's age and ability levels; and
2. The team finds that a child has a severe discrepancy between achievement and intellectual ability in one or more of the following areas:
(i) Oral expression;
(ii) Listening Comprehension;
(iii) Written expression;
(iv) Basic reading skill;
(v) Reading comprehension;
(vi) Mathematics calculation; or
(vii) Mathematics reasoning.

(b) The team may not identify a child as having a specific learning disability if the severe discrepancy between ability and achievement is primarily the result of:
1. A visual, hearing, or motor handicap;
2. Mental retardation;
3. Emotional Disturbance; or
4. Environmental, cultural, or economic disadvantage.
(Federal Register, Dec. 29, 1977, p. 65083)

This definition provided a basis to allow for the assessment of children with primary arithmetic disabilities.

Compared to the federal definition, DSM-III provided a very terse definition of developmental arithmetic disorder. This definition (APA, 1980) states the following:

> The essential feature is significant impairment in the development of arithmetic skills not accounted for by chronological age, mental age, or inadequate schooling. In addition, in school, the child's performance on tasks requiring arithmetic skills is significantly below his or her intellectual capacity. The diagnosis can be made only by individually administered IQ tests that yield a level or full-scale IQ, plus a variety of academic achievement tests containing arithmetic subtests. (p. 94)

This definition does not address the contribution of emotional and behavioral difficulties such as anxiety and/or concentration. It also states that, whereas reading and spelling disabilities may also be present, they are not as pronounced. Moreover, DSM-III states that "the disorder is apparently not common" (p. 95). However, Weinstein (1978) suggested that almost 6% of school children suffer from "developmental dyscalculia." This percentage would suggest that the disorder was not uncommon in the population. At the present time, the incidence of primary arithmetic disability is largely unknown. Badian (1983) found that 6.4% of elementary students in a small town showed poor mathematics achievement compared to 4.9% with reading delays. Moreover, Strang and Rourke (1985) found that many of a group of randomly selected children (ages 8–12 years) referred for reading and writing difficulties also showed calculation difficulties.

DSM-III-R expands and refines the definition. No longer is the prevalence seen as uncommon; rather, DSM-III-R states that "the disorder is less common than developmental reading disorder" (p. 41). No attempt is made to distinguish arithmetic disabilities from anxiety or other types of emotional disturbance. This omission is of concern given the findings of Slade and Russell (1971) and Strang and Rourke (1985) that emotionally disturbed children exhibit very poorly developed arithmetic skills. However, DSM-III-R does expand the original DSM-III definition by including deficits in linguistic, perceptual, attention, and mathematical skills. Such deficits involve problems in copying of problems, understanding and naming operations and concepts, recognizing numerical signs, sequencing, counting, and multiplying objects.

Given these three definitions as well as their differences, what empirical support is there for such a diagnosis? Further, how do arithmetic skills develop and differ with various developmental stages? These topics are further examined in the following sections so that some conclusions can be drawn with regard to the validity of DSM-III-R's conceptualization, definition, and criteria for diagnosis.

REVIEW OF THE LITERATURE

Development of Arithmetic Abilities

In the first months of an infant's life, the development of object permanence begins. At the same time, the central nervous system is developing, and dendritic arborization expands during the postnatal ages of 3, 15, and 24 months (Pribram, 1971). Piaget and Inhelder (1971) stressed the importance of object permanence for the development of an understanding of number or quantity. This knowledge develops during the ages of 4 to 7 years. Prior to this point, a child has not developed an understanding of the property of conservation of quantity. The child needs to learn that a number is a number regardless of the unit or shape it takes. This developmental process becomes established approximately by age 6 or 7 years (Piaget, 1941). A child is ready to master arithmetic concepts once he or she is able to understand quantity as well as to manipulate two pieces of information mentally at the same time (Gelman, 1969; Piaget, 1976). In the next stage, the child learns to write and compute using various arithmetical operations (addition, subtraction, multiplication, and division). The child is also developing an understanding of maps, time, space, and measurement. In middle school, the child learns to solve mathematical reasoning problems that involve analysis and synthesis of information.

Table 4.2 presents the various prerequisites for success in the preschool, intermediate, and adolescent periods. Piaget (1941) and Luria (1980) contributed greatly to our understanding of the development of number concepts and arithmetic abilities. They found that the young child (aged 3–6 years) gains experience with tangible objects that can be arranged and rearranged based on various properties. The child utilizes spatial and visual information in arranging these objects in various patterns, which in turn leads to understanding of number and quantity. In addition, the young child begins to understand basic concepts such as more than, less than, longer, shorter, and so on. These skills and concepts lead into the ability to write numbers and associate these numbers with the previously learned quantities. This development usually begins between the ages of 6 and 10 years, which corresponds to the elementary school years. At this time, the child's task is to automatize previous learning of numbers through writing and drill. In addition, practice is given in serial order and in writing problems in correct spatial alignment. At the ages of 11–13 years, the child learns to deductively reason. Problems are presented that involve operations not necessarily stated. For example, word problems become increasingly prevalent. It is entirely possible that a child may be proficient in arithmetic skills up to age 10 but show deficits once abilities in abstract and deductive reasoning are needed in order to solve word problems. The adolescent continues to utilize these basic skills and apply them to everyday living situations, such as reading maps and graphs and developing skills in computers, calculators, and estimation.

How, then, can arithmetic abilities become disordered? There are so many facets to the development of arithmetic abilities that problems in one phase may impact differently on the resulting skill pattern. A child with deficits in memorizing basic arithmetic facts evidences different needs than a child who is unable to understand how to apply arithmetic skills to an abstract problem. The following section seeks to describe areas of weakness as well as factors contributing to difficulties in arithmetic.

TABLE 4.2
Prerequisites for Arithmetic Success*

Preschool (*3–6 years*)	*Elementary* (*6–12 years*)	*Adolescence* (*12–17 years*)
Ability to:	Ability to:	Ability to:
understand same and different	group objects by 10s	use numbers in everyday life
match objects by size, color, shape	read and write 0–99	(i.e., measuring; using recipes,
sort objects by characteristics	tell time	using dewey decimal system,
understand concepts of:	solve problems with missing elements	using Roman numerals)
long few	understand halves and fourths	use calculators, adding
short some	measure objects	machines, computers
big less than	name value of coins	use estimation of cost, amount,
small more than	measure volume	in shopping
arrange objects by size	count by 2s, 5s, 10s	read charts, graphs, maps
understand 1-to-1 correspondence	solve addition and subtraction	understand directions
use objects for simple addition	use regrouping	utilize problem solving for home
recognize numbers 0–9	understand ordinal numbers	projects or building
count to 10	complete easy word problems	understanding probability
reproduce block designs	beginning map skills	develop flexible problem solving
copy numbers	judging time lapse	
group objects by named number	estimate solutions	
label shapes	perform basic arithmetic operations	
reproduce shapes and complex figures		

*Composite of skills mentioned by Piaget (1941) and Smith (1983).

Specific Arithmetic Disabilities

Three main underpinnings appear to be involved in arithmetic disability: deficits in language, reading, and/or spatial imagery (Gaddes, 1980). Children with cerebral palsy and mental retardation appear to be at greatest risk for arithmetic difficulties, not only due to brain damage but also because it is more difficult to compensate for lack of arithmetic skill by simple drill (Kinsbourne & Caplan, 1979). Kosc (1981) described several information-processing difficulties that negatively impinge on arithmetic ability. Difficulty in verbal expression of arithmetic terms and relations, inability to read numerical symbols, inability to correctly copy arithmetic problems, difficulty in carrying out arithmetic operations, disturbance in mental calculation, and a history of inability to manipulate objects physically were all found to be related to arithmetic difficulty. Moreover, visual memory difficulties have also been found to be related to learning disabilities in arithmetic (Strang & Rourke, 1985). Tuoko (1982) found that children who experienced visual memory difficulties had more difficulty mastering mechanical arithmetic skills. In other words, these children were unable to remember the steps needed in order to subtract, add, multiply, or divide as well as how to approach the problem initially. One can readily see the pervasive influence engendered by visual memory problems. Moreover, Rourke and Finlayson (1978) reported that children with severe verbal memory deficits not only had arithmetic problems but also significant delays in reading and writing skills. Children with arithmetic disabilities have also been found to experience difficulty with understanding place value, regrouping, and mental arithmetic (Kinsbourne & Caplan, 1979). Although difficulty is not present in understanding numerical concepts, number order is often reversed due to difficulty in memory for numbers read orally, and there is an inability to order information as well. In other words, the numbers can be recalled but are generally in the wrong order (134 instead of 143). In this instance, the number sense is not disordered, but rather the information is not accurately visually processed.

Children with spatial difficulties often experience difficulty copying problems as well as understanding the basic concept of "greater than" and "less than." In a study of children with learning disabilities, Greenstein and Strain (1977) found frequent errors in alignment of numerals or placement of figures on a page, misunderstanding of subtraction rules (such as subtracting the smaller number irrespective of its place in the problem, e.g., $43 - 27 = 24$), or failure to utilize problem information such as decimal points, dollar signs, or fractions. In these cases, difficulty is present in the child's ability to interpret information visually as well as a deficit in visual-spatial knowledge. Luria (1980) suggested that, when parts of the brain responsible for spatial imagery are damaged (i.e., temporal-parietal-occipital association areas), the child will be unable to represent spatial relationships mentally and thus be unable to calculate and perform arithmetic operations correctly. Consequently difficulties with spatial imagery can significantly impact on arithmetic competence. In support of this notion, Gaddes and Spellacy (1977) found significant correlations between arithmetic and sequential memory for light patterns. They concluded that, whereas "sequential memory may be related to arithmetical competence, it seems almost certain that the spatial skills demanded by the test are also strongly involved" (p. 291).

Children with arithmetic difficulties have also been found to experience difficulty with abstract reasoning and cognitive flexibility (Lee & Hudson, 1981; Strang & Rourke, 1983). For example, Strang and Rourke (1983) administered the category test to two groups of children with learning disabilities: one group with average arithmetic abilities but dysfunctional reading and spelling, and a second group with low arithmetic but intact reading and spelling skills. The group with deficient arithmetic skills scored one standard deviation below the mean on the category test, whereas the group with deficient reading ability scored at the mean. Thus, the children with arithmetic difficulties were found to have less ability to be flexible and adaptive in their thinking and to show poorer problem-solving abilities.

These problem-solving difficulties have also been found to impact negatively on the ability to understand interpersonal situations and interactions. Kirby and Asman (1984) found that children with deficient arithmetic skills also had difficulty learning appropriate social skills as well as generalizing previously learned interactional abilities to various situations. Children with arithmetic disabilities have also been found to show disturbances in body image and representation (Badian, 1983), social perception of self and others (Gaddes, 1985), and interpretation of emotional as well as spatial tasks (Wiig & Harris, 1974). Moreover, Badian and Ghublikian (1983) found that children who were good readers but did poorly in math computation tended to be more inattentive and disorganized, avoided responsibility, and were not as socially appropriate as children with only reading difficulties. For a more complete review of the relationship between arithmetic disability and social problems in learning disabled children, see Semrud-Clikeman and Hynd (1990).

Finally, poor reading skill acquisition can negatively impact on arithmetic performance. Cohn (1971) found that improvement in arithmetic skills is directly related to improvement in general neurological status, development of reading and writing skills, and visual recall. By fourth grade, arithmetic instruction moves from drills of number facts to the use of word problems. Children with reading difficulties may at this stage experience a concurrent arithmetic problem. It may also be that arithmetic and reading disability frequently coexist because of information-processing disturbances that negatively impact on math, especially at the more advanced levels, due to shared problems with language, attention, and basic conceptual skills. Compounded with these difficulties, core arithmetic skills tend to be interrelated with other subject areas, such as science and social studies; thus, core arithmetic skill deficits expand and impact on subjects that had previously been undisturbed. Children with arithmetic difficulties may show such deficiencies due to underlying reading and comprehension problems rather than a primary arithmetic disability (Stone & Michals, 1986). Syntactic complexity may also negatively impact on arithmetic performance (Bennett, 1981; Fafard, 1976; McCuller, 1981). Therefore, future studies need to control for reading level and complexity when examining arithmetic-based learning disability.

Subtypes of Arithmetically Disabled Children

Rourke and associates (Rourke & Finlayson, 1978; Strang & Rourke, 1983; Strang & Rourke, 1985) studied different subtypes of learning disabled children with arithmetic difficulties. They selected three groups of children, ages 9–14 years, based on their performance on the Wide Range Achievement Test (WRAT) reading, spelling, and arithmetic subtests. The first group evidenced difficulties in all academic areas; the second had stronger arithmetic than reading and spelling skills, although all areas were below age expectations; and the final group had average reading and spelling skills with deficient arithmetic ability. It should be remembered that the WRAT arithmetic subtest involves calculation and not necessarily math reasoning abilities; therefore, these subtypes reflect calculation difficulties without attention to the possibility of application and word problem deficits. Groups 2 and 3 were then compared on auditory-perceptual, verbal, and visual-perceptual tasks as well as visual-motor and tactile measures. The findings indicated that children in Group 3 (i.e., deficient arithmetic) had relatively intact auditory-perceptual and verbal skills with deficits in visual-perceptual and organizational skills, whereas Group 2 (i.e., deficient reading and spelling) showed difficulty in auditory perceptual skills. Group 2 had intact visual motor and tactile perception, whereas Group 3 evidenced bilateral impairment in tactile perceptual tasks as well as delayed psychomotor abilities. Moreover, Group 3 children experi-

TABLE 4.3
Comparisons of the Characteristics of Arithmetic Disability in Three Studies

Cohn (1971)	*Kosc (1981)*	*Rourke & Strang (1983)*
1. Malformed, reversed, outsized number writing	1. Disturbed writing of arithmetic symbols	1. Errors in spatial organization
2. Dyslexia	2. Disturbance in carrying out arithmetic operations	2. Misreading visual detail
3. Inability to perform simple addition	3. Low muscle tone in hands and fingers	3. Procedural errors
4. Inability to recognize and pay heed to changing operation signs	4. Inability to read arithmetic terms and symbols	4. Failure to shift operational set
5. Inability to read multidigit numbers accurately	5. Inability to read numbers accurately	5. Poor number writing
6. Poor memory for number facts	6. Inability to carry out arithmetic operations	6. Poor memory for number facts
7. Inability to use regrouping strategies correctly	6. Inability to carry out arithmetic operations	7. Poor number reasoning
8. Spatial misalignment of problems		

enced significant difficulties with spatial organization and analysis skills, whereas Group 2 performed within expectations. Therefore, the Luria (1980) hypothesis of the necessity for intact spatial imagery for arithmetic calculation processes would seem to be upheld by these findings.

Rourke's findings further support findings by Kosc (1981) when the errors made by Group 3 are analyzed. Table 4.3 outlines the errors proposed by these researchers. When Rourke and associates compared the types of errors made by Group 3 children with other children who had less dramatic arithmetic difficulties, three findings emerged: Group 3 children made more types of mechanical errors; they made more errors than the other group; and, finally, the errors by Group 3 did not appear to be related to impaired judgment and reasoning. These authors concluded that basic sensorimotor processes and development were disturbed for Group 3 children with the result being an inability to form the foundation for later arithmetic development. In other words, fundamental cognitive processes such as understanding the concepts of quantity, counting with one-to-one correspondence, and developing an understanding of basic additive and subtractive properties were deficient (see Table 4.2). It would appear that this hypothesis is directly testable and should be further investigated. A complete developmental history of the child's attainment of motor milestones as well as a longitudinal study of children with early motor difficulties would be helpful in isolating precursors to later arithmetic disability. In addition, a comparison between children with difficulties in all academic areas to Group 3 children would be instructive, particularly for the development of intervention techniques.

Neuroanatomical Basis for Arithmetic Ability

When normal subjects are compared to patients with documented brain damage and arithmetic deficits, significant psychophysiological differences are found. Children and adults with arithmetic difficulty have been found to have different evoked potentials with a slowing of alpha activity over the right hemisphere (John, Karmel, & Corning, 1977). Deterioration in calculation abilities has been found to be more pronounced in right hemispheric damaged patients (Querishi & Diamond, 1979). However, Giannitrapani (1982), in an electroencephalographic (EEG) factor analytic study of subjects while they listened to prose or performed mental calculation, concluded that mental arithmetic in children most likely involves bilateral hemispheric processes.

Further studies have found that arithmetic abilities are shared between the hemispheres with damage to either hemisphere compromising arithmetic ability (Diamond & Beaumont, 1972; Katz, 1980; Murphy, Darwin, & Murphy, 1977). For example, in patients with temporo-occipital lesions exhibiting alexia or agraphia for numbers, 84% showed left-sided cerebral damage (Boller & Grafman, 1985), whereas 89% of patients with spatial acalculia had right-sided lesions. Of particular interest is that arithmetic ability appears to be mediated by both hemispheres, particularly when linguistic material is required (Grafman, Passaflume, Faglioni, & Boller, 1982). Therefore, it is not unreasonable to speculate that arithmetic skills may well be affected when reading skills, usually thought to be mediated through left-hemispheric dysfunction, are also affected. Needless to say, however, arithmetic skills may be deficient in isolation as well.

Subcortical processes may also be significantly involved in arithmetic performance. For example, when left and right thalamic regions were electrically stimulated in normal subjects, differences were found in arithmetic function. Left thalamic stimulation produced an orienting response and an acceleration of mental arithmetic abilities, whereas right thalamic stimulation evoked a slowing in the rate of counting and number identification as well as an increase in calculation errors (Ojemann, 1974). The thalamus may well serve as an orienting mechanism for sensory input as well as integrating several necessary subprocesses for performance (Semrud-Clikeman & Hynd, 1990). Ojemann (1974) concluded from his studies that dysfunction in any brain region may decrease cognitive flexibility and adaptivity.

Summary

From the preceding review, it can be stated that arithmetic disability appears to be related to problems in spatial ability, problem solving and cognitive flexibility, deficient information processing skills, social imperception, disturbance in body image and representation as well as difficulty with interpersonal interactions in varying situations. Moreover, there is emergent electrophysiological evidence suggesting that arithmetic problems may be related to disturbances in posterior association regions Luria (1980) originally hypothesized and implicating the involvement of parieto-temporo-occipital regions in both hemispheres. Thalamic contributions to arithmetic calculation and reasoning have also been demonstrated, thus, implicating subcortical structures.

VALIDITY OF THE DSM-III, DSM-III-R, AND PL 94–142 DEFINITIONS

Complications arise in the measurement of arithmetic achievement not only in terms of definition but also in terms of what the obtained scores mean. The PL 94–142 definition requires a significant discrepancy between achievement and ability. How the discrepancy is arrived at is of utmost importance. Comparison of Table 4.1 with the preceding review leaves one with several questions as to omissions and/or vagueness endemic in the various definitions. Although most areas of concern are addressed by the PL 94–142 definition, several are not addressed by the DSM definitions. Each of these areas is discussed in the following paragraphs.

Concurrent Disabilities Allowed

The DSM-III-R and PL 94–142 definitions both allow for other disabilities without restriction. This allowance is important given the finding that significant numbers of children with both reading and written language problems also experience arithmetic disabilities (Torgesen, Rashotte, Greenstein, Houck, & Portes, 1987). The DSM-III and DSM-III-R definitions appear to address this issue adequately.

Specification of Type of Arithmetic Disability

The PL 94–142 definition appears to be the most consonant definition with the previously cited research through the inclusion of deficits in calculation and arithmetic reasoning. Although DSM-III does not address this issue, DSM-III-R includes attention, perception, counting, following directions, and learning multiplication tables. However, important skills such as arithmetic reasoning are not included. DSM-III-R, however, is the only definition that alludes to the importance of attentional variables. This is an important contribution as some studies report a concordance between attention deficit disorder without hyperactivity and learning disabilities, particularly in arithmetic (Brumback & Staton, 1982; Brumback, Staton, & Wilson, 1984; Douglas & Peters, 1979; Hynd et al., in press). PL 94–142 alludes to perceptual and attentional problems but addresses them indirectly by stating "one or more basic psychological processes" may be involved. It is not clear what is meant by this statement, but clearly attention could be subserved here.

As practitioners attempt to operationalize these definitions, problems become apparent as to what meanings can be attributed to the global statements included in these definitions. Although the DSM-III-R definition lists symptoms as well as types of disability, it may be helpful to include these areas separately. Symptoms may vary by age level but generally include difficulty recognizing, naming, and writing numbers; impaired spatial organization of numbers (misalignment, reversals, and inability to utilize the concept of quantity); calculation deficits; and arithmetic reasoning skills (Levin, 1979).

Language Involvement Addressed

This area is addressed by both DSM-III-R and PL 94–142 but not by DSM-III. The inclusion of language skills and the deficient development negatively impacting on arithmetic abilities is important. Arithmetic abilities are often seen as a form of language placed into numeral format (Kosc, 1981). Therefore, if difficulty is present in the ability to understand basic language concepts (i.e., smaller, larger, greater than, less than, between), it will be problematic for the child to understand arithmetic relationships as applied to the world at large.

Degree of Discrepancy from IQ

Unfortunately, the degree to which a discrepancy must exist between IQ and arithmetic achievement for diagnosis continues to trouble all the definitions. DSM-III and DSM-III-R do not specify what constitutes a significant discrepancy. Therefore, individual practitioners and researchers must utilize clinical judgment. This practice can be problematic as illustrated by Siegel and Heaven (1986). In their review, they discussed two studies; one used grade level discrepancies and percentiles less than 25 to determine the sample, whereas the other utilized grade scores below cognitive ability but not necessarily below average (Hall, Wilson, Humphreys, Tinzmann, & Bowyer, 1983). The Hall et al. (1983) study was not replicated, due to the problematic manner of the selection of subjects. Utilizing grade level retardation as a method for subject selection makes the false assumption that the same amount of discrepancy between achievement and grade levels is present at all levels. The same absolute amount of delay at a younger age represents a more significant problem than that same delay at an older age (Siegel & Heaven, 1986). In order to eliminate this problem and, thus, allow for more direct comparison of various samples, the use of standard scores and percentiles for comparisons is indicated.

PL 94–142 allows the individual states to determine the extent of discrepancy needed for a diagnosis of disability to be made. Suggestions have been made for establishing discrepancy through the use of regression equations (Thorndike, 1963; Yule, Rutter, Berger, & Thompson, 1974); utilizing a ratio of arithmetic age to mental age, expressive language, or chronological age with >.80 seen as significant (Benton, 1975); using the reliabilities of the tests in a formula to determine the magnitude of a discrepancy (Reynolds, 1981); or utilizing 15–20 standard score point discrepancy between ability and achievement. However, it must be concluded that none of these diagnostic procedures are uniquely more appropriate with regard to the diagnosis of arithmetic disorders.

Presumed Neurological Involvement

DSM-III and DSM-III-R do not address this issue. This omission is unfortunate given the rapidly expanding research available from electrophysiological studies (Ollo & Squires, 1986). PL 94–142 does implicate neurological involvement in the definition through its use of terms such as minimal brain dysfunction, developmental dysphasia, and dyslexia. In the past decade, there has been additional documentation as to the neurological foundation of learning disabilities through research employing MRI (Rumsey et al., 1986; Hynd, Semrud-Clikeman, Lorys, Novey, & Eliopulos, 1990), CT scans (Denckla, LeMay, & Chapman, 1985; Hynd & Semrud-Clikeman, 1989; Parkins, Roberts, Reinarz, & Varney, 1987), and postmortem studies (Galaburda, Sherman, Rosen, Aboitiz, & Geschwind, 1985). Although the utility of diagnosing learning disabilities through these neuroimaging procedures has not been established (Otto et al., 1984), the preceding studies have at least provided some documentation as to the neurological basis of these disorders.

Exclusions

PL 94–142 excludes children whose arithmetic disability is primarily the result of sensory handicaps, mental retardation, emotional disturbance, or environmental disadvantage. DSM-III-R excludes children with sensory handicaps and mental retardation. Several studies have found that anxiety can negatively impact on arithmetic performance (Kinsbourne & Caplan, 1979; Slade & Russell, 1971; Strang & Rourke, 1983). This is not to say that the two diagnoses cannot coexist, but rather, when a diagnosis of overanxious disorder is made, additional caution is needed in making the diagnosis of arithmetic disability. For instance, high levels of anxiety impact negatively on a child's ability to pay attention and process incoming stimuli. When anxiety levels continue to be elevated over a span of time, it becomes more and more difficult for a child to profit from instruction (Kinsbourne & Caplan, 1979). Similarly, a child with significant depression or dysthymia may experience difficulty with underarousal and, thus, not profit from arithmetic experiences designed to improve skill development.

Proposed Diagnostic Criteria

Based on our review, Table 4.4 presents proposed diagnostic criteria for developmental arithmetic disorders. These modifications incorporate the areas that were not addressed in the original DSM-III-R definitions as well as include findings provided by research. For instance, the specification of type of discrepancy from ability is addressed. Moreover, the various subtypes of arithmetic disability are included as well as deficits in language, perception, and/or attention.

Therefore, it would appear reasonable to include the basic elements delineated in PL 94–142 into the DSM-III-R definition for arithmetic

TABLE 4.4
Proposed Diagnostic Criteria for Developmental Arithmetic Disorders*

A. Arithmetic skills, as measured by a standardized, individually administered test, are significantly below the expected level (>1.5 standard deviations), given the person's schooling and intellectual capacity (as determined by an individually administered IQ test).
B. The disturbance significantly interferes with academic achievement or activities of daily living requiring arithmetic skills. This disturbance may manifest in either computational skills or mathematical reasoning abilities and may include associated linguistic, perceptual, and attentional difficulties.
C. Not due to a deficit in visual or hearing acuity or an acquired neurologic disorder. Although developmental arithmetic disorders are presumed to be due to central nervous system dysfunction, no direct evidence is required for a diagnosis.
D. Although developmental arithmetic disorder may frequently cooccur with mental retardation, depression, anxiety disorders, other developmental learning disorders, attention-deficit hyperactivity disorder, or significant difficulties in social perception and social skill development, it is not the direct result of these conditions or disorders.

*Underscored portions of this proposed definition represent additions to the DSM-III-R (APA, 1987) definition of developmental arithmetic disorder.

disability. Acknowledgment of neurological contributions to arithmetic dysfunction as well as the inclusion of mathematical reasoning skills as part of the DSM-III-R criteria appear to be warranted. This is particularly relevant, because the proposed revision of the PL 94–142 definition states that learning disabilities are presumed to be due to central nervous system dysfunction (Wyngaarden, 1987).

Moreover, recognition of the developmental contribution to arithmetic disability is important and could be operationalized to some extent through the use of standard scores rather than grade equivalents or percentiles in determination of disability. Rather than listing isolated symptoms, such as counting, following directions, and learning multiplication tables, DSM-III-R may be better served through either expanding this list to include more crucial subprocesses (for example, spatial orientation, knowledge of quantity, development of concepts, etc.) or deleting the list in favor of statements already included in DSM-III-R such as those noting linguistic, perceptual, and attentional difficulties. Although PL 94–142 includes perception and linguistic variables, attentional difficulties are not addressed. Moreover, no definition includes social imperception difficulties that often cooccur with arithmetic disabilities (Semrud-Clikeman & Hynd, 1990).

DIAGNOSIS OF ARITHMETIC DISABILITY

The assessment of arithmetic disability differs depending on the developmental age of the child. Arithmetic skill development also appears to be more related to native ability than to an ability to overachieve through hard work (Kinsbourne & Caplan, 1979). Although rote counting and memorization of mathematical facts may be overlearned, the flexibility needed to succeed in higher-level math problems is often deficient in children with arithmetic disability.

The development of arithmetic concepts and skills in learning disabled children has been found to follow a similar pattern to children with normal arithmetic ability albeit at a delayed pace (Weinstein, 1980). The pattern and type of errors made by such children are similar to children several years younger. Moreover, problem-solving skills appear to be similar to younger children. For example, counting on the fingers or making hatch marks to solve problems when utilized by a first grader are appropriate but are far less appropriate or efficient when a fourth grader uses these strategies. In addition, children with arithmetic disabilities have been found to show delay in Piaget's stages for the development of logical thought (Weinstein, 1978) that are predicated to be a necessary step for conceptual understanding in arithmetic (Piaget, 1941).

These findings suggest a developmental lag theory for arithmetic learning disabilities. However, this explanation may be too simplistic given the recent electrophysiological studies that indicate that the development of hemisphericity parallels cognitive development. Hemisphericity refers to the propensity for one hemisphere to specialize in certain cognitive functions (Butler & Glass, 1974; Dumas & Morgan, 1975). For instance, the left hemisphere is thought to be predominant in adults for calculations, whereas for children this dichotomy is not as well established (Semrud-Clikeman & Hynd, 1990). As a child develops, a shift from the use of right hemispheric processing of spatial information (i.e., holistic) to the analytic skills of the left hemisphere allows for more efficient arithmetic skill development. For children with arithmetic learning disabilities, this shift may not be as readily accomplished (Weinstein, 1975). In contrast, theories that stipulate that the right hemisphere is dysfunctional throughout the child's development suggest that there is not a developmental lag but rather a deficit in neurological development. The arithmetic disabled child is not seen as readily equipped to develop neurological traces during the early sensorimotor stages and utilize the holistic strategies thought to represent right hemisphere functioning. Therefore, the basic experiences of object and number constancy, conservation of number, and numeration needed to form the cornerstone for later formal operational thought are impaired from the start. Comparisons between children with normal arithmetic skills and those with deficient arithmetic ability may shed further light upon the deficit versus delay model. Some studies (Querishi & Diamond, 1979) show bilateral hemispheric involvement in math skills rather than solely a left hemisphere function as postulated by the delay hypothesis. Therefore, it is reasonable to speculate that some skills may be spared in a child with right hemisphere deficit, and some may develop more slowly.

Although the delay-versus-deficit argument cannot be resolved at the present time, children with arithmetic disabilities continue to present a challenge for diagnosis and remediation. It is important what tests are utilized in diagnosis, but also the performance within the test gives several clues as to type of dysfunction. Difficulties with understanding of number, place value, knowledge of math facts, ability to retain facts such as measurement and temporal concepts, and understanding of word problem operations can all be distinguished through analysis of a child's performance. Incorrect alignment of problems as well as procedural mistakes can be targets for remediation. Moreover, a child may have memorized the math facts needed to solve a word problem but does not know which operation to select. This problem may be related to an inability to understand key words in the problem, such as how many are left (subtract), how many in all (add), or how many did each receive (divide). Another difficulty often evidenced by such children is the lack of recognition of changing operations. In other words, if the first few problems ask for addition and the next few for subtraction, children who are inattentive to detail or perseverate in their response pattern may well add the entire page with little recognition or insight as to their mistake. These children may also use an incorrect procedure such as borrowing in addition problems and carrying in subtraction problems. It is often helpful in diagnosis to ask the child how they solved the problem. Often the use of classroom materials can elucidate particular difficulties in arithmetic skills. Such curriculum-based assessment can not only provide assistance with diagnosis but also can be of more practical value in the development of the individual educational plan as well as in the immediate remediation of difficulties.

There is relatively little research available as to the types of errors made by arithmetic disabled adolescents (Smith, 1983). Problems have generally been found to exist in spatial orientation, problems with reversals in subtraction, inattention to details, choice of mathematical rules, inability to maintain decimal and dollar signs, and inability to perform more than one mathematical operation involved in one problem (Cox, 1975; Greenstein & Strian, 1977). Assessment needs to take into consideration computation facility as well as mathematical understanding (Bartel, 1978).

In addition to informal testing and use of curriculum materials, it is often helpful to utilize diagnostic teaching when attempting to determine the scope and type of arithmetic dysfunction. The examiner can observe how the child makes use of the teaching, how he or she approaches the problem, his or her level of frustration, and the amount of review needed to master the problem. Moreover, the amount of retention of previously learned material can be assessed.

Informal measures are helpful for developing intervention strategies. However, formal measures are needed in order to ascertain the magnitude of disability as compared with peers as well as to provide objective baseline data. The test or tests selected will, of course, need adequate reliability and validity. Table 4.5 presents several tests that adequately evaluate mathematical skills. If a test is selected that utilizes only calculation or reasoning abilities, then additional information should be gathered.

Problems in Arithmetic Ability Measurement

The measurement of arithmetic abilities is fraught with problems. Some tests have timed calculation problems (Wechsler Intelligence Scale for Children–Revised [WISC–R] arithmetic subtest; Wechsler, 1974); some have time limits for completing the test (WRAT–R; Jastak & Wilkinson, 1984); some have recognition of the correct answer to a calculation problem (Peabody Individual Achievement Test–Revised; Dunn & Markwardt, 1989); some utilize a combination of calculation and reasoning (Basic Achievement Skills Individual Screener; Psychological Corporation, 1984; Key Math Diagnostic Test–Revised; Connolly, Nachtmann, & Prichett, 1981); and some have separate measures of calculation and reasoning skills (Diagnostic Achievement Battery; Newcomer & Bryant, 1986a; Diagnostic Achievement Test for Adolescents; Newcomer & Bryant, 1986b; Woodcock Achievement Test Battery–Revised; Woodcock, 1989). Given this discrepancy of approaches to assessment, it is not unreasonable to speculate that various scores indicate different skills, and although a child may have difficulties in arithmetic reasoning without calculation deficits (possibly contributed to by attentional and memory problems), if problems arise in computational abilities, arithmetic reasoning skills will be affected.

Scores are also reported in differing fashions depending on the instrument utilized. Grade-equivalent scores are often used to determine

TABLE 4.5
Representative Arithmetic Assessment Instruments

Measure	*Scores*		*Skills*						*Output*	
	Standard Scores	*Separate Subtest Scores*	*Readiness Included*	*Calculation*	*Reasoning*	*Basic Concepts*	*Measurement*	*Time*	*Written*	*Oral*
BASIS[1]	Y	N	Y	Y	Y	Y	N	N	Y	Y
Key Math–Revised	N	Y	Y	Y	Y	Y	Y	Y	Y	N
DAB[2]	Y	Y	Y	Y	Y	Y	Y	Y	Y	Y
DATA[3]	Y	Y	N	Y	Y	N	N	N	Y	N
K–ABC Achievement Test	Y	N	Y	Y	Y	Y	Y	Y	N	Y
Woodcock Psychoeducational Battery–Revised	Y	Y	Y	Y	Y	N	N	N	Y	Y
WRAT–R[4]	Y	—	Y	Y	N	N	N	N	Y	For younger children
PIAT–R[5]	Y	Y	Y	Y	Y	Y	Y	Y	N	Y
WISC–R Arithmetic	Y	N	Y	Y	Y	N	N	N	N	Y

[1]Basic Academic Skills Individual Screener
[2]Diagnostic Achievement Battery
[3]Diagnostic Achievement Test for Adolescents
[4]Wide Range Achievement Test–Revised
[5]Peabody Individual Achievement Test–Revised

the discrepancy between achievement and ability needed to qualify for learning disability support. Often a 1.5 to 2.0 grade-level discrepancy is utilized (Siegel & Heaven, 1986). Given the wide variance in the meaning of the grade-equivalent score as well as the statistical variance associated with grade equivalents, it is unfortunate that this practice is utilized (Benton, 1975; Burns, 1982; Erickson, 1975). Given the ordinal nature of grade-level equivalents, the same amount of delay for a younger child as for an older child may represent a more significant problem for the younger child. Moreover, the growth of most academic skills is a negatively accelerated function; the growth rate is not commensurate between all ages (Siegel & Heaven, 1986). The grade-equivalent score would not make such an assumption and, as such, is clearly invalid as a measure for severity of dysfunction. Recommendations from a number of sources indicate that the use of standard scores is most appropriate for interpretation of test performance (Sattler, 1988).

CONCLUSIONS

The study of arithmetic learning disabilities is in its infancy when compared with the research available on developmental reading disability or dyslexia. As computers and calculators become more critical in our daily lives, the importance of mathematical ability may well become a more essential skill to possess. Until recently, a fourth-grade arithmetic level was found to be adequate for functioning in everyday life (Chandler, 1978). In the future, the work place will not be as amenable to poorly developed mathematical skills. Rather than the current stress given to mastery of basic skills, the use of calculators to solve arithmetic problems will need to be directly taught to children with arithmetic learning disabilities, with the emphasis in curriculum shifted to problem solving and an understanding of math relationships (Capps & Hatfield, 1977; Caravella, 1977; Gawronski & Coblentz, 1976; Tietelbaum, 1978).

However, paramount problems exist in the recognition of arithmetic disabilities at early ages. Studies by Rourke (1985) and associates (Rourke & Strang, 1983) stated that children with arithmetic disabilities were generally identified in the early to middle adolescent period. Early identification is important to begin treatment, especially in the development of basic concepts. Of equally great concern is the tendency for these children to be identified as depressed and as behaving in a socially inappropriate manner without attention to associated cognitive underpinnings, (e.g., spatial-perceptual deficits) that may impact on social imperception disabilities and arithmetic skills. Children are generally screened for disability prior to kindergarten entrance. This screening often consists of speech and language, gross and fine motor abilities, and reading readiness skills. Additional attention to basic concept development, understanding of number, counting abilities, and shape, size, and color recognition need to be addressed as well as the child's social skill level.

Additional difficulty is found in the definitions of arithmetic learning disabilities. Subtypes may well exist, but research is just now addressing this issue. Although DSM-III-R and PL 94–142 recognize arithmetic learning disabilities, the definitions are global and imprecise, leaving several areas either unaddressed or poorly delineated. Part of the problem lies in the lack of knowledge as to how an arithmetic disability manifests. Another issue is a history of benign neglect as to the articulation of possible cooccurring disorders such as attention deficit disorder without hyperactivity (not recognized as a separate entity by DSM-III-R; Hynd et al., in press). Clearly, the cooccurrence of psychiatric disturbance in some of these children needs to be further documented.

It would appear that there is much need for research regarding the nature of arithmetic disabilities. Noninvasive techniques, such as evoked potentials, CT scans, and MRI scans, may allow for further exploration of concomitant neurological involvement in arithmetic disabled children. Interest in arithmetic disabilities has increased in recent years, and research is beginning to identify possible neuropsychological subtypes as well as appropriate intervention techniques (Rourke, 1985). Clinicians using DSM-III-R will need to be cognizant of the research advances and use them to flesh out the now sketchy definition provided by this manual. Although it is recognized that DSM-III-R was not designed to be utilized as a tool for educational diagnosis and remediation, it is clear that available research findings were not employed in developing the definition or diagnostic statement. Thus, use of the DSM-III-R criteria for sole diagnosis of arithmetic learning disability would appear to be inappropriate. However, the DSM-III-R criteria coupled with PL 94–142 and the available research can be helpful in developing a more current perspective and diagnostic criteria. The proposed criteria outlined in this regard in Table 4.4 may be helpful, if paired with reliable and valid clinical measures assessing components of arithmetic skill. We hope that the revision of DSM-III-R will address these issues and provide a more insightful and complete definition as supported by the existing literature. Inclusion of neurological underpinnings, the possibility of comorbidity, use of a more clearly defined "significant discrepancy," and expanding the subtypes of arithmetic disability ultimately will broaden the definition and make it more clinically useful and meaningful in research endeavors.

REFERENCES

American Psychiatric Association. (1980). *Diagnostic and statistical manual of mental disorders* (3rd ed.). Washington, DC: Author.

American Psychiatric Association. (1987). *Diagnostic and statistical manual of mental disorders* (3rd ed., rev.). Washington, DC: Author.

Badian, N. A. (1983). Dyscalculia and nonverbal disorders of learning. In H. R. Mykelbust (Ed.), *Progress in learning disabilities* (Vol. 5, pp. 235–264). New York: Grune & Stratton.

Badian, N. A., & Ghublikian, M. (1983). The personal-social characteristics of children with poor mathematical computation skills. *Journal of Learning Disabilities, 16*, 154–157.

Bartel, N. R. (1978). Problems in mathematics achievement. In D. D. Hammill & N. R. Bartel (Eds.), *Teaching children with learning and behavioral problems* (pp. 271–294). Boston: Allyn & Bacon.

Bennett, K. K. E. (1981). The effects of syntax and verbal mediation on learning disabled students' verbal mathematical problem scores. *Dissertation Abstracts International, 42*, 1093A.

Benton, A. L. (1975). Developmental dyslexia: Neurological aspects. In W. J. Friedlander (Ed.), *Advances in neurology* (Vol. 7, pp. 93–135). New York: Raven.

Boller, F., & Grafman, J. (1985). Acalculia. In P. J. Vinkin, G. W. Bruyn, & H. L. Klavans (Eds.), *Handbook of clinical neurology* (pp. 315–345). Amsterdam: North-Holland.

Brumback, R. A., & Staton, R. D. (1982). A hypothesis regarding the commonality of right hemisphere involvement in learning disability, attentional disorder, and childhood major depressive disorder. *Perceptual and Motor Skills, 55*, 1091–1097.

Brumback, R. A., Staton, R. D., & Wilson, H. (1984). Right cerebral hemisphere dysfunction. *Archives of Neurology, 41*, 248–249.

Burns, E. (1982). The use of standardized grade equivalents. *Journal of Learning Disabilities, 15*, 17–18.

Buswell, G., & John, L. (1926). Diagnostic studies in arithmetic. *Supplementary Educational Monograph, 30*.

Butler, S., & Glass, A. (1974). Asymmetries in the EEG associated with cerebral dominance. *EEG and Clinical Neurophysiology, 36*, 481–491.

Capps, L. R., & Hatfield, M. M. (1977). Mathematical concepts and skills: Diagnosis, prescription, and correction of deficiencies. *Focus on Exceptional Children, 8*, 1–8.

Carvella, J. R. (1977). *Minicalculators in the classroom*. Washington, DC: National Education Association.

Chandler, H. N. (1978). Confusion confounded: A teacher tries to use research results to teach math. *Journal of Learning Disabilities, 11*, 361–369.

Cohn, R. (1971). Arithmetic and learning disabilities. In H. R. Myklebust (Ed.), *Progress in learning disabilities* (Vol. 2, pp. 176–194). New York: Grune & Stratton.

Connolly, A. J., Nachtmann, W., & Pritchett, E. M. (1981). *The Key Math Diagnostic Arithmetic Test–Revised*. Circle Pines, MN: American Guidance Service.

Cox, L. S. (1975). Diagnosing and remediating systematic errors in addition and subtraction computations. *The Arithmetic Teacher, 22*, 151–157.

Denckla, M. B., LeMay, M., & Chapman, C. A. (1985). Few CT scan abnormalities found even in neurologically impaired learning disabled children. *Journal of Learning Disabilities, 18,* 132–135.

Diamond, S. J., & Beaumont, J. G. (1972). A right hemisphere basis for calculation in the human brain. *Psychonomic Science, 26,* 137–138.

Douglas, V. I., & Peters, K. G. (1979). Toward a clear definition of the attentional deficit of hyperactive children. In G. A. Hale & M. Lewis (Eds.), *Attention and cognitive development* (pp. 173–248). New York: Plenum.

Dumas, R., & Morgan, A. (1975). EEG asymmetry as a function of occupation, task, and task difficulty. *Neuropsychologia, 13,* 219–228.

Dunn, L. W., & Markwardt, F. C. (1989). *Peabody Individual Achievement Test–Revised.* Circle Pines, MN: American Guidance Service.

Eliasberg, W., & Feuchtwanger, E. (1922). Zur psychologischen und psychopathologischen. Untersuching and theorie des erworben Schwachsinns. Z. ges. *Neurologic Psychiat., 75,* 516–595.

Erickson, M. T. (1975). The z–score discrepancy method for identifying learning disabled children. *Journal of Learning Disabilities, 8,* 308–312.

Fafard, M. (1976). The effects of instructions on verbal problem solving in learning disabled children. *Dissertation Abstracts International, 37,* 5741A.

Federal Register. (1977). *Assistance to states for education for handicapped children: Procedures for evaluating specific learning disabilities.* Bethesda, MD: U.S. Department of Health, Education, and Welfare, *42,* 62082–62085.

Gaddes, W. H. (1980). *Learning disabilities and brain function: A neuropsychological approach.* New York: Springer-Verlag.

Gaddes, W. H. (1985). *Learning disabilities and brain function.* New York: Springer-Verlag.

Gaddes, W. H., & Spellacy, F. J. (1977). *Serial order perceptual and motor performances in children and their relation to academic achievement.* Research Monograph No. 35. Victoria, BC: Department of Psychology, University of Victoria.

Galaburda, A. M., Sherman, G. F., Rosen, G. D., Aboitiz, F., & Geschwind, N. (1985). Developmental dyslexia: Four consecutive patients with cortical anomalies. *Annals of Neurology, 18,* 222–233.

Gawronski, J. D., & Coblentz, D. (1976). Calculators and the mathematics curriculum. *Arithmetic Teacher, 23,* 510–512.

Gelman, R. (1969). Conservation acquisition, a problem of learning to attend to relevant attributes. *Journal of Experimental Child Psychology, 7,* 167–187.

Giannitrapani, D. (1982). Localization of language and arithmetic functions via EEG factor analysis. *Research Communications in Psychology, Psychiatry, and Behavior, 7,* 39–55.

Grafman, J., Passafiume, D., Faglioni, P., & Boller, F. (1982). Calculation disturbances in adults with focal hemispheric damage. *Cortex, 18,* 37–50.

Greenstein, J., & Strain, P. S. (1977). The utility of the Key Math Diagnostic Arithmetic Test for adolescent learning disabled students. *Psychology in the Schools, 14,* 275–282.

Groen, G. J., & Parkman, J. M. (1972). A chronometric analysis of simple addition. *Psychological Review, 79,* 329–343.

Hall, J. W., Wilson, K. P., Humphreys, M. S., Tinzmann, M. B., & Bowyer, P. M. (1983). Phonemic similarity effects in good vs. poor readers. *Memory and Cognition, 11,* 520–527.

Homan, D. (1970). The child with a learning disability in arithmetic. *Arithmetic Teacher, 17,* 199–203.

Hynd, G. W., Lorys, A., Semrud-Clikeman, M., Nieves, N., Huettner, M., & Lahey, B. B. (in press). Attention deficit disorder without hyperactivity (ADD/WO): A distinct behavioral and neurocognitive syndrome. *Journal of Child Neurology.*

Hynd, G. W., & Semrud-Clikeman, M. (1989). Dyslexia and neurodevelopmental pathology: Relationships to cognition, intelligence, and reading skill acquisition. *Journal of Learning Disabilities, 22,* 204–220.

Hynd, G. W., Semrud-Clikeman, M., Lorys, A. R., Novey, E. S., & Eliopulos, D. (1990). Brain morphology in developmental dyslexia and attention deficit disorder/hyperactivity. *Archives of Neurology, 47,* 919–926.

Jastak, S., & Wilkinson, G. S. (1984). *The Wide Range Achievement Test–Revised.* Wilmington, DE: Jastak.

John, E. R., Karmel, B. Z., & Corning, W. C. (1977). Neurometrics. *Science, 196,* 1392–1410.

Johnson, D., & Myklebust, H. (1967). *Learning disabilities: Educational principles and practices.* New York: Grune & Stratton.

Kaliski, J. (1962). Arithmetic and the brain-injured child. *Arithmetic Teacher, 9,* 245–251.

Katz, A. (1980). Cognitive arithmetic: Evidence for right hemispheric mediation in an elementary component stage. *Quarterly Journal of Experimental Psychology, 32,* 69–84.

Kinsbourne, M., & Caplan, P. J. (1979). *Children' learning and attention problems.* Boston: Little, Brown.

Kirby, J. R., & Asman, A. F. (1984). Planning skills and mathematics achievement: Implications regarding learning disability. *Journal of Psychoeducational Assessment, 2,* 9–22.

Kosc, L. (1981). Neuropsychological implications of diagnosis and treatment of mathematical learning disabilities. *Topic in Learning and Learning Disabilities, 1,* 19–30.

Larsen, S. C., & Hammill, D. D. (1975). The relationship of selected visual-perceptual abilities to school learning. *Journal of Special Education, 9,* 281–291.

Larsen, S. C., Parker, R., & Trenholme, B. (1978). The effects of syntactic complexity upon arithmetic performance. *Learning Disability Quarterly, 1,* 80–85.

Lee, W. M., & Hudson, F. G. (1981). *A comparison of verbal problem-solving in arithmetic of learning disabled and non-learning disabled seventh grade males* (Research Report No. 42). Lawrence, KS: University of Kansas, Institute for Research in Learning Disabilities.

Levin, H. S. (1979). The alcalculias. In K. M. Heilman & E. Valenstein (Eds.), *Clinical neuropsychology* (pp. 128–140). New York: Oxford.

Luria, A. (1980). *Higher cortical functions in man.* New York: Basic.

McCuller, C. C. (1981). The effects of varying the syntactic complexity of mathematics word problems on the performance of learning disabled students. *Dissertation Abstracts International, 42,* 1099A.

McLeod, T. M., & Crump, W. D. (1978). The relationship of visuo-spatial skills and verbal ability to learning disabilities in mathematics. *Journal of Learning Disabilities, 11,* 237–241.

Michael, W. B., Guilford, J. P., Fruchter, B., & Zimmerman, W. S. (1957). The description of spatial-visualization abilities. *Educational and Psychological Measurement, 17,* 185–199.

Murphy, P., Darwin, J., & Murphy, D. (1977). EEG feedback training for cerebral dysfunction: A research program with learning disabled adolescents. *Biofeedback and Self Regulation, 2,* 288–295.

Newcomer, P. L., & Bryant, B. R. (1986a). *Diagnostic achievement battery.* Austin, TX: Pro-Ed.

Newcomer, P. L., & Bryant, B. R. (1986b). *Diagnostic achievement test for adolescents.* Austin, TX: Pro-Ed.

Ojemann, G. (1974). Mental arithmetic during human thalamic stimulation. *Neuropsychologia, 12,* 1–10.

Ollo, C., & Squires, N. (1986). Event-related potentials in learning disabilities. *Evoked potentials* (pp. 497–512). New York: Elsiver.

Otto, R., Karrer, R., Halliday, R., Horst, R. L., Klorman, R., Squires, N., Thatcher, R. W., Fenelon, B., & Lelord, G. (1984). Developmental aspects of event-related potentials. *Annals of the New York Academy of Sciences, 425,* 319–337.

Parkins, R., Roberts, R. J., Reinarz, S. J., & Varney, N. R. (1987, January). *CT asymmetries in adult developmental dyslexics.* Paper presented at the Annual Convention of the International Neuropsychological Society, Washington, DC.

Piaget, J. (1941). *The child's conception of number.* London: Routledge & Kegan Paul.

Piaget, J. (1976). Piaget's theory. In B. Inhelder, H. H. Chipman, & C. Zwingmann (Eds.), *Piaget and his school: A reader in developmental psychology* (pp. 14–36). New York: Springer-Verlag.

Piaget, J., & Inhelder, B. (1971). *Mental imagery in the child.* London: Routledge & Kegan Paul.

Pribram, C. H. (1971). *Languages of the brain.* Englewood Cliffs, NJ: Prentice-Hall.

Psychological Corporation. (1984). *Basic Achievement Skills Individual Screener.* NY: Author.

Querishi, R., & Diamond, S. J. (1979). Calculation and the right hemisphere. *Lancet, 1,* 322–323.

Reid, D. K., & Hresko, W. P. (1981). A developmental study of the language and early reading in learning disabled and normally achieving children. *Learning Disabilities Quarterly, 3,* 54–61.

Reynolds, C. R. (1981). The fallacy of "two years below grade level for age" as a diagnostic criterion for dyslexia. *Journal of School Psychology, 19,* 26–51.

Rosner, J. (1973). Language arts and arithmetic achievement, and specifically related perceptual skills. *American Educational Research Journal, 10,* 59–68.

Rourke, B. P. (Ed.). (1985). *Neuropsychology of learning disabilities.* New York: Guilford.

Rourke, B. P., & Finlayson, M. A. J. (1978). Neuropsychological significance of variations in patterns of academic performance: Verbal and visual-spatial abilities. *Journal of Abnormal Child Psychology, 6,* 121–133.

Rourke, B. P., & Strang, J. D. (1983). Subtypes of reading and arithmetical disabilities: A neuropsychological analysis. In M. Rutter (Ed.), *Developmental neuropsychiatry* (pp. 323–359). New York: Guilford.

Rumsey, J. M., Dorwart, R., Vermess, M., Denckla, M. B., Kruesi, M. J. P., & Rapoport, J. L. (1986). Magnetic resonance imaging of brain anatomy in severe developmental dyslexia. *Archives of Neurology, 43,* 1045–1046.

Sattler, J. (1988). *Assessment of children.* (3rd ed.) San Diego, CA: Jerome M. Sattler.

Semrud-Clikeman, M., & Hynd, G. W. (1990). Right hemispheric dysfunction in nonverbal learning disabilities. Social, academic, and adaptive functioning in adults and children. *Psychological Bulletin, 107,* 196–209.

Siegel, L. S., & Heaven, R. K. (1986). Categorization of learning disabilities. In S. J. Ceci (Ed.), *Handbook of cognitive, social, and neuropsychological aspects of learning disability* (pp. 95–121). Hillsdale, NJ: Lawrence Erlbaum Associates.

Slade, P. D., & Russell, G. F. M. (1971). Developmental dyscalculia: A brief report on four cases. *Psychological Medicine, 1,* 292–298.

Smith, C. R. (1983). *Learning disabilities: the interaction of learner, task, and setting.* Boston: Little, Brown.

Stone, A., & Michals, D. (1986). Problem solving skills in learning-disabled children. In S. J. Ceci (Ed.), *Handbook in cognitive, social, and neuropsychological aspects of learning disabilities* (pp. 291–315). Hillsdale, NJ: Lawrence Erlbaum Associates.

Strang, J. D., & Rourke, B. P. (1983). Concept-formation/nonverbal reasoning abilities of children who exhibit specific academic problems with arithmetic. *Journal of Clinical Child Psychology, 12,* 33–39.

Strang, J. D., & Rourke, B. P. (1985). Arithmetic disability subtypes: The neuropsychological significance of specific arithmetical impairment in childhood. In B. P. Rourke (Ed.), *The neuropsychology of learning disabilities* (pp. 167–183). New York: Guilford.

Svenson, O. (1975a). Analysis of time required by children for simple additions. *Acta Psychologia, 39,* 289–302.

Svenson, O. (1975b). Strategies for solving simple addition problems. *Scandinavian Journal of Psychology, 16,* 143–151.

Svenson, O., Hedenbord, M., & Lingman, L. (1976). On children's heuristic for solving simple additions. *Scandinavian Journal of Educational Research, 20,* 161–173.

Thorndike, R. L. (1963). The measurement of creativity. *Teachers College Record, 64,* 422–424.

Teitelbaum, E. (1978). Calculators for classroom use? *Arithmetic Teacher, 26,* 18–20.

Torgesen, J. K., Rashotte, C. A., Greenstein, J., Houck, G., & Portes, P. (1987). Academic difficulties of learning disabled children who perform poorly on memory span tasks. In H. L. Swanson (Ed.), *Memory and learning disabilities* (pp. 305–333). Greenwich, CT: JAI Press.

Tuoko, H. (1982). *Cognitive correlates of arithmetic performance in clinic referred children.* Unpublished doctoral dissertation, University of Victoria.

Wechsler, D. (1974). *Wechsler Intelligence Scale for Children–Revised.* New York: Psychological Corporation.

Weinstein, M. L. (1978). *Dyscalculia: A psychological and neurological approach to learning disabilities in mathematics in school children.* Unpublished doctoral dissertation, University of Pennsylvania.

Weinstein, M. L. (1980). A neuropsychological approach to math disability. *New York University Education Quarterly, 11,* 22–28.

Wiig, E., & Harris, S. (1974). Perception in interpretation of nonverbally expressed emotions by adolescents with learning disabilities. *Perceptual and Motor Skills, 38,* 239–245.

Woodcock, R. W. (1989). *Woodcock–Johnson Psychoeducational Battery–Revised.* Boston: Teaching Resources.

Wyngaarden, J. B. (Ed.). (1987). *Learning disabilities: A report to the U.S. Congress.* Washington, DC: National Institute of Health, Interagency Committee on Learning Disabilities.

Yule, W., Rutter, M., Berger, M., & Thompson, J. (1974). Over and under-achievement in reading: Distribution in the general population. *British Journal of Educational Psychology, 44,* 1–12.

CHAPTER 5

Expressive Writing Disorders

Noel Gregg
University of Georgia

The written expressive abilities of both children and adults have been a concern of psychologists, teachers, and clinicians for many years. Educators, observing a growing number of students underachieving in writing, have turned to the psychological community for guidance. However, despite the fact that writing remains one of man's primary tools for communication to the world as well as for structuring internal thought schemes, there has been limited empirical research in the area of written language. A major responsibility and challenge to educators and psychologists is the development of more efficient and effective procedures to investigate and assess written discourse.

Writing is a complex form of communication requiring various underlying processing abilities. Johnson and Myklebust (1967) and Myklebust (1965, 1973) discussed a hierarchy of language abilities and stated that the ability to write follows developmentally the abilities to listen, speak, and read. Litowitz (1981) pointed out that subsumed under the need for those basic language capacities, both receptive and expressive, are cognitive and psychological abilities (e.g., selective attention, perception, categorization, memory, problem solving).

Written language requires that the writer simultaneously deal with a subject, text, and reader; breakdowns in the system may occur for several different reasons. Problems in oral language and reading are often precursors to difficulties with the writing process (Johnson & Myklebust, 1967). Underlying information-processing abilities of attention, perception, memory, and reasoning are, therefore, necessary for the art of writing. Difficulty with the integration of subject, text, and reader, however, is not always due to an underlying processing problem. It may be the result of poor instruction, lack of adequate experience in manipulating language structures, or the result of overall intellectual or emotional functioning.

The *Diagnostic and Statistical Manual of Mental Disorders–Third Edition Revised* (DSM-III-R) (American Psychological Association, 1987) added the category of developmental expressive writing disorder (315.80) as one of the subclassifications within developmental disorders. The Axis II developmental disorders characteristically have an onset in childhood or adolescence and persist (without periods of remission or exacerbation) into adult life. A developmental expressive writing disorder is identified by an individual demonstrating "writing skills markedly below the expected level, given the person's schooling and instructional capacity" (DSM-III-R, p. 44). However, this category should be reserved only for individuals demonstrating specific cognitive processing deficits impacting on written language. As is discussed herein, it should not be applied to individuals demonstrating instructional deficits and/or whose primary disorder is not in written language (e.g., developmental expressive or receptive language disorder, a developmental reading disorder, thought disorder, affective disorder). It is shown in this chapter that there is little empirical research available to understand the cause(s) and/or characteristics of a developmental expressive writing disorder; therefore, a clinician's effectiveness in assessment and/or instruction is limited.

HISTORY OF THE CLINICAL SYNDROME

The term *agraphia* has historically been applied to disorders of written language without a clear description of the term and/or application to assessment. Unfortunately a majority of the research regarding agraphia that has been conducted has been very narrowly focused and gathered mainly from clinical observations of patients with acquired neurological disorders. A review of recent literature concerning the written language characteristics of children and adolescents demonstrating developmental written expression disorders reveals a need for extensive research in this area (Newcomer, Nodine, & Barenbaum, 1988).

Working from a somewhat simplistic definition of written discourse, researchers have investigated the relationship of agraphia to aphasia and dyslexia and have attempted to classify the various types of agraphia. Although one can not assume that conclusions drawn from acquired disorders will apply to developmental disorders, an understanding of these investigations is important as the foundation for grasping the impact of cognitive processing deficits on written discourse among the developmental disorders.

Classifications of agraphia (Benson, 1979; Kaplan & Goodglass, 1981; Leischner, 1969; Roeltgen, 1985) have resulted in five well-defined groups. These subgroups, derived from clinical observations of acquired agraphic patients, include the following: pure agraphia, aphasic agraphia, agraphia with alexia (parietal agraphia), apraxic agraphia, and spatial agraphia. This classification system is described in Table 5.1. Recently, Roeltgen (1985) suggested that, rather than defining the subgroups on clinical descriptions of agraphic patients, a better alternative would be to classify by the neuropsychological mechanism within the writing system that appears disordered. Such a model is similar to those proposed by cognitive psychologists (Ellis, 1983). Roeltgen described a model of writing that looks at the linguistic, motor, and interaction between motor components and visual-spatial skills. This method of classification certainly offers a promising method that might prove useful as more research is conducted on the cognitive processes impacting on written discourse. However, at this time,

TABLE 5.1
Classifications of Agraphia

Type	*Definition*	*Cause*	*Literature*
Pure agraphia	Agraphia without aphasia alexia, acalculia	Focal lesion or acute confusional state	(Laine & Marttila,1981; Auerbach & Alexander, 1981; Baxter & Warrington, 1986; Levine et al., 1988)
Aphasic agraphia	Agraphia with aphasia	Lesions associated with aphasia	(Marcie & Hecaen,1979; Kaplan & Goodglass, 1981)
Agraphia with alexia (parietal agraphia)	Agraphia with alexia	Lesions associated with alexia	(Kaplan & Goodglass, 1981; Roeltgen, 1985)
Apraxic agraphia	Difficulty forming graphemes	Parietal lobe opposite preferred hand	(Leischner, 1969; Hecaen & Albert, 1978; Marcie & Hecaen, 1979)
Spatial agraphia	Spatial disorders	Lesions in nondominate	(Hecaen & Albert, 1978; Marcie & Hecaen, 1979; Roeltgen, 1985)

such a method appears speculative in nature and does not account for the impact of cognitive processing deficits on a developing system (Bryant & Bradley, 1985) nor the social-cultural influences (Scribner & Cole, 1981; Vygotsky, 1962) on literacy.

The subgroup described as pure agraphia appears to be the most relevant to the category of developmental expressive writing disorder (DSM-III-R, 1987). Research suggests that loss of ability to write may occur in isolation after brain damage (Pitres, 1884; Wernicke, 1903) and be accompanied by little or no aphasia, alexia, or limb apraxia (Auerbach & Alexander, 1981; Baxter & Warrington, 1986; Laine & Marttila, 1981; Kapur & Lawton, 1983; Rosati & deBastiani, 1979). Researchers disagree on a theory accounting for isolated agraphia. Levine, Mani, and Calvanio (1988) grouped the debate along two dimensions. They stated that some theories propose the loss of specific skills (center theories), and others postulated a dissociation of skills (dissociation theories). Within their review of the literature, they also concluded that some theories are narrow in scope (i.e., agraphia is isolated), whereas other theories suggest that agraphia is only one manifestation of a more general disorder. Based on a single case study of a right-handed man who suffered a left parieto-occipital cerebral infarction, Levine et al. (1988) concluded that pure agraphia and Gerstmann's syndrome are due to a dissociation of language skills and visuospatial skills caused by a dominant parieto-occipital lesion. However, they cautioned that their theory of isolated agraphia is limited to those cases with parieto-occipital lesions where the agraphia is generally accompanied by other elements of the Gerstmann syndrome, particularly because isolated agraphia has been reported in association with other lesions of the dominant hemisphere (Exner, 1881). They stated that no single theory will probably be able to account for all of the types of agraphias that can result from lesions.

Research from the field of cognitive psychology has also contributed to a better understanding of the cognitive processes involved in producing written text. Ellis (1982) proposed one of the most sophisticated cognitive models describing the processes involved in writing. His schemata draws heavily on the work of Morton (1980), Morton and Patterson, (1980) Seymour (1979), and Marshall and Newcombe (1980). However, Ellis' model, like most cognitive models of written language, basically focuses on the area of spelling, ignoring grammatism, form, and sense of audience (Ellis, 1983). Ellis (1982) admitted to the limitations of his model and suggested that it is only the broad architecture of the system. The identification of specific cognitive processes impacting on other areas of written language besides spelling (e.g., syntax, form, sense of audience) has received very little attention by researchers.

Myklebust's (1965) seminal publication of the theory and application of the Picture Story Language Test described several cognitive processes impacting on the ability to express written discourse. Although the cognitive processes he identified as disordered and resulting in dysgraphia have been questioned by current research, the contribution of Myklebust's work in the area of developmental writing disorders can not be underestimated today. While terminology and etiology continue to be debated, the impact of cognitive processes on written language among populations of handicapping condition (i.e., hearing impaired, emotionally disturbed, learning disabled, mentally retarded) was thoroughly addressed by Myklebust in light of available research. Furthermore, Myklebust explored areas of written language in addition to handwriting and spelling. Although it is easy to critique this research as simplistic, it provided a solid foundation for future investigation of written language disorders among developmentally disordered populations. He very clearly defined dysgraphia from a developmental perspective, differentiating it from ataxia and paralytic disorders. Myklebust used the term *dysgraphia* to apply only to disorders that are "symbolic in nature." In such cases, he felt that a breakdown occurred between the mental image of the word and the motor system. Therefore, Myklebust described dysgraphia as a type of apraxia. He was one of the first researchers to define the "dysgraphic" child as one who presents no other obvious symptoms except the inability to write (dysgraphia without dyslexia). Such a child would be classified as having a developmental expressive writing disorder under the DSM-III-R (1987) classification taxonomy.

Unlike the work of early pioneers (Myklebust, 1965), recent research (Rapp & Caramazza, 1989) has recognized the importance of identifying neurological functional proximity rather than neural proximity when trying to understand the association of deficits in written expression. This has encouraged a move away from the old medical model (Lyon & Moat, 1989) where identifying single lesions or deficits was the priority. It will be advantageous for researchers to investigate common processing mechanisms that support or share the input and output of processes involved in writing. Kossyln (1981) outlined an account that included the sharing of resources/cognitive skills between perceptual and productive processes. Although specific resources might be shared in the writing processes, investigation of functionally distinct processes will also need to continue in order to understand the components of the system. Current research (Caramazza, 1988; Hill & Caramazza, 1989; Rapp & Caramazza, 1989) supports a distinction between the processes that compute visual spatial information from abstract levels of letter representation and the reverse operation. Therefore, theories of cognitive architecture must take into account the impact of linguistic and visual spatial processes.

As writing disorders have taken on political importance in the United States, the neurological causation of the disorders has begun to take some precedence to the research areas of aphasia and alexia. However, the ability to explain how the brain integrates all the visual, linguistic, and motor skills required to produce written text remains in the infancy stage. Cognitive psychologists continue to debate appropriate models to describe the processes involved in writing within the normally achieving population and the social-cultural influences on literacy. Therefore, the impact of processing disorders on models of written language development has received very little attention. However, the neglect of systematic research to define the characteristics of written expression disorders has not deterred publication of articles that make suggestions for assessment and teaching strategies. Many of these articles are based on little knowledge of theories supported by research in the area of written language. Therefore, it is imperative the research from the fields of neuropsychology, cognitive psychology, English, and special education be integrated to construct valid and reliable assessment tools for the diagnosis of developmental expressive writing disorders.

Debate continues on defining the specific cognitive processing abilities required in the ability to write; therefore, the areas of written language most impacted on by cognitive breakdowns are presented in this chapter with hypothesis as to the causation of the problem and the development of more accurate diagnostic criteria. These areas include the following: spelling, syntax, text organization, and sense of audience. An understanding of the processing problems often impacting on these areas and the relationship of this to assessment techniques are defined. First, however, the category of developmental expressive written language disorder as defined by the new DSM-III-R (1987) is reviewed critically in relation to diagnostic criteria.

DEVELOPMENTAL EXPRESSIVE WRITING DISORDER: A GENERAL CRITIQUE

The DSM-III-R includes the new category of expressive writing disorder under the rubric of developmental disorders. It was added to identify those individuals who demonstrate underachievement in written language when compared to their instructional opportunities and intellectual capacity. The diagnostic criteria for 315.80 developmental expressive writing disorders are listed in Table 5.2. Several problems

TABLE 5.2
Diagnostic Criteria for Developmental Expressive Writing Disorder (315.80)

A. Writing skills, as measured by a standardized, individually administered test, are markedly below the expected level given the person's schooling and intellectual capacity (as determined by an individually administered IQ test).
B. The disturbance in A significantly interferes with academic achievement or activities of daily living requiring the composition of written texts (spelling words and expressing thoughts in grammatically correct sentences and organized paragraphs).
C. Not due to defect in visual or hearing acuity or a neurologic disorder.

arise when reviewing the implications of these criteria if translated literally by professionals.

The first criterion states that a standardized instrument must be individually administered, and a marked discrepancy between achievement in written language and overall intellectual ability should be observed. However, a developmental expressive writing disorder should not be synonymous with underachievement in written language, discrepancy being only a symptom of the condition. As Kavale (1987) stated, "discrepancy as a concept cannot be validated in a theoretical sense and remains what Hempel (1952) termed a 'fictious concept' " (p. 13). Several researchers have discussed in great detail the theoretical (Kavale, 1987), psychometric (Wilson, 1987), developmental (Parrill, 1987), and educational (Hessler, 1987) fallacies of utilizing discrepancy as the sole criterion of eligibility for developmental learning disorders. If such a model is allowed to dominate the identification process, it is hypothesized that the chance for misdiagnosis of developmental expressive writing disorders will likely occur.

An individual who qualifies for classification under developmental expressive writing disorder (315.80) should demonstrate a cognitive processing (i.e., attention, perception, memory) deficit as well as a discrepancy between achievement in written language and intellectual potential. The individual would be unable to perform written language tasks due to disorders in one or several cognitive processing abilities. Table 5.3 lists a few of the characteristic written language behaviors resulting from linguistic and visual-spatial processing deficits. However, these behaviors are only speculative, because weak empirical support exists for most cognitive models of written text. Further research is needed to identify how other processing deficits and/or lack of integration of cognitive processes impact on the ability to write (Rapp & Caramazza, 1989). The cognitive processing deficits that are impacting on written language can also effect other academic and social areas. Therefore, commonly associated disorders are developmental reading disor-

TABLE 5.3
Characteristic Behaviors of Selected Cognitive Deficits on Written Text

Grammatical Disorders	*Phonological Disorders*	*Visual-Spatial*
Substitution, ommission, or addition of single noun, verb, adjective, or adverb	Morpheme substitution	Letter of confusion
	Morpheme ommission	Rate of visual perception Slow
	Phoneme substitutttion	Letter inversions
	Phoneme ommission	Internal detail errors
Substitution, omission, or addition of single preposition, pronoun, determiner, quantifier, conjunction, or adverb	Syllable substitution	Letter transpositions (blame>blame)
	Syllable ommission	Letter deletions
	Transpositions of morphemes phonemes syllables	Letter substitutions
	Sound-symbol conversion	
Substitution, omission, or addition of an inflexional affix.		
Distorted order of words.		

der, developmental expressive and receptive language disorders, developmental arithmetic disorders, and developmental coordination disorders. As discussed previously, the category Developmental Writing Disorder should be used only to identify an individual whose primary disorder is in the ability to produce written text, not the individual whose writing problems are secondary to other disorders.

Difficulty with writing is not always due to an underlying cognitive processing problem. It may be the result of poor instruction or lack of adequate experience in manipulating language structures. Students experiencing such a problem have been labeled "basic writers" by Shaugnessy (1977). Although they do not experience an underlying processing deficit, they lack the ability to integrate subject, text, and reader in order to compose a written text. Such a problem is likely an instructional deficit and requires the attention of regular education. Thought disorders (i.e., schizophrenia), depression, anxiety, and self-concept problems also can lead to poor performance in written language. However, as with instructional deficits, these should not be labeled developmental expressive writing disorders.

A second problem with the current DSM-III-R (1987) criteria for development expressive writing disorders is the requirement of the use of standardized tools. The validity and reliability for most tools currently utilized for assessment of written language are very weak. Isaacson (1988) stated that, lacking better methods, psychologists often use measures that are too subjective or too narrowly focused. Such evaluators often consider only the student's written product during the assessment and not the cognitive processing abilities required to perform such a task. Lacking tools with content validity, psychologists must utilize clinical judgment through the use of standardized and qualitative data in assessing written language. The DSM-III-R (1987) perpetuates the utilization of invalid, unreliable, or inappropriate diagnostic tools by restricting the user to only standardized assessment tools. However, behaviors observed may be more appropriate than standard scores.

Another problem with the DSM-III-R (1987) discussion of developmental expressive writing disorders is the age of onset. The DSM-III-R (1987) states that, in severe cases, the disorder may not be apparent until age 7 (i.e., about second grade), and, in less severe cases, the disorder may not be apparent until age 10 (fifth grade). Such a statement is in direct conflict with current theory on written language development. The development of many writing skills begins prior to picking up a pencil to record symbols on paper. Vygotsky (1978) discussed the development of representation that leads up to symbolization in writing and the importance of the development of gesture, play, and drawing to the ability to write. The cognitive processing deficits that will later impact on written text will show up early in these other modes of representation. Therefore, an examiner should be able to identify high-risk children for written language disorders before such children are required to use pen and paper. State eligibility criteria might not allow for children to qualify for services despite the fact that detection of the disorder can be made much earlier by trained diagnosticians.

In summary, diagnosticians utilizing the DSM-III-R (1987) criteria for identifying developmental expressive writing disorders must, therefore, be cautious of four important problem areas. These include: (a) the issue of discrepancy; (b) the identification of cognitive processing deficits; (c) the problem with reliability, validity, and appropriateness of standardized assessment tools; and (d) the age of onset of the disorder.

TESTING FOR DEVELOPMENTAL EXPRESSIVE WRITING DISORDERS: GENERAL IMPLICATIONS

A single informal measure or a single standardized assessment tool should never be the sole means of determining if an individual meets the DSM-III-R (1987) criteria for developmental expressive writing disorders. Although Luria (1980) was biased in favor of the phonocentric

(Scinto, 1986) theory of written language, he did describe a series of tasks designed to analyze the "state of the various elementary components and levels of writing" (p. 537) that would be useful to professionals developing assessment batteries in the area of written language. In order to evaluate the different areas of written language, Luria (1980) suggested dividing the tasks into three formats: copying, dictation, and spontaneous writing. Such measures allow the evaluator to examine the different cognitive processing abilities of the individual confronted with very different task demands. Each of the areas of written language (e.g., spelling, syntax, text organization, and sense of audience) should be examined on each of these types of tasks. Although a thorough investigation is presented on the cognitive processing demands required for each of the areas on the individual tasks, several general points must first be highlighted regarding copying, dictation, and spontaneous writing tasks.

Copying

Luria (1980) pointed out several ways to vary the copying task to observe different cognitive processing abilities of the writer. The first suggestion is to have the individual, if developmentally appropriate, copy individual letters, single words, isolated sentences, and paragraphs. Spelling, syntax, and organizational deficits can be noted as the task demands increase the need for the integration of specific cognitive processing abilities. Luria (1980) also recommended varying the type of script (size, density, and type) to assess specific motor and visual processing abilities accurately. The copying of nonsense figures (Roeltgen, 1985) is also valuable, because it is more difficult to transcribe figures with no apparent symbolic meaning. Luria (1980) also encouraged varying the distance between the presentation of the stimuli and when the individual is allowed to reproduce the information as a means to evaluate specific motor, memory, and spatial abilities. Monitoring the strategies an individual uses (auditorization, tracing, etc.) as well as the amount of time necessary to complete the task is significant information in drawing conclusions regarding a disorder versus a cognitive style difference.

Dictation

The ability to complete a dictation task requires the individual to integrate linguistic, visual-spatial, and motor skills in an automatic manner. Therefore, dictation tasks should begin by requesting the person to write individual letters, syllables, and sentences. In order to distinguish between linguistic and motor disorders, the examiner should vary the client's response by the use of anagrams (blocks with single letters written on them).

Spontaneous Writing

A spontaneous writing sample would require the individual to write either a sentence, paragraph, or story on a familiar topic. Many researchers have indicated that the complexity of such a task requires extensive integration of several cognitive processing abilities (Cromer, 1981; Luria, 1980; Roeltgen 1985). It is suggested that the topic or picture be standard from client to client in order to compare performance. However, the type of task (sentence starter, picture, or topic) significantly influences the individual's performance. Ideally, the evaluator should collect samples of the individual's writing across different audiences (i.e., peer, teacher) and types of genre (i.e., letter, narrative, descriptive).

AREAS IMPACTED BY A DEVELOPMENTAL EXPRESSIVE WRITING DISORDER

A developmental expressive writing disorder can impact on one or several areas of written language depending on the number, type, and severity of the cognitive processing deficits. The assessment of written language requires an understanding of each of these areas and the cognitive processing abilities most necessary to complete task demands required in each of the areas. No single, standardized assessment tool is currently available that will alone provide an evaluator the necessary information to investigate all of these areas. However, an understanding of the areas will allow a trained diagnostician to utilize the tools on the market in combination with qualitative measures to make a more accurate assessment of this disorder. The areas of written language most commonly impacted on by a development expressive writing disorder include the following: spelling, syntax, text organization, sense of audience, and ideation (Gregg, Hoy, & Sabol, 1988; Gregg & McAlexander, 1989; Myklebust, 1973). Each of these areas is discussed next by reviewing the current literature, relating known research to the DSM-III-R criteria, and, finally, by proposing more effective diagnostic criteria for the identification of developmental expressive writing disorders.

Spelling

Recent evidence suggests that visual-orthographic and phonological strategies may serve different functions in reading as compared to complex spelling (Barron, 1980; Rapp & Caramazza, 1989). Frith (1980) also pointed out that the strategies used in spelling are affected by several variables, including: stage of learning, individual preferences, demands of the task, and neurological impairment. Thus, individuals' spelling strategies change as a function of their knowledge and experience.

The terminology for the two syndromes most often discussed in analyzing spelling disorders are *surface dysgraphia* and *deep dysgraphia,* as Marshall and Newcombe (1973) identified. Individuals with surface dysgraphia usually can read and write nonwords that conform to English phonology better than irregular English words (Deloche, Andreesky, & Desi, 1982). Their construction of words on tasks of spelling frequently are phonologically similar to the target words (Gregg, Hoy, & Sabol, 1988). Individuals who demonstrate the deep dysgraphia syndrome have greater difficulty reading and writing nonsense words, and their errors on reading and writing tasks include semantic or derivational errors but not phonemic errors (Shallice & Warrington, 1975).

Surface dysgraphia is thought to stem from the disruption of the lexical route (Delouche et al., 1982; Marshall & Newcombe, 1973). However, Kremin (1982) cautioned that the typical error patterns of surface dysgraphia might not be the result of a single component syndrome. He proposed the following two possible causes: (a) disruption at the level of visual word recognition or (b) disruption at the level of postlexical phonological output (Kremin, 1985). Shallice and Warrington (1980) also cautioned against the single-cause error with respect to surface dysgraphia. They noted that reliance on the attribution of phonology to subword orthographic units can be initiated by any one of a myriad of deficits in the process for assigning the output phonology of whole familiar written words. On the other hand, deep dysgraphia is said to be caused by disruption of the phonologic route, resulting in dependence on the limits of the lexical route for writing functions (Patterson & Marcel, 1977; Saffron & Marin, 1977; Shallice & Warrington, 1975).

Measures of Spelling Ability

Two of the most commonly used spelling tasks are those that investigate the recall and recognition of single words. The ability to recognize a correctly spelled word puts very different demands on an individual's cognitive processing system than a recall task.

Smith (1980) suggested that a recognition spelling task is largely a measure of reading ability, whereas a recall task is more a measure of spelling, because it involves a substantial amount of phonological overgeneralization. Frith (1979) also pointed out the significance of a phonological component in spelling recall that is not present in recognition tasks. Frith (1980) noted that poor spellers and young children tend to rely on visual processing skills in reading and phonetic processing in spelling, lacking flexibility in automatically changing from phonetic to visual-spatial processing strategies. This finding is consistent with the research of Gregg, Hoy, and Sabol (1988). Older students and good spellers rely more on visual-spatial processing skills for spelling and phonetic processes for reading. Research also appears to be suggesting that poor spellers have difficulty going directly from memory to the motor output (Simon, 1976; Smith, 1973).

Recall spelling tasks also appear to require the writer to overgeneralize graphemic rules on a phonological basis (Smith, 1980). Gregg, Hoy, and Sabol (1988) found that a recall spelling task (Wide Range Achievement Test–Revised, spelling; Jastak & Wilkinson, 1984) was a better discriminator of spelling deficits among adults with developmental expressive writing disorders when compared to performance on a standardized recognition spelling test.

The ability to copy text either from a paper or a chart also provides significant information about the transcription abilities of individuals. Gregg (1982) found that, even when adults demonstrating developmental written language disorders copied correctly spelled words from one paper to another, they made significantly more spelling errors as compared to basic writers or normally achieving writers. Therefore, as described previously, the use of oral spelling tasks and anagrams help the evaluator determine whether the spelling disorder appears to be more linguistic, visual-spatial, or motor related.

Unfortunately, many evaluators examine spelling achievement simply by giving the client a standardized spelling measure that usually requires the examiner to prompt the individual orally. However, Cromer (1980) stressed the necessity also to examine the spontaneous spelling of individuals in order to observer spelling strategies that might not be utilized on single-word dictation or recognition spelling tasks. Gregg, Hoy, and Sabol (1988) adapted the Cromer analysis when investigating the spontaneous writing of adults demonstrating developmental expressive writing disorders. These are listed in Table 5.4. They found that the majority of spelling errors of this population were phonographical.

Cromer (1980) found that receptive language disordered and deaf children made more visual errors, and the expressive aphasic or speech-disordered children made more phonographical errors. The

TABLE 5.4
Analysis of Spelling Errors of Individuals with Developmental Writing Disorders

Type of Error	*Description*
Phonological	As Cromer (1980) described, these spelling errors resemble in some respect the sound of the target word when pronounced.
Visual	Cromer (1980) described these errors as preserving the general shape or look of the word, but pronunciation would not lead to a similar sound.
Morphological	These errors Cromer (1980) described are problems with the morphological form of the word.
Spelling rule	Cromer (1980) classified errors in this category to include suffix-adding rules, doubling rules for final letters, the end rule, and other spelling-rule errors.
Segmentation	Cromer (1980) described incorrect segmentation errors usually as one word written as two (e.g., away/a way).
Context	This category was not included in Cromer's spelling analysis. The investigators included words in this category that were spelled correctly when oberved in isolation; however, in the context of the text, they were incorrect. These errors included homonyms and other word substitutions.

overreliance on phonetic strategies also resembles the dysgraphic spellers in the Frith (1980) and Rourke (1981) studies. However, Frith (1983) and Luria (1980) cautioned that the problem is not simply visual memory but rather graphemic memory. This refers to a letter-by-letter meaning of words that enables a differentiation over their sound. The individual demonstrating a developmental expressive writing disorder appears to pay attention to some letters but not to others (Frith, 1983).

Diagnostically, this research indicates that a recognition recall, copying, and spontaneous writing sample provides the clinician a valid examination of the different spelling processes. It also emphasizes that no single standardized spelling assessment can adequately identify a developmental expressive writing disorder that impacts on the area of spelling.

DSM-III-R Criteria Related to Spelling Disorders

The DSM-III-R (1987) requires, as reviewed previously, three criteria in order for a developmental expressive writing disorder to be determined. First, an individual must demonstrate a discrepancy between achievement and potential based on standardized assessment tools. Second, the individual must demonstrate significant underachievement in the ability to compose written text. Last, the major cause of the underachievement must not be acuity-based or neurologically based.

The major problem with these criteria, as mentioned in the first part of the chapter, is that no mention is made as to documenting the cognitive processing deficits that cause the underachievement in spelling. Many emotionally disturbed children, for instance, could be underachieving in spelling based on achievement test scores. However, their inability to spell is more likely to be due to lack of instruction or concentration than to a specific cognitive processing deficit. It is inappropriate to categorize this population as development expressive writing disordered. A clinician must be able to identify the cognitive processing deficits that impact on an individual's ability to perform spelling tasks either on individually dictated words, words copied, or in a spontaneous writing sample.

Proposed Criteria for Expressive Writing Disorders Impacting on Spelling

The evaluator investigating the spelling abilities of an individual should begin by administering various types of task formats (e.g., recall, recognition, spontaneous). Second, by an analysis of the spelling errors (e.g., Table 5.4), hypotheses can be generated as to the type of cognitive processing deficits involved. By comparing this information to data gathered throughout the evaluation on intellectual, linguistic, motor, and cognitive processing tools, the specific deficits can be determined. Further informal evaluations might need to be completed to confirm or disconfirm these hypotheses. Table 5.5 lists the proposed criteria to be utilized in determining if spelling errors are the result of a developmental expressive writing disorder.

TABLE 5.5
Proposed Criteria for Developmental Writing Disorder Impacting on Spelling Performance

1. Cognitive processing deficits, as measured by standardized, individually administered tests or subtests, are markedly below the expected level given the person's schooling and intellectual capacity (as determined by an individually administered IQ test).
2. Performance on at least *two* of the *four tasks* listed below are markedly below the expected level given the person's schooling and intellectual capacity (as determined by an individually administered IQ test).
   a. copy task
   b. recall spelling measure
   c. recognition spelling measure
   d. oral spelling measure
   e. spontaneous writing sample
3. Not due to a defect in visual, hearing, or lack of instruction.

Written Syntax

Early studies in written language were mainly concerned with evaluating writers' productivity and syntactic abilities. Such a focus led to research investigating number of words and words per sentence (Myklebust, 1965, 1973), *t* units (Hunt, 1965, 1970), and other indices of syntactic maturity and difficulty (Harris, 1977; O'Donnell, 1976). Many of these early studies were influenced by the theory of transformational generative grammar. Hunt (1965, 1970) was the first researcher to report the application of transformational grammar to the analysis of written discourse.

Transformational grammar also led to the research investigating the sentence combining skills of students (Mellon, 1969; O'Hare, 1973). Sentence combining provided a limited, patterned manipulation that allowed researchers to view a writer addressing only a few syntactic options at one time (Tomlinson, 1980). O'Donnell and Hunt (1975) attempted to assess syntax through the use of a rewrite paragraph, controlling for verbosity and topic selection. Following their research, the use of a controlled rewrite paragraph became a popular method of measuring syntactic complexity of students' written language. Despite differences of opinions on how best to assess an individual's syntactic maturity, researchers agree the selection and utilization of syntactic structures is an essential component of any writer's ability to reach specific rhetorical goals.

The literature pertaining to the written syntax characteristics of developmental expressive writing disorders is inadequate in both quantity and quality of research studies. First, a clear distinction between syntax disorders that are secondary to an oral language disorder and syntax disorders that are specific to written language production is seldom, if ever, made for the reader. Myklebust (1965, 1973) clearly differentiated the individual demonstrating specific dysgraphia from those for whom dysgraphia was secondary to dysphasia or dyslexia. However, he did not clearly define the syntax errors that are specific to the written language of such a population. Coltheart (1980) discussed the research from acquired disorders and concluded that agrammatism of speech is not always associated with an individual being agrammatic in their reading or writing. The implications of these findings have not been explored in relation to developmental disorders.

Future research is also needed to further investigate the written syntax errors commonly associated with a developmental expressive written language disorder. Research (Gregg, 1982; Johnson, 1987; Myklebust, 1965, 1973) finds that the following behaviors demonstrated in written text can be indicative of written syntax disorders: word omissions, word substitutions, verb and pronoun use, omission and substitutions of word endings, lack of punctuation, and a discrepancy between oral and written syntax.

The *t* unit has also been a method utilized to investigate written syntax. Hunt (1970) defined a *t* unit as "one main clause plus any subordinate clause or nonclausal structure that is attached to or embedded in it" (p. 44). Several researchers (Gregg, 1982; Vogel & Moran, 1982), when comparing the written syntax of learning disabled individuals to normally achieving writers, felt that *t* unit analysis was not a sensitive enough tool for identifying error patterns.

The syntactic density scoring (SDS) index that Golub & Kidder (1974) developed provided professionals with a more sophisticated measure of oral and written syntax. Belanger (1978) reported serious problems in the scoring procedures that resulted in the SDS index. He proposed a mathematical correction (CSDS) to remediate this problem. Simms & Crump (1983) later used this CSDS index to compare the written syntax of learning disabled (LD) to normally achieving children. Vogel (1986) used the CSDS to compare the written language of 33 college students demonstrating a learning disability to 33 nonLD college writers. Both researchers felt that the use of the CSDS appears to be more applicable for research purposes than as as assessment tool at this time.

Research (Blalock, 1981; Gregg, 1982; Poplin, Gray, Larsen, Banikowski, & Mehring, 1980; Poteet, 1980) consistently reports that many disabled writers also have significant deficits in the knowledge and application of the rules of capitalization and punctuation. The relationship of syntactic disorders to punctuation errors that Shaugnessy (1977), Johnson (1987), and Vogel (1986) have discussed. Specific cognitive processing deficits affect an individual's ability to acquire and/or produce the rules applicable to the formulation of ideas within sentences and the assigning of appropriate punctuation. Mechanical errors can often be a red flag for a syntax disorder rather than simply the result of lack of instruction. However, many current written language assessment tools require a knowledge of both capitalization and punctuation rules in determining syntactic maturity.

The diagnostician can not ignore the mechanical errors resulting from the breakdown in the writing process. It is quite obvious that mechanical errors limit the overall semantic, syntactic, or pragmatic coherence of a piece of writing. However, the diagnostician must identify the source of a mechanical error. As Shaugnessy (1977) pointed out, error-laden work must be viewed in light of its intentional structure, as errors evidence systematic, rule-governed behavior. Gregg (1982) found that the mechanical errors made by learning disabled college writers differed from those made by basic writers. The learning disabled writers' errors appeared to be the result of cognitive processing deficits and the basic writers' the result of instructions.

Measuring Written Syntax

The assessment of written syntax should include tasks measuring both receptive (identification) and expressive syntax. Standardized measures of written language can supplement informal measures in obtaining this information. An evaluator should begin by comparing the individual's oral syntax to their written syntax to identify receptive and expressive oral language disorders impacting on written language. There are a variety of syntax formats useful for assessment. These include the following.

Sentence Combining. Within the last several years, research efforts have been devoted to the impact of sentence combining as an evaluation measure of syntactic maturity. Sentence combining is an approach designed to allow students to manipulate language in order to improve the maturity of the syntactic structures. A student is given sets of kernal sentences that they must combine and write out as a single, complex statement. Exercises may contain grammatical cues (Mellon, 1969), nongrammatical cues (O'Hare, 1973), or no guiding cues (Strong, 1973). Gregg (1986) developed a sentence combining task and administered it to a group of learning disabled college females, nonLD writers, and basic writers. She found, at the level of the sentence, that punctuation continued to be a problem for LD writers. Second, on a set of sentences where the subjects were given a nongrammatical guiding cue, the LD group performed the poorest. Tasks such as this were labeled complementization (Clark & Clark, 1977; Phelps-Gunn & Phelps-Teraski, 1982). Phelps-Gunn and Phelps-Teraski (1982) wrote, "Complementization involves combining propositions into more efficient units by cognitive arrangement" (p. 128).

The Test of Written Language–2 (TOWL–2) (Larsen & Hammill, 1988) added a sentence combining subtest that was not in the original Test of Written Language (Larsen & Hammill, 1983). However, on the TOWL–2 subtest, only sentence combining without cues is provided to the client. Gregg (1986) found that a sentence-combining task using substitution cues can provide significantly more diagnostic information to distinguish between instructional deficits and cognitive processing deficits impacting on syntax.

Controlled Stimulus Passage. A controlled stimulus passage (O'Donnell & Hunt, 1975) attempts to measure syntactic complexity of a student's writing by controlling such writing variables as topic, verbosity, and lack of information. A controlled stimulus passage is designed to study a student's ability to rewrite material that contains short, choppy sentences. It is similar to sentence combining except the student is provided a total paragraph and must make decisions as to which sentences to combine or eliminate. Syntax, mechanical, organization, and spelling errors should be noted by the evaluator and compared to the types of errors made on a spontaneous writing sample.

Spontaneous Writing. An evaluation of a student's syntactic maturity should always include a sample of spontaneous writing. On the TOWL–2 (Larsen & Hammill, 1986), syntactic maturity is measured by the number of words in the spontaneous writing sample that are used to form "grammatically acceptable sentences" (p. 6). This syntactic maturity scale appears to be attempting to combine a *t* unit index with Myklebust's (1965, 1973) Syntax Quotient, although it falls short on both accounts. In addition, the determination of "grammatically acceptable" is determined with few criteria as guidance. Finally, no consideration is given as to whether the student's word choice and order are appropriate to the task given.

A diagnostician in determining whether an individual's spontaneous writing displays syntactic errors should evaluate mechanics and syntax together in searching for the cognitive processing deficits impacting on written language. Therefore, an examination of punctuation, word usage, patterns, and transformations is more important than a number count of words, sentences, or *t* units. Figures 5.1 and 5.2 provide some of the syntactic structures an evaluator would want to observe on both receptive and expressive tasks.

Identification Tasks. An evaluator will want to know whether an individual understands the syntactic rules by identification as well as on production tasks. Multiple-choice word usage, punctuation, and style subtests on many of the achievement tests provide such a score. This information should be compared to the individual's spontaneous writing sample. The TOWL (Larsen & Hammill, 1983) provides two subtests (Word Usage and Style) that yield excellent identification measures. However, the TOWL–2 (Larsen & Hammill, 1986) Style subtest adds in dictation of the sentences rather than correction, therefore changing the cognitive demands.

DSM-III-R Criteria Related to Syntax Disorders

The DSM-III-R (1987) requires a standardized assessment of written language to be utilized in determining a developmental expressive writing disorder. As discussed previously, a standardized tool is not

I. Mechanical Components	Accurate	Omission	Substitutes	Addition	Rarely Used	Comments
A. Punctuation						
1. Periods						
2. Question Marks						
3. Exclamation Marks						
4. Commas						
5. Semi-colons						
6. Colons						
7. Apostrophes						
8. Capitalization						
9. Quotes						
10. Others						
B. Word Usage						
1. Determiners						
2. Nouns						
3. Verbs (Tense)						
4. Adjectives						
5. Adverbs						
6. Prepositions						
7. Pronouns						
8. Conjunctions						
9. Functors						
10. Others						

FIG. 5.1. Mechanics assessment. Adapted from syntax scale from *Picture Story Language Test* by H. Myklebust, 1965, in *Mechanics assessment—an informal* (unpublished manuscript) by M. Habiger, 1988, Atlanta Area School for the Deaf, Atlanta, GA.

always available to measure specific areas of written language. Second, although many tests might state that they measure syntax (i.e., PSLT, TOWL–2), the validity of this claim is questionable. Third, the major problem with the current DSM-III-R (1987) criteria is that no mention is made as to documenting the cognitive processing deficits that cause the underachievement in syntax.

The oral and written language structures common to the individual's school, work, and home must always be investigated prior to evaluating the individual's written syntactic patterns. An examiner must be sure

II. Syntactical Components

	WRITTEN SAMPLE			ORAL SAMPLE			
	Errors	Errors	Rarely Never	Errors	Errors	Rarely Never	Comments
A. Patterns							
1. $N_1 + V$							
2. $V + N_2$							
3. $N_1 + V_{be} + Adj$							
4. $N_1 + V_{be} + N_1$							
5. $N_1 + V_{be}$ + When							
6. $N_1 + V + N_2$							
7. $N_1 + V_{be}$ + Where							
8. $N_1 + V$ + Where							
9. $N_1 + V$ + When							
10. $N_1 + V + N_3 + N_2$							
11. $N_1 + V$ + Where + When							
12. Others							
B. Transformations							
1. T/Yes-No							
2. T/Do							
3. T/Wh							
4. T/V_{tto}							
5. T/Negative							
6. Others							

FIG. 5.2. Syntax assessment. Adapted from *Developmental Language System Guide* by L. Laughton, 1980, cited in *Written Syntax Assessment—An Informal* (unpublished manuscript) by M. Habiger, 1988, Atlanta Area School for the Deaf, Atlanta, GA.

to eliminate oral language differences (e.g., dialect) from language disorders that impact on syntactic maturity. Further, an examiner must distinguish whether the written syntax disorder is, in reality, a secondary problem to a developmental receptive/expressive language disorder or a developmental reading disorder. In such cases, the written syntax deficit is not the result of a developmental expressive writing disorder.

Proposed Criteria for Expressive Writing Disorders Impacting on Syntax

A diagnostician investigating the written syntax of an individual should begin by administering various types of tasks that would require the integration of different cognitive processing systems. Such formats might include the following: sentence combining, controlled stimulus passage, identification, and spontaneous writing. Specific syntactic structures should be examined to determine if any were used appropriately, omitted, substituted, added, or rarely used in the written text (Myklebust, 1965, 1973). Therefore, by presenting the client different task formats, an examination of syntactic development and punctuation can be made both in a contrived and spontaneous manner. Figures 5.1 and 5.2 are adaptations of Myklebust's (1965) Syntax Subtest from the PSLT and Laughton's (1980) developmental language guide (Habiger, 1988). Such behavior checklists in combination with an examiner's knowledge of normal syntactic development are more appropriate assessment tools for the identification of written syntactic disorders than taking standard scores off of individual subtests of current written language assessment tasks.

Proposed criteria for identifying a developmental expressive writing disorder impacting on written syntax are listed in Table 5.6. Again, as in other areas of written language, the determination of specific cognitive processing deficit(s) is the key diagnostic component of the disorder. Therefore, intellectual, psychological, linguistic, and motor assessment performance should be compared to any written language tasks completed by an individual. In addition, performance on at least two of the four tasks previously described (i.e., sentence combining,

TABLE 5.6
Proposed Criteria for Developmental Writing Disorder Impacting on Syntax Performance

1. Cognitive processing deficits, as measured by standardized, individually administered tests or subtests, are markedly below the expected level given the person's schooling or intellectual capacity (as determined by an individually administered IQ test).
2. Performance on at least *two* of the *four tasks* listed below are markedly below the expected level given the person's schooling and intellectual capacity (as determined by an individually administered IQ test).
   a. sentence combining task
   b. controlled stimulus passage
   c. identification task
   d. spontaneous writing sample
3. Not due to a deficit in visual, hearing, or lack of instruction.

controlled stimulus passage, identification tasks, and spontaneous writing) should be markedly below the expected level given the person's schooling and intellectual capacity (as determined by an individually administered IQ test). Finally, the disorder should not be due to a deficit in visual or hearing acuity, or to lack of instruction.

Text Organization

The concept of text discourse and its relationship to thought and language is currently at the forefront of research in cognitive psychology. Discourse theories have attempted to integrate linguistic, psychological, and cultural research to heighten the understanding of the processes involved in the production of written language, particularly the development of form. Until very recently, studies investigating the structure of written language have dealt more with identifying grammatical types and numbers of sentences produced rather than with the text as a whole.

A goal, therefore, of the process of composition is the production of coherent form (Scinto, 1982). Much of form is developed in written language by cohesion and coherence, the terms not being synonymous. Witte and Faigley (1981) differentiated the terms in the following manner: "Cohesion defines those underlying semantic relations that allow a text to be understood and used and coherence conditions are governed by the writer's purpose, the audience's knowledge and expectations, and information to be conveyed" (p. 201). Therefore, in an evaluation of text organization, an assessment of the relationships between sentences (cohesion) and the total form (coherence) must be investigated.

Cohesion

It has been suggested that syntagmatic approaches to studies of written language do not address the issue of the meaning of sentences. Halliday's (1978) theory suggests a fruitful way of conceptualizing the evaluation of writing. He proposed that the textual function in writing depends on cohesion both within and between sentences. Halliday and Hasan (1976) defined cohesion as the structures beyond the sentence level that establish language as a particular type of text.

Cohesion implies that a written text is more than a mere collection of statements no matter how syntactically mature or grammatically correct such sentences appear. In an attempt to help evaluators analyze individual writing samples when cohesion is questioned, Gregg (1985) developed a cohesive scale by adapting the work of Halliday and Hasan (1976) and Strong (personal communication, 1981). This scale can be

TABLE 5.7
Cohesive Scale of Written Language

I. *GRAMMAR TIE*: Whenever a pronoun refers to a noun in another sentence or paragraph. Also when certain words are substituted for others or when words are omitted so that meaning depends on a previous statement (William strong, personal communication, 1981).

  A. *Reference*: "Relation between an element of the text and something else by reference to which it is interpreted in the greater instance (relationship between meaning)" (Halliday & Hasan, 1976, p. 308).
    1. *pronominal*: he, she, it, they.
    2. *demonstrative and definite article*: this, these, here, that, those, then, the.
    3. *comparative*: same, identical, similarly, such, different, other, else, additional, more, less, as many, ordinals.

  B. *Substitution*: "Words used as direct substitutes for other words (sometimes more precise ones) and repetition of the first term is avoided. There are nominal, verbal, and clausal substitutes" (Halliday & Hasan, 1976, p. 314). Example: one, the same, so do, be, have, do the same, likewise, do so, so that, not.

  C. *Ellipsis*: "A deletion of a word, phrase, or clause" (Halliday & Hassan, 1976, p. 144).
  Example: "<u>Some,</u> in his opinion, hold real promise for writing."

II. *TRANSITIONAL TIE*: Words and phrases showing relationships between statements (Strong, personal communication, 1981).

  A. *Consequence*: therefore, then, thus, hence, accordingly, as a result.
  B. *Likeness*: likewise, similarly.
  C. *Contrast*: but, however, nevertheless, on the contrary, on the other hand, yet.
  D. *Amplification*: and, again, in addition, further, furthermore, moreover, also, too.
  E. *Example*: for instance, for example.
  F. *Concession*: to be sure, granted, of course, it is true.
  G. *Insistence*: anyway, indeed, in fact, yes, no.
  H. *Sequence*: first, second, finally.
  I. *Restatement*: that is, in other words, in simpler terms, to put it differently.
  J. *Recapitulation*: in conclusion, all in all, to summarize, altogether.
  K. *Time or place*: afterward, later, earlier, formerly, elsewhere, here, there, hitherto, subsequently, at the same time, simultaneously, above, below, further on, this time, so far, until now (Crew, 1976, p. 176).

III. *LEXICAL TIE*: Cohesive effect achieved by the selection of vocabulary. (Halliday & Hasan, 1981, p. 318).

  A. *Lexical reiteration*: Lexical cohesion that is achieved by the use of repeated words, synonyms, near synonyms, and superordinates (Halliday & Hasan, 1976, p. 318).
    1. same item.
    2. synonym.
    3. superordinate.

  B. *Collocation*: Lexical items that regularly co-occur (Halliday & Hasan, 1976, p. 284).

Note: From "College learning disabled, normal, and basic writers: A comparison of frequency and accuracy of cohesive ties" by N. Gregg, 1985, *Journal of Psychoeducational Assessment, 3*, 223–231.

seen in Table 5.7 and is further discussed hereafter. In addition, there are a variety of cohesive ties that are important. These include the following.

Grammar Ties. Grammatical ties include any pronominal, demonstrative, or comparative that refers to a noun in another sentence or paragraph. This category of ties also includes words substituted for others or words that are dependent on the previous sentences for their meaning.

Research by Wiig and Semel (1976) notes that many learning disabled individuals have difficulty understanding and utilizing demonstratives and pronominals in oral language. Gregg (1982) found that an adult population demonstrating expressive writing disorders utilized fewer demonstratives in their expository and rewrite tasks. She also found that they appeared to use a high percentage of substitution and ellipsis ties. This could have been the result of the developmental expressive writing disordered population applying more nonspecific "general item" words rather than specific synonyms. Difficulty with word finding also could have caused the excessive use of general terms. Frequency and accuracy of cohesive ties can be affected by a number of different processing deficits.

Transitional Ties. It was felt that the description that Halliday and Hasan (1976) provided for conjunctive ties would not be specific enough for the classroom teacher. Therefore, Strong's (personal communication, 1981) label of *transitional ties* was used along with Frederic Crew's (1976) 11 labels for transitional links. Transition ties are words and phrases showing relationships between statements.

Researchers (Herbert & Czerniejewski, 1976; Johnson, 1987; Wiig & Semel, 1976) noted the difficulty that learning disabled writers have understanding and utilizing cohesive ties between sentences. Gregg (1982) found that an adult population of writers demonstrating developmental expressive writing disorders often used more transitional ties than basic or normal college writers. It was noted, however, that "amplification" and "contrast ties" were the primary reasons for this inflated total. The developmental expressive writing disordered writers used more *ands* and *buts* to connect thoughts. Such a pattern is typical of the writing of much younger children.

Lexical Ties. The first subclass of the lexical ties (reiteration) contains same item, synonyms, and superordinate ties. The second subclass of the lexical ties category consists of collocation that are ties "that were achieved through the association of lexical items that regularly co-occur" (Halliday & Hasan, 1976, p. 284).

The inability to utilize lexical ties frequently and accurately occurs for many different reasons. For instance, the student's conceptual organization strategies might be interfering with the understanding of word meaning. Wiig and Semel (1976) discussed the problems that learning disabled students have understanding different types of word relationships. Word finding and auditory discrimination problems can also lead to inaccurate use of lexical ties. Gregg (1982) found that, in an adult population demonstrating developmental expressive writing disorders, difficulty with morphological endings and omissions of words in written language resulted in a significant number of accuracy errors.

Coherence

Coherence is the communication relationship between reader and writer; as Phelps (1985) wrote, a "relationship which takes form as the intendedness of that integration or wholeness" (p. 21). The National Assessment of Educational Progress (NAEP) reported in 1975 that students' writing skills were declining due to a lack of essay coherence. In 1980, the NAEP constructed a primary-trait coherence measure for use in the Third National Writing Assessment. Such a holistic measure allowed for an investigation of the degree to which coherence affects quality in written language. Using the NAEP coherence scale, McCulley (1985) found that general coherence was an important element of writing quality.

Yet the study of the production of written form has not been thoroughly investigated. Charolles (1978) attempted to develop a model for the understanding of coherence and for a text-level approach to the study of children's writing. Scinto (1982) investigated the development of written text, concluding that the development of coherence is neither related to age nor reading ability but rather to the use of combinatory logic.

Gregg and Hoy (1989) investigated the ability of a population of adults demonstrating developmental expressive writing disorders to comprehend and produce text form as compared to normally achieving writers. They found that this population was more like the normally achieving writers on an assessment of comprehension of text structure. On the production of a coherent text, normally achieving writers performed significantly better than the developmental expressive writing disordered population. These results suggest that individuals demonstrating a developmental expressive writing disorder often will show discrepancies between their comprehension and production of written form. This more than likely would not be true of writers whose primary disorder is receptive language and/or reading-based. However, future research is needed to address the specific cognitive processes involved in the actual comprehension and production of text structure.

The concern for finding a measure that would investigate writing from the point of view of the total discourse lead to the development of the holistic score. The Educational Testing Service first coined the term "holistic" to define the type of scoring of written language that can be rated in a quick and reliable manner (Vogel & Moran, 1982). It is an impressionistic evaluation of writing that looks at the total text and sorts and ranks ratings in a guided procedure. The underlying philosophy of the approach to evaluation is that writing is communication; that is, meaning and intent of the writer must not be fragmentized such that the whole is diminished.

Holistic scoring reliably rank orders essays according to a set of specified criteria. This procedure is utilized when large numbers of essays must be evaluated quickly (i.e., placement, minimal competency testing). However, the information gained from holistic scoring would not provide a diagnostician specific information on an individual's organizational strategies information needed for documenting a developmental expressive writing disorder.

The ability to organize ideas and arrange them to create an organizational plan involves many linguistic and cognitive skills (Scardamalia, Bereiter, & Goelman, 1982). Close examination of the writer's cognitive development has been found to have strong correspondence with the ability to create literary patterns (Applebee, 1978). This has led to the use of story grammars as a diagnostic tool. However, current research (Englert & Thomas, 1987; Newcomer, Barebaum, & Nodine, 1988) investigating story grammar of handicapped populations often fails to report the cognitive stages of the subjects. Therefore, reporting story scheme performance is similar to comparing to apples to oranges. In addition, Weaver and Dickinson (1982) reported the limited diagnostic utility of story grammars due to their relative insensitivity to developmental and individual differences.

Research in the area of story scheme has not been validated thoroughly with respect to its application to written production of text form. Therefore, at present there are no reliable and valid standardized assessment measures an evaluator can utilize to individually investigate a writer's cohesion or coherence of text. The Test of Written Language–2 (TOWL–2) (Larsen & Hammill, 1988) does address some aspects of coherence under their Thematic Maturity subtest. However, this subtest also investigates some issue of audience awareness. Therefore, actually what this subtest measures is questionable. Graham and Harris (1986) devised a scale for assessing the schematic structure of written stories. Yet the validity of the scale was demonstrated by cor-

TABLE 5.8
Components of Text Organization that Should be Used in Assessment

Area	*Not Noted*	*Frequently Noted*	*Significantly Noted*
Sentence organization			
1. Fragments			
2. Run-ons			
3. Tense shifts			
4. Subject/predicate disagreements			
5. Word omissions			
6. Word additions			
7. Repetitious structure			
8. Semantic breakdowns			
Intra paragraph organization			
1. Unformatted			
2. Omission of topic sentence			
3. Related details insufficiently developed			
4. Sequencing errors			
5. Addition of unrelated details			
Inter paragraph organization			
1. Omission of topic sentence			
2. Beginning insufficiently developed			
3. Middle insufficiently developed			
4. Ending insufficiently developed			
5. Closure not established			
6. Sequencing errors			
7. Ideas organized by:			
a. Inference			
b. Generalization			
c. Senses			

Note: From *Text Organization—Informal Behaviors to Observe* by N. Henry, 1988 (unpublished manuscript), University of Georgia, Athens.

relation with the Thematic Maturity Score on the TOWL (.40) and holistic rating of the story quality (.71 to .86). The application of this scale to genre other than narrative has also not been investigated.

An evaluator investigating an individual's ability to produce written text must first consider the genre (narrative, expository, persuasive, descriptive), the age of the individual (logical reasoning development), and the experience with the topic asked to write. Then a comparison should be made of syntax, intraparagraph organization, and interparagraph organization. Table 5.8 lists some of the aspects of text organization that should be rated by the evaluator.

DSM-III-R Criteria Related to Text Organization Disorders

The DSM-III-R requires that a standardized assessment of written language be utilized in determining a developmental expressive writing disorder. However, there are no standardized tests measuring coherence and/or cohesion that are both valid and reliable. Therefore, an evaluator will need to develop some checklists to utilize in evaluating the frequency and accuracy of between sentence relationship as well as text structure or form. Again, the evaluator should be attempting to identify the cognitive processing deficits that are causing the difficulty in producing organized text.

Gregg and McAlexander (1989) discussed the case of a developmental expressive writing disabled adult who had significant difficulty producing organized text. The adult's organizational deficits, that is, his inability to subordinate and coordinate external data, affected his ability to produce written text. Although he had an excellent vocabulary and understanding of the meaning of individual words, he demonstrated significant difficulty in grasping the meaning of the whole-text structure. Understanding the relationship between logical-grammatical structures at the sentence and text level was also difficult for him. Luria (1973) described such characteristics in individuals with left parietal lesions.

Proposed Criteria for Expressive Writing Disorders Impacting on Organization

Table 5.9 lists the proposed criteria related to text organization disorders. First, an examiner should identify the cognitive processing deficits impacting on the ability to organize text. Second, the intersentence and intrasentence structures of the text should be evaluated. Finally, the examiner should be sure the text structure problem is not secondary to other developmental disorders.

TABLE 5.9
Proposed Criteria for Developmental Writing Disorders Impacting on Text Structure

1. Cognitive processing deficits, as measured by standardized, individually administered tests or subtests, are markedly below the expected level given the person's schooling and intellectual capacity (as determined by an individually administered IQ test).
2. Performance on a spontaneous writing sample must indicate at least 4/5 intraparagraph errors and 5/7 interparagraph organizational errors (see Table 5.8).
3. Not due to a deficit in visual, hearing, or lack of instruction.

Sense of Audience

A sense of audience requires the writer to make the necessary adjustment and choices in writing that take into account the intended reader(s). A writer's ability to identify and remain sensitive to a specific audience influences almost every aspect of written communication, including: syntax (Flower, 1979; Shaugnessy, 1977), organization (Berkenkotter, 1981; Crowhurst & Picke, 1979; Flower, 1979; Rubin, 1982; Smith & Swan, 1978), and pronoun reference (Shaugnessy, 1977).

Gregg (1982) reported that many learning disabled adult writers often exhibit pragmatic errors in their written language. Such individuals do not elaborate and/or provide the reader with enough information to understand the meaning of the written text. Their writing often appears disorganized, lacking cohesion and coherence. Researchers (Rubin, 1984) are just beginning to explore the relationship between an individual's social cognitive skills and sense of audience in written language. In a search for a better understanding of sense of audience, Gregg, Hoy, and Wahlers (1988) explored college-able learning disabled writers' social cognitive complexity compared to normally achieving writers. They found that learning disabled students were deficient not only in the total number of constructs they could produce but also in the abstractness of the constructs. Individuals who have developed a greater number of abstract constructs should be better able to adapt a message to an audience's perspective. Gregg, Hoy, and Wahlers (1988) concluded that many learning disabled writers would have a difficult time adjusting their word choice and style to different audience demands.

This research on learning disabled writers is informative in terms of defining characteristics of this population as a group, but it does not address how specific types of cognitive processing deficits impact on a writer's sense of audience. Therefore, Gregg and McAlexander (1989) explored the relationship between a writer's unique social and cognitive profiles and the sense of audience in their written text by examining representative writing samples of two developmentally disordered writers. The first student had a developmental expressive writing disorder; the second had a developmental expressive language disorder. Although the writing of both students was affected by their disabilities, this research suggests that the quality of their audience awareness was indeed related to the nature of their cognitive deficits.

Social Cognitive Skills

In the past, audience awareness was treated as a "monolithic" rather than a "multidimensional" construct (Rubin, 1984, p. 239), that is, as one skill rather than a combination of several. In recent years, however, researchers have identified several social cognition subskills required in developing sensitivity to audience in written language. It must be noted, however, that whereas these subskills have been identified in theory as separate, some are so interrelated that one written statement can illustrate more than one subskill. Gregg and McAlexander (1989) identified six social cognitive subskills as good predicators of sense of audience ability (see Table 5.10).

Gregg and McAlexander's (1989) research has three major implications for the identification of cognitive processing deficits impacting on sense of audience. First, professionals must recognize when a student's mechanical and grammatical problems do not arise simply from a lack of concern for readers. Second, certain specific cognitive processing deficits are more likely to cause problems with other sense of audience skills. More research, however, is needed to determine the impact of various specific deficits. Finally, professionals need to recognize the significant relationship between self-concept and audience awareness. Certainly, the relationship between an individual's cognitive weaknesses or deficits, self-concept, and sense of audience is an area of research that must receive further exploration (Gregg & McAlexander, 1989).

Ideation/Abstraction

Integral with sense of audience and text organization is the concept of abstraction. Moffett and Wagner (1983) defined abstractness as the umbrella structure of discourse in that it is the set of relations among speaker–listener–subject that are involved in referential and rhetorical communication. They feel referential relations are how someone abstracts from raw phenomena, and rhetorical relations are how someone abstracts from an audience. Moffett and Wagner (1983) concluded that abstraction reduces the stress of an individual by selecting and ranking the elements of experiences. According to Moffett and Wagner, increased consciousness of abstracting has as much to do with developmental growth as progress up the abstraction ladder.

Ideation also is a very difficult area to assess and should not be evaluated in isolation from sense of audience. Myklebust's (1965, 1973) Abstraction Quotient is one of the best standardized measurements of ideation/abstraction in written language. He developed an abstract–concrete scale to be used with the PSLT. Problems with the scale involve the type of task used with the scale, lack of sensitivity to audience, and validity and reliability issues. The Test of Written Language–2 (Larsen & Hammill, 1988) purports to evaluate ideation and sense of audience in the Thematic Maturity subtest. However, this subtest is a mixture of unrelated characteristics based on little theoretical foundation. Combin-

TABLE 5.10
Aspects of Ideation/Abstractness

Assessment Area	*Components*
I. Type of discourse	
	Drama/recording
	Description
	Exposition
	Argument/persuasive
II. Temporal aspect	
	What is happening
	What happened (or will happen)
	What happens
	What might/shall happen
III. Degree of decentering	
	Egocentric
	Well-known audience
	Unknown audience
	Emotions projected
	Emotions focused
IV. Basis of abstraction	
	Sensations
	Perceptual
	Memory
	Generalization
	Inferences
V. Communication mode (ideas)	
	Implicit
	Explicit
VI. Communication mode (vocabulary)	
	Implicit
	Explicit

Note: From *Ideational Abstractness—Assessment Areas* by N. Henry, 1988 (unpublished manuscript), University of Georgia, Athens.

ing Mykelbust's (1965, 1973) Abstraction Quotient and Moffett and Wagner's (1983) Abstraction Scale, an evaluator will need to develop questions to use in probing a writer's ideational level. Table 5.10 provides some aspects of ideation an examiner must keep in mind while probing an individual's written language abstraction skills (Henry, 1988a).

Measurement of Sense of Audience

Cognitive and developmental psychology has developed many measurement techniques in order to assess social cognition and/or audience awareness. However, many of these instruments are of questionable psychometric adequacy (Enright & Lapsley, 1980; Kurdek, 1978). Rubin

(1984) stated that the problem has been that social cognition measures fail to correlate with indices of communication adequacy, because the wrong instrument was selected to tap the social awareness of a particular task. Rubin (1984) and Gregg and McAlexander (1989) reported that the awareness of audience requires the use of many social cognitive subskills. Therefore, it is not as simple as "keep your audience in mind."

Britton, Burgess, Martin, McLeod, and Rosen (1975) also advocated that students need experience in communicating directly to specified audiences for different purposes in writing. He claimed that the student "must have a lively representation of this audience in mind—or, if he does not, he will fail in his intent" (p. 59). Therefore, it is also important for the evaluator to assess an individual's writing across different audiences. Some writers can produce much more elaborate and better organized text when they feel comfortable with the audience.

In addition to evaluating an individual's ability to deal with different audiences, the subskills involved in producing a sensitivity to the audience must be observed and probed. Table 5.11 lists the six subskills that appear to be predictive of sense of audience (Rubin, 1984). In addition, types of problems possibly impacting on each of these subskills are described.

DSM-III-R Criteria Related to Sense of Audience

The DSM-III-R (1987) does not list sense of audience as an area possibly impacted by a Developmental Expressive Writing Disorder. However, research indicates that this area is vital to assess when evaluating the written language of individuals.

Gregg and McAlexander (1989) demonstrated that the quality of an individual's audience awareness is indeed related to the nature of their specific cognitive processing profile. However, many other thought and affective disorders could also impact on a writer's sensitivity to their audience (e.g., depression). Therefore, an examiner will need to identify specific cognitive processing deficits (i.e., visual-perceptual, nonverbal organization) that might impact on an individual's adapting the text to their audience's needs. Again, no standardized instrument is available to assess sense of guidance. Table 5.11 can provide the examiner with behaviors to identify in evaluating levels of sense of audience in written text.

Proposed Criteria for Expressive Writing Disorders Impacting on Sense of Audience

Table 5.12 lists the proposed criteria for a developmental expressive writing disorder impacting on sense of audience. The first criteria, as with all other areas, must be the identification of a cognitive processing

TABLE 5.11
Audience Awareness Subskills Important for Assessment

Writer Ability for Successful Use of Sense of Audience	*Influencing Use of Writer Ability*
Content	
Knowledge/value base "I know something about what I'm going to write about."	Language deficits Reading deficits Long-term memory deficits Pragmatic deficits Conceptual deficits Nonverbal deficits Limited experience
Execution	
Actual linguistic resources. "I have the skills to express my ideas to meet the demands of this audience."	Word finding deficits Oral receptive syntax deficits Oral express syntax deficits Semantic deficits Written syntax deficits Phonological deficits Spelling disorders Text structure deficits
Construct Differentiation	
Transferring the mental image of the audience into communication strategies aiding message delivery. "In order to communicate my ideas, I need to do it this way using these strategies."	Distractibility Pragmatic deficits Oral language deficits Conceptual deficits Organizational deficits Low self-esteem Nonverbal deficits Limited experience
Perspective	
Realizing that the reader has a perspective different form the writer's. "She may look at this problem in a different way than I do."	Egocentricity Pragmatic deficits Conceptual deficits Cognitive flexibility Nonverbal deficits Limited experience
Maintenance of Perspective	
Interaction between text and reader. "How well I say what I know depends on the words I can have to choose and who continues to read my product."	Egocentricity Rigidity of thought Pragmatic deficits Oral deficits Word finding Syntax Semantics Organizational deficits Nonverbal Limited experience

(*continued*)

TABLE 5.11
(Continued)

Writer Ability for Successful Use of Sense of Audience	*Influencing Use of Writer Ability*
Role-Taking	
Engaging in social inference. "I have something to say, but my reader is thinking a certain way. I have to consider this."	Pragmatics Conceptual deficits Rigidity Egocentricity Nonverbal deficits Limited experience

Note: Adapted from "Social Cognition and Written Communication" by D. Rubin, 1984, *Written Communications, 1*(2), pp. 211–243, in *Audience Awareness: Subskills for Assessment—An Informal* by N. Gregg & B. McCarty, 1989 (unpublished manuscript), University of Georgia, Athens.

deficit impacting on an individual's ability to remain sensitive to his or her audience. Second, underachievement must be observed across different audiences. Third, at least three out of the five subskills necessary for an appropriate sensitivity to an audience (Table 5.11) must be identified as inadequate in two writing samples.

Handwriting (Manual Writing)

Individuals demonstrating a developmental expressive writing disorder have at times been identified by poor handwriting performance (Brenner, Gillmann, Zangwill, & Farrell, 1967; Sovik, 1984b; Sovik,

TABLE 5.12
Proposed Criteria for Writing Disorders Impacting on Sense of Audience

1. Cognitive processing deficits, as measured by standardized, individually administered tests or subtests, are markedly below the person's schooling or intellectual capacity (as determined by an individually administered IQ test).
2. Performance must be compared across at least two writing samples which require writing to different types of audiences.
3. Performance on three out of the five following subskills required for audience awareness must be significantly below average on both writing samples (Criteria 2).
    1. Content
    2. Execution
    3. Construct differentiation
    4. Role taking
    5. Perspective taking.
4. Not due to a deficit in acuity, thought disorder, attention disorder, or lack of instruction.

Arntzen, & Thygesen, 1986). Distinguishing between individuals demonstrating a developmental expressive writing disorder, a developmental coordination disorder, or a developmental reading disorder is, to say the least, quite difficult. Certainly, research indicates that heterogeneity characterizes the dyslexic and dysgraphic populations (Solheim, Nygaard, & Aasved, 1984; Sovik, 1984; Sovik & Thygesen, 1985). Currently, under the DSM-III-R (1987) classification system, deficits impacting on manual writing would be classified under developmental coordination disorders. This would be consistent with Myklebust's (1965) definition of dysgraphia as "symbolic in nature."

Yet the relationship between the DSM-III-R developmental expressive writing disorder and developmental coordination disorder might be more significant than is currently recognized. Research indicates a distinction between normal and dysgraphic children concerning motor control in their writing hand and their ability to exploit sensory feedback during the act of copying and writing tasks (Sovik, Arntzen, & Thygesen, 1986). There is also evidence that contextual factors and motion patterns used in producing letters, words, and sentences will affect children's as well as adult's writing behavior (Ellis, 1979; Sovik et al., 1986; Teulings, Thomassen, & van Galen, 1983; Wing, Nimmo-Smith, & Eldridge, 1983). Contextual factors influencing the spelling and writing process were reviewed in the literature (Ellis, 1979; Slovik et al., 1986; Teulings et al., 1983; Wing et al., 1983). Recently, Sovik et al. (1986) found that both semantic and physiological (movement) factors may influence the writing process. Further research is needed to examine the associations between graphemic memory (Ellis, 1983), motor coordination, and linguistic processes.

SUMMARY OF PROPOSED DIAGNOSTIC CRITERIA FOR A DEVELOPMENTAL EXPRESSIVE WRITING DISORDER

The DSM-III-R (1987) criteria for determining a developmental expressive writing disorder are extremely vague, providing professionals with little direction for identifying the population. As discussed throughout the chapter, a major flaw with the current criteria rests on the requirement for documenting the developmental writing disorder with a standardized, individually administered test. The lack of empirical research defining the written language characteristics of individuals demonstrating a developmental expressive writing disorder as well as a lack of valid and reliable measurement tools to assess spelling, written syntax, text organization, and sense of audience make such criteria inappropriate. Qualitative indices and clinical judgment should be incorporated as diagnostic supplements when a valid, standardized measurement of a specific writing task is not available.

The age of onset for a developmental expressive writing disorder set by the DSM-III-R (1987) certainly needs modification as well. Vygotsky (1978) and Luria (1980) both documented the development of representation that leads up to symbolization in writing. In particular, Vygotsky (1978) discussed the importance of the development of gesture, play, and drawing to the ability to write. Some of the cognitive processing deficits that will later impact on written text (e.g., visual-motor skills, symbolization) can be identified early in these other modes of representation.

Further, the lack of criteria specifying the identification of cognitive processing deficits impacting on the production of written text remains the most glaring problem with this DSM-III-R category. Professionals must recognize that a developmental expressive writing disorder is not synonymous with underachievement. The cause for a discrepancy between achievement in written language and general overall intelligence can stem from various sources (i.e., thought disorder, affective disorder, lack of instruction). However, the DSM-III-R (1987) category for a developmental expressive writing disorder should be reserved for those individuals (Myklebust, 1965, 1973) described as dysgraphic. Although research (Rapp & Caramazza, 1989) is still unclear as to which linguistic or visual-spatial cognitive processing deficits contribute to deficits in the different stages of written language acquisition, the fact that a written language disorder is the result of a breakdown in the cognitive system is well supported in the literature (Rapp & Caramazza, 1989; Roeltgen, 1985). In addition, the concept of dysgraphia without dyslexia or dysaphasia has been supported in the literature (Baxter & Warrington, 1986; Levine et al., 1988). The DSM-III-R category for developmental expressive writing disorders should be used for the "pure" dysgraphia population. Although writing disorders are secondary to oral language, reading, and personality disorders, the DSM-III-R category for developmental expressive writing disorders should not be used in identifying these other populations. The relationship between primary and secondary writing disorders, although not well examined in the literature, must be recognized by professionals. Proper intervention rests on accurate identification of such populations.

The current DSM-III-R (1987) criteria for developmental expressive writing disorders states that the disturbance impacts on written composition in spelling words, expressing thoughts in grammatically correct sentences and organized paragraphs. However, no criteria for indi-

TABLE 5.13
General Guidelines for Proposed Criteria for Developmental Writing Disorder

1. Performance measured on three different formats including copying, dictation, and spontaneous writing tasks.
2. Selected cognitive processing deficits, (e.g., visual-motor) as measured by standardized, individually administered tests or subtests, are markedly below the expected level given the person's schooling and intellectual capacity (as determined by an individually administered IQ test).
3. Criteria for each of the areas of written language expression (i.e., spelling, syntax, organization, and sense of audience) must be met in at least one of those areas.
4. Not due to a deficit in visual or hearing, but tactile deficits may be contributors.
5. Not due to lack of instruction.
6. Not a secondary problem to any other developmental disorder.

cating what constitutes a "significant" problem is described, and there is no description of the types of task that should be used in exploring those areas. In addition, no mention of the impact of the disorder on sense of audience or ideation is provided in the DSM-III-R, despite the fact that current research clearly documents the importance of these areas to the writing process.

Professionals required to identify developmental expressive writing disorders could very easily misdiagnose individuals demonstrating this disorder by following the criteria specified in the DSM-III-R. Therefore, it is suggested that modification of the current criteria be provided for clinicians and researchers. Table 5.13 lists these proposed recommendations. First, no decision should be made regarding an individual's writing ability without at least the presentation of three different task formats. Luria's (1980) suggested copy, dictation, and spontaneous writing tasks appear to be the most appropriate variety of tasks that would best identify specific cognitive processing deficits impacting on written language. Each of the areas of written language (spelling, syntax, organization, sense of audience) can be observed effectively by these three tasks.

The second modification suggested refers to the type of tools accepted as appropriate assessments of written language. It is suggested that qualitative indices should be used to supplement and/or replace the need for standardized instruments in the area of written language. Third, specific criteria for identification of a disorder have been defined for each of the following areas of written language: spelling, syntax, organization, and sense of audience. It is suggested that, for an individual to be classified under this DSM-III-R diagnosis, he or she must meet the standards described for at least one area. Therefore, the proposed guidelines would make it mandatory to identify the cognitive processing deficits impacting on the specific area of written language. Finally, this diagnosis would be reserved for those individuals whose primary disorder is in the production of written text and not a secondary problem to an oral or reading developmental disorder.

FUTURE DIRECTIONS

Vygotsky (1962) discussed the complex process of writing and postulated that learning to write involves the mastering of cognitive skills within the development of new social understanding. He felt that we categorize and synthesize our lives through inner speech—the language of thought. To transform the inner language to written text requires one step outside of thought to enter the social context of the reader. Therefore, writing involves not only knowledge of the topic, rhetorical knowledge, and metacommunication skills (Burleson, 1984) but an awareness of an audience (reader) and a sensitivity to the reader's needs. Indeed, coherence and cohesion in writing result in a large part from intentions that are appropriate to a specified audience. Thus, audience awareness must be coupled with genre competence to allow a writer to be effective in communicating ideas.

The empirical research investigating the characteristics of writing disorders and the development of appropriate written language assessment tools is lacking. Therefore, professionals attempting to assess the different areas of an individual's written product have little guidance theoretically and clinically. The DSM-III-R (1987) criteria for developmental expressive writing disorders does little to help in specifying the population, and in fact it may lead to the misdiagnosis of the population.

Criteria for each of the areas of written language most commonly impacted on by a written language disorder have been described. Professionals are encouraged always to investigate the writing abilities of an individual using copying, dictation, and spontaneous writing tasks. Each of those tasks investigates the different cognitive demands required for the writing process. Finally, the key component for the diagnosis of a developmental expressive writing disorder rests in the ability of the examiner to identify the cognitive processing deficit(s) impacting on specific areas of written language (e.g., spelling, syntax, organization, and sense of audience). Professionals need to recognize that certain specific cognitive processing deficits are more likely to impact on different areas of the writing process and product. However, more research is certainly needed to determine the impact of various specific cognitive deficits on the writing process.

REFERENCES

American Psychiatric Association. (1987). *Diagnostic and statistical manual of mental disorders* (3rd ed.–rev.). Washington, DC: American Psychiatric Association.

Applebee, A. N. (1978). *The child's concept of story: Ages two to seventeen.* Chicago: University of Chicago Press.

Auerbach, S. H., & Alexander, M. P. (1981). Pure agraphia and unilateral optic ataxia associated with a left superior lobale lesion. *Journal of Neurology, Neurosurgery and Psychiatry, 44,* 430–432.

Barron, R. W. (1980). Visual and phonological strategies in reading and spelling. In U. Frith (Ed.), *Cognitive processes in spelling* (pp. 195–214). London: Academic Press.

Baxter, D. M., & Warrington, E. K. (1986). Ideational agraphia: A single case study. *Journal of Neurology, Neurosurgery and Psychiatry, 49,* 369–374.

Belanger, J. F. (1978). Calculating the syntactic density score: a mathematical problem. *Research in the Teaching of English, 12,* 149–153.

Benson, D. F. (1979). *Aphasia, alexia and agraphia.* New York: Churchill Livingston.

Berkenkotter, C. (1981). Understanding a writer's awareness of audience. *College Composition and Communication, 32,* 388–399.

Blalock, J. (1981). Persistent problems and concerns of young adults with learning disabilities. In W. Cruickshank & L. Silvers (Eds.), *Bridges to tomorrow. The Best of ACLD* (Vol. 2, pp. 45–62). Syracuse: Syracuse University Press.

Brenner, M. V., Gillmann, S., Zangwill, D. L., & Farrell, M. (1967). Visuo-motor disability in school children. *British Medical Journal, 4,* 259–262.

Britton, J., Burgess, T., Martin, N., McLeod, A., Rosen, H. (1975). *The development of writing abilities (11–18).* London: Macmillan.

Bryant, P., & Bradley, L. (1985). *Children's reading problems.* New York: Basil Blackwell.

Burleson, B. R. (1984). The affective perspective taking process: A test of Turiel's role-taking model. In M. Burleson (Ed.), *Communication yearbook II* (pp. 473–488). Beverly Hills, CA: Sage.

Caramazza, A. C. (1988). Some aspects of language processing revealed through the analysis of acquired aphasia: The lexical system. *Annual Review of Neurosciences, 11,* 395–421.

Charolles, M. (1978). Introduction aux problems de la coherence des textes [Introduction to problems with the coherence of text]. *Langue Fracaisc, 38,* 7–41.

Clark, H., & Clark, E. (1977). *Psychology and language: An introduction to psycholinguistics.* New York: Harcourt Brace Jovanovich.

Coltheart, M. (1980). Deep dyslexia: A review of the syndrome. *Deep Dyslexia.* Boston: Routledge & Kegan Paul.

Crew, F. (1976). *The Random House handbook* (2nd ed.). New York: Random House.

Cromer, R. F. (1980). Spontaneous spelling by language-disordered children. In U. Frith (Ed.), *Cognitive processes in spelling* (pp. 402–422). London: Academic Press.

Crowhurst, M., & Piche, G. L. (1979). Audience and mode of discourse effects on syntactic complexity at two grade levels. *Research in the Teaching of English, 13,* 101–109.

Deloche, G., Andreesky, E., & Desi, M. (1982). Surface alexis: A case report and some theoretical implications to reading models. *Brain and Language, 15,* 12–31.

Ellis, A. W. (1979). Slips of the pen. *Visible Language, 13*, 265–282.

Ellis, A. W. (1982). Spelling and writing (and reading and speaking). In A. W. Ellis (Eds.), *Normality and pathology in cognitive functions* (pp. 113–146). London: Academic Press.

Ellis, A. (1983). *Reading, writing and dyslexia: A cognitive analysis.* Hillsdale, NJ: Lawrence Erlbaum Associates.

Englert, C. S., & Thomas, C. C. (1987). Sensitivity to text structure in reading and writing: A comparison of learning disabled and non-learning disabled students. *Learning Disability Quarterly, 10*, 93–105.

Enright, R. D., & Lapsley, D. K. (1980). Social role-taking: A review of the constructs, measures, and measurement properties. *Review of Education Research, 50*, 647–674.

Exner, S. (1881). *Untersuchungen uber die Lokalisation der Funktionen in der Grosshirnrinde des Menschen.* Vienna: Wilhelm Braumuller.

Flower, L. (1979). Writer-based prose: A cognitive basis for problems in writing. *College English, 41*, 19–37.

Frith, U. (1979). Reading by ear and writing by ear. In P. A. Kilers, M. Wrolstad, & H. Bouma (Eds.), *Processing of vidible language* (pp. 24–49). New York: Plenum.

Frith, U. (1980). Unexpected spelling problems. In U. Frith (Ed.), *Cognitive processes in spelling* (pp. 495–516). London: Academic Press.

Frith, U. (1983). The similarities and differences between reading and spelling problems. In M. Rutter (Ed.), *Developmental neuropsychiatry* (pp. 453–472). New York: Guilford.

Golub, L. S., & Kidder, C. (1974). Syntactic density and the computer. *Elementary English, 51*, 128–131.

Graham, S., & Harris, K. (1986, April). *Improving learning disabled students composition a story grammar training: A component analysis of self-control strategy training.* Paper presented at the Annual Meeting of the American Educational Research Association, San Francisco.

Gregg, N. (1982). *An investigation of the breakdown in certain aspects of the writing process with college age learning disabled, normal, and basic writers.* Unpublished doctoral dissertation, Northwestern University.

Gregg, N. (1985). College learning disabled, normal, and basic writers: A comparison of frequency and accuracy of cohesive ties. *Journal of Psychoeducational Assessment, 3*, 223–231.

Gregg, N. (1986). College learning disabled, normal, and basic writer's sentence combining abilities. *B.C. Journal of Special Education, 10*, 153–166.

Gregg, N., & Hoy, C. (1989). Coherence: The comprehension and production abilities of college normally achieving, learning disabled and underprepared writers. *Journal of Learning Disabilities, 22*, 370–373.

Gregg, N., Hoy, C., & Sabol, R. (1988). Spelling error patterns of normal, learning-disabled, and underprepared college writers. *Journal of Psychoeducational Assessment, 6*, 14–23.

Gregg, N., Hoy, C., & Wahlers, D. (1988). *Social cognitive ability as a predictor of the quality of expository writing among college learning disabled and normal writers.* Unpublished manuscript, University of Georgia.

Gregg, N., & McAlexander, P. A. (1989). The relation between sense of audience and specific learning disabilities: An exploration. *Annals of Dyslexia, 39*, 206–226.

Gregg, N., & McCarty, B. (1989). *Audience awareness: Subskills for assessment—an informal.* Unpublished manuscript, University of Georgia.

Habiger, M. (1988a). *Mechanics assessment—an informal.* Unpublished manuscript, Atlanta Area School for the Deaf, GA.

Habiger, M. (1988b). *Written syntax assessment—an informal.* Unpublished manuscript, Atlanta Area School for the Deaf, GA.
Halliday, M. A. K. (1978). *Language as social semiotics: The social interpretation of language and meaning.* London: University Park Press.
Halliday, M. A. K., & Hasan, R. (1976). *Cohesion in English.* London: Longman.
Hammill, D. D., & Larsen, S. C. (1988). *The Test of Written Language–2.* Austin, TX: Pro-Ed.
Harris, M. M. (1977). Oral and written syntax attainment of second graders. *Research in the Teaching of English, 11,* 117–132.
Hecaen, H., & Albert, M. L. (1978). *Human neuropsychology.* New York: Wiley.
Hempel, C. G. (1952). *Fundamentals of concept formation in empirical science.* Chicago: University of Chicago Press.
Henry, N. (1988a). *Ideation/abstractness—Assessment Areas,* Unpublished manuscript, University of Georgia, Department of Special Education, Athens.
Henry, N. (1988b). *Text organization—Informal behaviors to observe.* Unpublished manuscript, University of Georgia, Department of Special Education, Athens.
Herbert, M. A., & Czerniejewski, C. (1976). Language and learning therapy in a community college. *Bulletin of the Orton Society, 26,* 96–106.
Hessler, G. (1987). Educational issues surrounding severe discrepancy. *Learning Disabilities Research,* 3(1), 43–49.
Hill, A. E., & Caramazza, A. (1989). The graphemic buffer and attentional mechanisms. *Brain and Language, 36,* 208–235.
Hunt, K. W. (1965). *Grammatical structures written at three grade levels.* (Research Report, No. 3). Urbana, IL: National Council of Teachers of English.
Hunt, K. W. (1970). Syntactic maturity in school children and adults. *Monographs of the Society for Research in Child Development, 35,* pp. 18–31.
Isaacson, S. (1988). Assessing the writing product: Qualitative and quantitative measures. *Exceptional Child, 45,* 528–535.
Jastak, S., & Wilkinson, G. S. (1984). *Wide Range Achievement Text–Revised.* Wilmington, DE: Jastak.
Johnson, D. (1987). Disorders of written language. In D. J. Johnson & J. W. Blalock (Eds.), *Adults with learning disabilities: Clinical studies* (pp. 173–204). Orlando, FL: Grune & Stratton.
Johnson, D., & Myklebust, H. (1967). *Learning disabilities: Educational principles and practices.* New York: Grune & Stratton.
Kaplan, E., & Goodglass, H. (1981). Aphasia-related disorders. In M. T. Sarno (Ed.), *Acquired aphasia,* (pp. 256–272). New York: Academic Press.
Kapur, N., & Lawton, N. F. (1983). Dysgraphia for letters: A form of motor memory deficit? *Journal of Neurology, Neurosurgery and Psychiatry, 46,* 573–575.
Kavale, K. (1987). Theoretical issues surrounding severe discrepancy. *Learning Disabilities Research,* 3(1), 12–20.
Kinsbourne, M., & Warrington, E. K. (1962). A study of finger agnosia. *Brain, 1,* 215–225.
Kosslyn, S. M. (1981). The medium and the message in mental imagery: A theory. *Psychological Review, 88,* 46–66.
Kremin, H. (1982). Alexia: Theory and research. In R. N. Malatesha & P. G. Aaron (Eds.), *Reading disorders: Varieties and treatments* (pp. 25–52). New York: Academic Press.
Kremin, H. (1985). Routes and strategies in surface dyslexia and dysgraphia. In K. E. Patterson, J. C. Marshall, & M. Coltheart (Eds.), *Surface dyslexia: Neuropsychological and cognitive studies of phonological reading* (pp. 105–138). Hillsdale, NJ: Lawrence Erlbaum Associates.
Kurdek, L. (1978). Perspective taking as the cognitive basis of children's moral development: A review of the literature. *Merrill–Palmer Quarterly, 24,* 3–28.

Laine, T., & Marttila, R. J. (1981). Pure agraphia: a case study. *Neuropsychologia, 19,* 311–316.

Larsen, S. C., & Hammill, D. D. (1983). *The Test of Written Language.* Austin, TX: Pro-Ed.

Larsen, S. C., & Hammill, D. D. (1988). *The Test of Written Language-2.* Austin, TX: Pro-Ed.

Laughton, T. (1980). *Developmental language system guide.* Atlanta: Department of Education.

Leischner, A. (1969). The agraphias. In P. J. Vinken & G. W. Bruyn (Eds.), *Disorders of speech, perception and symbolic behavior* (pp. 31–56). Amsterdam: North-Holland.

Levine, D. N., Mani, R. B., & Calvanio, R. (1988). Pure agraphia and Gerstmann's syndrome as a visuospatial-language dissociation: An experimental case study. *Brain and Language, 35,* 172–196.

Litowitz, B. (1981). Developmental issues in written language. *Topics in Language Disorders, 1*(2), 73–89.

Luria, A. R. (1973). *The working brain: An introduction to neuropsychology.* New York: Basic.

Luria, A. R. (1980). *Higher cortical functions in man.* New York: Basic.

Lyon, R., & Moat, L. (1989). Critical issues in the instruction of the learning disabled. *Journal of Consulting and Clinical Psychology, 56,* 830–855.

Marcie, P., & Hecaen, H. (1979). Agraphia. In K. M. Heilman & E. Valenstein (Eds.), *Clinical neuropsychology* (1st ed. pp. 105–131). New York: Oxford University Press.

Marshall, T., & Newcombe, F. (1973). Patterns of paralexia: A psycholinguistic approach. *Journal of Psycholinguistic Research, 2,* 1975–2000.

Marshall, T. C., & Newcombe, F. (1980). The conceptual status of deep dyslexia: An historical perspective. In M. Coltheart, K. E. Patterson, & J. C. Marshall (Eds.), *Deep dyslexia* (pp. 1–21). London: Routledge & Kegan Paul.

McCulley, G. A. (1985). Writing quality, cohesion, and coherence. *Research in the Teaching of English, 19,* 269–282.

Mellon, J. C. (1969). *Transformational sentence combining: A method for enhancing the development of syntactic fluency in English composition.* (Research Report, No. 10). Urbana, IL: National Council of Teachers of English.

Moffett, J. M., & Wagner, B. J. (1983). *Student-centered language arts and reading, K–13: A handbook for teachers.* Boston: Houghton Mifflin.

Morton, T. (1980). The logogen model and orthographic structure. In U. Frith (Ed.), *Cognitive processes in spelling* (pp. 117–134). London: Academic Press.

Morton, T., & Patterson, K. E. (1980). A new attempt at an interpretation, or, an attempt at a new interpretation. In M. Coltheart, K. E. Patterson, & T. C. Marshall (Eds.), *Deep dyslexia* (pp. 91–118). London: Routledge & Kegan Paul.

Myklebust, H. (1965). *Development and disorders of written language: Picture Story Language Test* New York: Grune & Stratton.

Myklebust, H. (1973). *Development and disorders of written language: Studies of normal and exceptional children.* New York: Grune & Stratton.

National Assessment of Educational Progress (1980). *Writing achievement, 1969–79: Results from the third national writing assessment, volume 1–17 years-olds.* (Report No. 110-W-01). Denver, CO: National Assessment of Educational Progress.

National Assessment of Educational Progress (1975). *Writing Mechanics, 1969–1974.* Writing Report No. 05-W-01. Denver, CO: National Assessment of Educational Progress.

Newcomer, P., Nodine, B., & Barenbaum, E. (1988). Teaching writing to exceptional children: Reaction and recommendations. *Exceptional Children, 54,* 559–541.

O'Donnell, R. C. (1976). A critique of some indices of syntactic maturity. *Research in the Teaching of English, 10*, 31–38.
O'Donnell, R. C., & Hunt, K. (1975). Syntactic maturity test. In W. T. Fagan, C. R. Cooper, & J. M. Jensen (Eds.), *Measures for research and evaluation in the English language arts* (pp. 16–32). Urbana, IL: National Council of Teachers of English.
O'Hare, F. (1973). *Sentence combining: Improving student writing without formal grammar instruction.* (Research Report No. 15). Urbana, IL: National Council of Teachers of English.
Parrill, M. (1987). Developmental issues surrounding severe discrepancy. *Learning Disabilities Research, 3*, 32–42.
Patterson, K. E., & Marcel, A. L. (1977). Aphasia, dyslexia, and the phonological coding of written words. *Quarterly Journal of Experimental Psychology, 29*, 307–318.
Phelps-Gunn, T., & Phelps-Terasaki, L. W. (1982). *Written language instruction.* Rockville, MD: Aspen Systems.
Phelps, L. W. (1985). Dialects of coherence: Toward an integrative theory. *College English, 47*(1), 12–29.
Pitres, A. (1884). Considerations sur l'agraphie [Considerations of surface agraphic]. *Revue de Medicine (Paris), 4*, 873–885.
Poplin, M., Gray, R., Larsen, S., Banikowski, A., & Mehring, T. (1980). A comparison of components of written expression abilities in learning disabled and non-learning disabled students at three grade levels. *Learning Disability Quarterly, 3*, 46–53.
Poteet, J. (1980). Informal assessment of learning disabilities. *Learning Disability Quarterly, 3*, 88–98.
Rapp, B. C., & Caramazza, A. (1989). Letter processing in reading and spelling: Some dissociations. *Reading and Writing: An Interdisciplinary Journal, 1*, 3–23.
Roeltgen, D. (1985). Agraphia. In K. M. Heilman & E. Valenstein (Eds.), *Clinical neuropsychology* (pp. 75–96). New York: Oxford University Press.
Rosati, G., & deBastiani, P. (1979). Pure agraphia: A discrete form of aphasia. *Journal of Neurology, Neurosurgery and Psychiatry, 42*, 266–269.
Rourke, B. P. (1981). Reading and spelling disabilities: A developmental neuropsychological perspective. In U. Kirk (Ed.), *Neuropsychology of language, reading and spelling* (pp. 85–106). New York: Academic Press.
Rubin, D. (1982). Adapting syntax in writing to varying audiences as a function of age and social cognitive ability. *Journal of Child Language, 9*, 497–510.
Rubin, D. (1984). Social cognition and written communication. *Written Communication, 1*, 211–243.
Saffron, E. M., & Marin, O. S. M. (1977). Reading without phonology: Evidence from aphasia. *Quarterly Journal of Experimental Psychology, 29*, 515–525.
Scardamalia, M., Bereiter, C., & Goelman, H. (1982). The role of production factors in writing ability. In M. Nystranl (Ed.), *What writers know: The language process, and structure of written language* (pp. 173–210). New York: Academic Press.
Scardamalia, M., Bereiter, C., & McDonald, J. D. S. (1977, April). *Role-taking in written communication investigated by manipulating anticipatory knowledge.* Paper presented at the annual meeting of the American Educational Research Association, Chicago, IL.
Scinto, L. F. M. (1982). *The acquisition of functional composition strategies for text.* Hamburg: Bushe.
Scinto, L. F. M. (1986). *Written language and psychological development.* New York: Academic Press.
Scribner, S., & Cole, M. (1981). *The psychology of literacy.* Cambridge, MA: Harvard University Press.

Seymour, P. (1979). *Human visual cognition*. West Drayton, UK: Collier-Macmillan.

Shallice, T., & Warrington, E. K. (1975). Word recognition in a phonemic dyslexic patient. *Quarterly Journal of Experimental Psychology, 27*, 187–199.

Shallice, T., & Warrington, E. K. (1980). Single and multiple component central dyslexia syndromes. In M. Coltheart, K. Patterson, & J. C. Marshall (Eds.), *Deep dyslexia* (pp. 119–145). London: Routledge and Kegan Paul.

Shaugnessy, M. P. (1977). *Errors and expectations: A guide for the teacher of basic writing*. New York: Oxford University Press.

Simms, R. B., & Crump, D. W. (1983). Syntactic development in the oral language of learning disabled and normal students at the intermediate and secondary level. *Learning Disability Quarterly, 6*, 155–165.

Simon, D. P. (1976). Spelling a task analysis. *Instructional Sciences, 5*, 277–302.

Smith, F. (Ed.). (1973). *Psycholinguistics and reading*. New York: Holt, Rinehart & Winston.

Smith, P. T. (1980). Linguistic information in spelling. In U. Frith (Ed.), *Cognitive processes in spelling* (pp. 67–82). London: Academic Press.

Smith, W. L., & Swan, M. B. (1978). Adjusting syntactic structures to varied levels of audience. *Journal of Experimental Education, 46*, 29–34.

Solheim, R., & Nygaard, H. D., & Aasved, H. M. (Eds.). (1984). *Bergen-prosjektet, II Sokelys paa smaaskolealderen*. Bergen: Universitetsforlaget.

Sovik, N. (1984a). Utvikling av skrivedugleik. Teoriar, forsking og pedagogiske retnignsliner. *Det Kgl. Norske Videnskabers Selskab Forhandlinger*, 69–81.

Sovik, N. (1984b). The effects of a remedial tracking program on writing performance of dysgraphic children. *Scandinavian Journal of Educational Research, 28*, 129–147.

Sovik, N., Arntzen, O., & Thygesen, R. (1986). Effects of feedback training on "normal" and dysgraphic students. In H. R. S. Kao (Eds.), *Graphonomics: Contemporary research in handwriting* (pp. 17–29). Amsterdam: North-Holland.

Sovik, N., & Thygesen, R. (1985). *Learning disabilities in reading, spelling and writing. Report no. 2 in the research project: The relationship between dysgraphia and dyslexia in primary school* (Mimeograph): Department of Education, University of Trondheim.

Strong, W. (1973). *Sentence combining: A composing book*. New York: Random House.

Teulings, H. L., Thomassen, A. J. W. M., & van Galen, G. P. (1983). Preparation of partly precured handwriting movements: The size of movement units in handwriting. *Acta Psychologica, 54*, 165–177.

Tomlinson, B. M. (1980). The influence of sentence combining instruction on the syntactic maturity and writing quality of minority college freshman in a summer preentry preparation program. (Doctoral dissertation, University of California, Riverside). *Dissertation Abstracts International, 42*, 5440A (Order No. 8020623).

Vogel, S. A. (1986). Syntactic complexity in written expression of LD college writers. *Annals of Dyslexia, 35*, 137–157.

Vogel, S. A., & Moran, M. (1982). Written language disorders in learning disabled college students: a preliminary report. In W. Cruickshank & J. Lerner (Eds.), *Coming of age: The best of ACLD* (Vol. 3, pp. 32–49). Syracuse: Syracuse University Press.

Vygotsky, L. S. (1962). *Thought and language*. Cambridge, MA: MIT Press.

Vygotsky, L. S. (1978). *Mind and society: The development of higher psychological processes*. Cambridge, MA: Harvard University Press.

Weaver, P. A., & Dickinson, K. P. (1982). Scratching below the surface structure: Exploring the usefulness of story grammars. *Discourse Processes, 5*, 225–243.

Wernicke, C. (1903). Ein Fall von isolierter Agraphie. *Monatschrift Fur Psychiatrie und Neurologie, 13*, 242–265.

Wiig, E. H., & Semel, E. M. (1976). *Language disabilities in children and adolescents.* Columbus, OH: Merrill.

Wilson, V. L. (1987). Statistical and psychometric issues surrounding severe discrepancy. *Learning Disabilities Research, 8*(1), 24–28.

Wing, A. M., Nimmo-Smith, I., & Eldridge, M. A. (1983). The consistency of cursive letter information as a function of position in the word. *Acta Psycholgica, 54*, 197–204.

Witte, S. P., & Faigley, L. (1981). Coherence, cohesion, and writing ability. *College Composition and Communication, 32*, 189–204.

CHAPTER 6

Developmental Reading Disorder

Keith E. Stanovich
Ontario Institute for Studies in Education

The learning disability (LD) now identified variously as *developmental reading disorder, dyslexia,* or *specific reading disability* (Hynd & Cohen, 1983) was brought to general attention by the turn-of-the-century clinical reports of Pringle Morgan (1896) and James Hinshelwood (1895) in Great Britain, among others, under the guise of the term *congenital word blindness.* Hinshelwood summarized many of these cases in his 1917 monograph of that name. Subsequent to Hinshelwood's work, the most influential single investigator was Samuel Orton (1925, 1928, 1937), who contributed numerous reports of clinical cases, theoretical speculation about proximal causation, and suggestions for treatment techniques. Although Orton coined the term *strephosymbolia* to describe the condition, he also often used the term *specific reading disability* (Duane, 1985).

The introduction of the generic term *learning disabilities* in the early 1960s (Kirk, 1963; Lerner, 1985) and the subsequent concerns with providing services for learning disabled (LD) children, the bulk of whom suffered from reading difficulties of varying degrees of severity (Bateman, 1979; Gaskins, 1982; Kirk & Elkins, 1975; Lerner, 1975), ushered in a period of great concern about definitional issues that has continued until the present. Several definitions had a considerable influence both on research on developmental reading disorder and in service delivery debates. The definition of the World Federation of Neurology had many features that became canonical for many researchers and practitioners. Specific developmental dyslexia was characterized by Critchley (1970) as "a disorder manifested by difficulty in learning to read despite conventional instruction, adequate intelligence, and socio-cultural opportunity. It is dependent upon fundamental cognitive abilities which are frequently of constitutional origin" (p. 11). This particular definition highlighted the well-known "exclusionary criteria" that subsequently caused much dispute in discussions of dyslexia (e.g., Applebee, 1971; Ceci, 1986; Doehring, 1978; Eisenberg, 1978; Rutter, 1978). In particular, this definition excluded from the dyslexia classification children of low intelligence, those who suffered from inadequate environments, and those who underachieved due to lack of educational opportunity.

The exclusionary criteria were carried over into the definition of learning disability employed in the landmark Education for All Handicapped Children Act (1975):

> Specific learning disability means a disorder in one or more of the basic psychological processes involved in understanding or in using language spoken or written, which may manifest itself in an imperfect ability to listen, think, speak, read, write, spell, or to do mathematical calculations. The term includes such conditions as perceptual handicaps, brain injury, minimal brain dysfunction, dyslexia, and developmental aphasia. The term does not include children who have learning problems which are primarily the result of visual, hearing, or motor handicaps, of mental retardation, of emotional disturbance, or of environmental, cultural, or economic disadvantage. (Federal Register, 1977, p. 65, 083)

The National Joint Committee for Learning Disabilities (NJCLD) responded (in Hammill, Leigh, McNutt, & Larsen, 1981) to criticisms of the exclusionary criteria by proposing the following:

> These disorders are intrinsic to the individual and presumed to be due to central nervous dysfunction. Even though a learning disability may occur concomitantly with other handicapping conditions (e.g., sensory impairment, mental retardation, social and emotional disturbance) or environmental influences (e.g., cultural differences, or inappropriate instruction, psycholinguistic factors), it is not the direct result of those conditions or influences. (p. 336)

The Committee thus emphasized that the mere presence of other impairments or of environmental deprivation should not exclude children from the LD categorization. The Interagency Committee on Learning Disabilities, established by the U.S. Health Research Extension Act of 1985 (Kavanagh & Truss, 1988), accepted the essentials of the NJCLD

TABLE 6.1
315.00 Developmental Reading Disorder

The essential feature of this disorder is marked impairment in the development of word recognition skills and reading comprehension that is not explainable by mental retardation or inadequate schooling and that is not due to a visual or hearing defect or a neurological disorder. The diagnosis is made only if this impairment significantly interferes with academic achievement or with activities of daily living that require reading skills.

Oral reading is characterized by omissions, distortions, and substitutions of words and by slow, halting reading. Reading comprehension is also affected. This disorder has been referred to as "dyslexia."

Associated features. Deficits in expressive language and speech discrimination are usually present, and may be severe enough to warrant the additional diagnosis of developmental expressive or receptive language disorder. Developmental expressive writing disorder is often present. In some cases, there is a discrepancy between verbal and performance intelligence scores. Visual perceptual deficits are seen in only about 10% of cases. Disruptive behavior disorders may also be present, particularly in older children and adolescents.

Age at onset. The disorder is usually apparent by age seven (second grade). In severe cases, evidence of reading difficulty may be apparent as early as age six (first grade). Sometimes developmental reading disorder may be compensated for in the early elementary grades, particularly when it is associated with high scores on intelligence tests. In this case, the disorder may not be apparent until age nine (fourth grade) or later.

Course. With reading therapy, if the disorder is mild, there are often no signs of the disorder in adulthood. If the disorder is severe, even with treatment many signs of the disorder remain for life.

Prevalence. Estimates of the prevalence of the disorder in school-age children have ranged from 2% to 8%.

Familial pattern. The disorder is more common among first-degree biologic relatives than in the general population.

Differential diagnosis. In *mental retardation,* reading difficulty is commensurate with the general impairment in intellectual functioning. However, in some cases of mild mental retardation, the reading level is significantly below the expected level given the person's schooling and level of mental retardation. In such cases, the additional diagnosis of developmental reading disorder should be made, because treatment of the reading difficulties can be particularly helpful to the child's chances for employment in adulthood.

Inadequate schooling can result in poor performance on standardized reading test. In such cases, however, there is likely to be a history of many school changes or absences, or most other children in the school are likely to have similar difficulty.

Impaired vision or hearing may affect reading ability and can be ruled out through audiometric or visual screening test.

Diagnostic Criteria for 315.00 Developmental Reading Disorder

---

A. Reading achievement, as measured by a standardized, individually administered test, is markedly below the expected level, given the person's schooling and intellectual capacity (as determined by an individually administered IQ test).
B. The disturbance in A significantly interferes with academic achievement or activities of daily living requiring reading skills.
C. Not due to a defect in visual or hearing acuity or a neurologic disorder.

---

definition but included disorders of social skills to the listing of learning disabilities and added that learning disabilities may also co-occur with attention deficit disorder.

All of these organizational and legal definitions are vague regarding the specific behavioral criteria that should be used to classify children on an individual basis (Reynolds, 1984–1985). In dealing with learning disabilities in the domain of reading, state and local educational authorities have most often operationalized these verbal definitions in terms of quantitative discrepancies between individually administered reading tests and intelligence tests (Frankenberger & Harper, 1987; Kavale, 1987; McKinney, 1987).

It is clear from a perusal of Table 6.1 that various aspects of developmental reading disorder in DSM-III-R (American Psychiatric Association, 1987) are consistent with the history of classification in the field of reading disability. The heart of the definition (Diagnostic Criterion A) is a measured discrepancy between reading achievement and intelligence; although, like the legal and professional definitions given previously, no quantitative discrepancy criterion is listed. Some exclusionary criteria (sensory defects or neurologic disorder) are listed in Diagnostic Criterion C, and others (mental retardation, inadequate schooling) are listed under Differential Diagnosis. However, the guidelines for differential diagnosis clearly indicate that developmental reading disorder can co-occur with mental retardation in some children, and thus DSM-III-R is consistent with the National Joint Committee for Learning Disabilities and the Interagency Committee on Learning Disabilities in this respect.

In summary, DSM-III-R is largely consistent with definitional practice as developed by various governmental commissions and professional organizations. As such, the DSM-III-R diagnostic criteria will share the strengths of these other definitional efforts but will additionally be plagued by the same theoretical, educational, and diagnostic disputes that have been spawned by the other definitions.

DEFINITIONAL ISSUES

Although numerous side issues and points of contention surround the classification of children as reading disabled, the core disputes concern the operationalization and theoretical justification of the discrepancy criterion: the fact that a dyslexic child has an "unexpected" disability in the domain of reading, one not predicted by their general intellectual competence and socioeducational opportunities. Practically, this has meant a statistical assessment of the difference between their objectively measured reading ability and general intelligence (Frankenberger & Harper, 1987; Kavale, 1987; Kavale & Nye, 1981; Reynolds, 1985; Shepard, 1980). Typically, very little effort is expended in ascertaining whether adequate instruction has been provided or whether the child suffers from sociocultural disadvantage—in short, in ascertaining whether the disability is "intrinsic to the individual." So much conceptual confusion has surrounded the more operational discrepancy criterion that researchers and theoreticians have been reluctant to take on the potential additional complications of the other criteria.

The disputes surrounding the discrepancy criterion can be classified into two types. First, there are statistical and measurement complications involved in the operationalization of severe discrepancy. Many early attempts at defining severe discrepancy for purposes of educational classification were psychometrically naive. Several of these early classification practices served to undermine the definitional clarification that various agencies and professional bodies were striving to achieve. For example, discrepancy formulas that ignore regression artifacts subvert the attempt not to exclude the possible cooccurrence of reading disability and low IQ. With the utilization of appropriate statistical knowledge, many of the difficulties encountered in applying a severe discrepancy criterion are, at least in theory, easily remediable.

Much more problematic have been challenges to the conceptual assumptions that underlie the motivation to create a category of reading disorder in the manner DSM-III-R and the other definitions outlined previously. The vast majority of poor readers in the schools are of course not characterized by severe discrepancies between their reading ability and assessed intelligence (Eisenberg, 1979). Their below-average reading performance is predictable from their general cognitive abilities. They are what Gough and Tunmer (1986) termed "garden-variety" poor readers who are, by mathematical necessity, much more numerous than discrepancy-defined poor readers.

One of the fundamental premises that has provided the impetus for the development of classifications, such as dyslexia or developmental reading disorder, has been the assumption that such children are qualitatively different from garden-variety poor readers. Classifications such as dyslexia or developmental reading disorder have been employed for purposes of educational treatment and for purposes of educational funding; thus, it would be desirable for qualitative differentiation to occur in a variety of domains. That is, strong justification for a discrepancy-based classification would derive from studies showing that these different types of poor readers employ somewhat different information-processing operations while reading, that they have differential educational prognoses, and that they show differential responses to treatment than do garden-variety poor readers. Lack of such an empirical differentiation would certainly question the wisdom of expending such great effort to differentiate poorly achieving children along the lines previously outlined.

Part of the reason for the malaise and soul searching that periodically overtakes researchers and practitioners in the area of reading disabilities (Coles, 1978, 1987; Lyon, 1987; Senf, 1986; Stanovich, 1989; Vaughn & Bos, 1987; Vellutino, 1979) is that the field plunged ahead into the domains of educational practice and diagnosis without setting itself on a firm empirical foundation by first unequivocally demonstrating the empirical differentiability that would establish construct validity for the reading disability concept. We review next the evidence that does exist.

Our review of the literature focuses on these two classes of controversy surrounding the construct of developmental reading disorder. First, what is the best way to operationalize severe discrepancy? Second, what is the empirical evidence that would justify the need to assess achievement/aptitude discrepancies for the purposes of educational classification?

SELECTIVE REVIEW OF THE LITERATURE

Defining Severe Discrepancy

Despite repeated admonitions that the diagnosis of reading disability should be multidimensional (Hooper & Willis, 1989; Johnson, 1988; McKinney, 1987; Senf, 1986; Tindal & Marston, 1986), in actual educational and clinical practice, it is the assessment of a discrepancy between aptitude and reading achievement that is the key defining feature (Frankenberger & Harper, 1987). Reynolds (1985) noted that, across the numerous legal, professional, and research definitions of learning disabilities, five major components recur with great frequency: (a) failure to achieve, (b) psychological process disorders, (c) exclusionary criteria, (d) etiology, and (e) severe discrepancy. However, Reynolds (1985) wrote that "the severe discrepancy criterion is the most widely applied across the states," probably because "severe discrepancy is easily measured relative to other components of the definition of LD" (pp. 38–39). As Shepard (1980) argued, "all LD definitions, either by connotation or denotation, rest on this discrepancy between achievement and ability" (p. 80). Referring to the expert testimony and public discussion during the development of Public Law 94–142, Reynolds (1984–1985) concluded that "the only consensus regarding the characteristics of this 'thing' called learning disability, was that it resulted in a major discrepancy between what you would expect academically of learning disabled children and the level at which they are actually achieving" (p. 452).

In short, the reading disabilities field seems wedded to the discrepancy notion, if, as has been the case, it insists on differentiating between types of poor readers. Attempts to discourage the use of discrepancy formulas (Board of Trustees of the Council for Learning Disabilities, 1987) will only result in the reliance on much less reliable clinical judgments (Reynolds, 1985) and will almost inevitably result in samples of "dyslexic" children who are not differentiable from other poor readers who are not so labeled. As is discussed in the next section, such a lack of differentiation will fuel criticism of the use of a reading disability category in research and in practice. Although advances in neuropsychology (Hynd & Hynd, 1984; Hynd & Semrud-Clikeman, 1989) and in theories of human abilities (Ceci, 1990; Gardner, 1983; Sternberg, 1985) may one day change this situation, at present we are wedded to a definition that involves the use of quantitative data obtained from standardized psychometric instruments.

DSM-III-R is consistent with this emphasis in that it presents a version of the severe discrepancy notion in Diagnostic Criterion A: "Reading achievement, as measured by a standardized, individually administered test, is markedly below the expected level, given the person's schooling and intellectual capacity (as determined by an individually administered IQ test)" (APA, 1987, p. 44). Like many such definitions (Reynolds, 1984–1985), DSM-III-R is vague concerning what is meant by "markedly below the expected level." Part of the vagueness in such definitions stems from an implicit recognition that the bivariate distribution of reading and intelligence is continuous in nature (Olson, Kliegl, Davidson, & Foltz, 1985; Scarborough, 1984; Share, McGee, McKenzie, Williams, & Silva, 1987; Silva, McGee, & Williams, 1985) and that any criterion for discrepancy will be, of necessity, a somewhat arbitrary choice (Ellis, 1985; Hynd & Cohen, 1983). Correspondingly, estimates of prevalence for this particular disorder are also, to some extent, arbitrary. As Ellis (1985) noted, trying to give estimates of the prevalence of dyslexia is analogous to giving prevalence estimates for a condition such as obesity.

In the history of research and diagnosis of reading disabilities, the assessment of "achievement below expectancy" has been operationalized in four different ways (Cone & Wilson, 1981; Reynolds, 1984–1985, 1985; Shepard, 1980; Wilson & Cone, 1984). The first, defining anyone more than two grade levels below current grade placement as reading disabled, involves no discrepancy from aptitude and so is inconsistent with almost all current conceptions of reading disability. A more popular method is to define reading disability as an achievement discrepancy from an expected grade equivalent, the latter based on some formula containing a measure of mental age or IQ. There are two fundamental problems with such formulas. First, they often ignore the fact that the standard deviation of grade equivalent distributions increases with age, leading to greater numbers of older children being classified as disabled, if a constant criterion is used (Cone & Wilson, 1981; Shepard, 1980). Second, they do not take into account regression effects, because they assume a perfect correlation between IQ and achievement (Cone & Wilson, 1981; Reynolds, 1985; Shepard, 1980; Wilson & Cone, 1984).

A third technique, the z-score discrepancy method (Cone & Wilson, 1981; Shepard, 1980), obviates the problems involved in assessing discrepancies in grade equivalent units (which have a number of undesirable properties; see Reynolds, 1981) but still fails to correct for the imperfect correlation between ability and achievement. In contrast, the fourth class of technique, the regression discrepancy method (Reynolds, 1984–1985, 1985; Rutter & Yule, 1975; Shepard, 1980), is psychometrically the most justifiable. Here, the discrepancy is calculated from an expected achievement level based on the regression of reading achievement on the aptitude measure, thus taking into account the imperfect correlation between achievement and aptitude. McKinney (1987; see also Yule, 1984) demonstrated how, in an actual data set, the regression discrepancy method identifies a more uniform distribution of children across the IQ range than does the z-score method, a desirable characteristic given the logic of what a learning disability is supposed to be, and also given that most recent definitions allow for the cooccurence of a reading disability with low intelligence. The z-score discrepancy method, in contrast to the regression method, overidentifies high-IQ children and underidentifies low-IQ children, a pattern of identification that has fueled social criticism of the learning disabilities concept (Coles, 1987; Senf, 1986). As Yule (1984) noted, "Where scarce remedial resources are provided according to the degree of a child's underachievement, the less able child will be doubly disadvantaged unless regression techniques are used" (p. 234).

Despite its many advantages, the regression method does have some drawbacks. It can require some sophistication to apply, and several precautions regarding test standardization are critical (Reynolds, 1984–1985, 1985; Shepard, 1980). However, the existence of some helpful computer programs (e.g., Reynolds & Stowe, 1985) serves to ease the conceptual load on practitioners. The current state of knowledge in brain science and in the psychology of individual differences in combination with the logical constraints imposed by our previous conceptualizations of what a reading disability was (see following and Stanovich, 1986a, 1988b) guarantees that we will have to live with discrepancy formulas for some time to come. It is thus worthwhile for clinicians to become conversant with this technique for measuring discrepancy and for researchers to continue to clarify the complications involved in its application (Cone & Wilson, 1981; Reynolds, 1984–1985; Shepard, 1980). The issue of how to measure the discrepancy is actually a relatively straightforward concern when compared with the conceptual complications involved in trying to establish construct validity for the concept of reading disability itself. We now turn to this complicated and contentious issue.

Construct Validity of the Developmental Reading Disorder Classification

From the beginning, what has both fueled theoretical interest in dyslexia and justified differential educational treatment has been the assumption that the reading difficulties of the dyslexic stem from problems that are different from those characterizing the garden-variety poor reader (to use Gough & Tunmer's [1986] term) or, alternatively, if they stem from the same factors, that the degree of severity is so extreme for the dyslexic that it constitutes, in effect, a qualitative difference.

The experimental contrasts that have operationalized the idea of qualitative difference and/or differential causation in the literature have been dominated by two different designs. One is the reading-level match design (see Bradley & Bryant, 1978, 1985; Bryant & Goswami, 1986), where an older group of dyslexic children is matched on reading level with a younger group of nondyslexic children. The cognitive characteristics and reading subskills of the two groups are then compared. The logic here is fairly straightforward. If the reading subskills and cognitive characteristics of the two groups do not match, then it would seem that they are arriving at their similar reading levels via different routes, and this would support the idea of a qualitatively different developmental model for dyslexic readers, although there are numerous conceptual and methodological complications surrounding this inference (see Bryant & Goswami, 1986; Goswami & Bryant, 1989; Jackson & Butterfield, 1989). In contrast, if the reading subskill profiles of the two groups are identical, this would seem to undermine the rationale for the differential educational treatment of dyslexic children and their theoretical differentiation. If dyslexic children are reading just like any other child who happens to be at their reading level, and are using the same cognitive skills to do so, why should we consider their reading behavior to be so special?

The second major design, one pertinent not only to theoretical issues but also to the educational politics of reading disability, is to compare dyslexic children with children of the same age who are reading at the same level but who are not labeled dyslexic. Adapting the terminology of Gough and Tunmer (1986), this design is termed the "garden-variety control" design. Again, the inferences drawn are relatively straightforward. If the reading subskills and cognitive characteristics of the two groups do not match, then it would seem that the two groups are arriving at their similar reading levels via different routes. In contrast, if the reading subskill profiles of the two groups are identical, this would certainly undermine the rationale for the differential educational treatment of dyslexic children, and this in turn would make dyslexic children considerably less interesting theoretically. As Fredman and Stevenson (1988) argued, "if there is no clear distinction between the groups in terms of how they read, then the practice of identifying a special group of poor readers for special attention may no longer be necessary" (p. 105).

Shockingly to an outsider, an applied field devoted to diagnosing and treating dyslexic children based on assumptions of differential causation, prognosis, and treatment flourished entirely in the *absence* of the critical empirical data relevant to its foundational assumptions. It was not until the mid-1970s that we had the data from the groundbreaking epidemiological comparison of dyslexic and garden-variety poor readers that Rutter and Yule (1975) conducted, and only in the last 10 years or so has their data been supplemented by other garden-vareity control investigations. Additionally, only recently have enough studies employing reading-level matches been accumulated so that patterns were discernible.

The absence of empirical support for its most fundamental assumptions is the reason that the concept of reading disability has endured such widespread skepticism and criticism throughout its history. Typical of this criticism is an article Ellis (1985) wrote in which he asked, in an emperor-has-no-clothes fashion, "is it worth studying dyslexia?" (p. 199), and he further pressed the following point:

> Does applying all the exclusionary tests . . . to a group of poor readers in order to obtain a sample of high-grade, refined dyslexics actually yield a sample whose reading problems are qualitatively different from those of non-dyslexic rejects? Surprisingly this question seems to have received hardly any attention at all. . . . No one, it seems, has ever shown that the initial laborious screening is necessary in the sense that it produces a population of individuals whose reading characteristics are different from the great mass of poor readers. (pp. 199–200)

Ellis' use of the word "surprisingly" alludes to the point that the reading disabilities field expanded and grew in virtual absence of the critical data needed to test its foundational assumptions. Similarly, Seidenberg, Bruck, Fornarolo, and Backman (1986) lamented the absence of the necessary empirical demonstrations of qualitative differences based on discrepancy criteria:

> If the dyslexic readers differ from poor readers along the same dimensions that differentiate poor readers from good, it cannot be concluded that the dyslexic readers' performance is due to decoding processes specific to this group. Hence the results fail to provide evidence for the kind of qualitative differences between groups entailed by the standard view. (pp. 79–80)

This situation has only recently begun to be remedied by researchers employing the two designs that I described previously. When we did begin to accumulate the necessary empirical data in the last few years, however, the data were not always convergent. Empirically there are reading-level match studies that reveal similar processing profiles (Baddeley, Logie, & Ellis, 1988; Beech & Harding, 1984; Treiman & Hirsh-Pasek, 1985) and those that have identified differences (Baddeley, Ellis, Miles, & Lewis, 1982; Bradley & Bryant, 1978; Kochnower, Richardson, & DiBenedetto, 1983; Olson et al., 1985; Olson, Wise, Conners, Rack, & Fulker, 1989; Snowling, 1980; Snowling, Stackhouse, & Rack, 1986). Similarly, garden-variety comparisons have supported qualitative similarities (Fredman & Stevenson, 1988; Siegel, 1988; Taylor, Satz, & Friel, 1979) and differences (Jorm, Share, Maclean, & Matthews, 1986; Rutter & Yule, 1975; Silva et al., 1985).

The Phonological-Core Variable Difference Model

These mixed results have troubled many in the field, because they relate to some of the foundational assumptions of the concept of dyslexia as it is used in both research investigations and in educational practice. Nevertheless, if we step back from the minutia of the experimental details and look for patterns, there are trends discernible in the admittedly somewhat confusing data base. I described these trends within the context of what I termed the phonological-core variable-difference model (Stanovich, 1988a)—actually perhaps more of a framework than a model. The model rests on a clear understanding of the assumption of specificity in definitions of dyslexia (see Hall & Humphreys, 1982; Stanovich, 1986a, 1986b). This assumption underlies all discussions of the concept of dyslexia, even if it is not explicitly stated. It is the idea that a child with this type of learning disability has a brain/cognitive deficit that is reasonably specific to the reading task. That is, the concept of dyslexia requires that the deficits displayed by such children not extend too far into other domains of cognitive functioning. If they did, this would depress the constellation of abilities we call intelligence and thus reduce the reading/intelligence discrepancy that is central to all the definitions we reviewed previously.

In short, the key deficit in dyslexia must be a vertical faculty rather than a horizontal faculty (see Fodor, 1983); that is, a domain-specific process (Cossu & Marshall, 1986) rather than a process that operates across a wide variety of domains. For this and other reasons, many investigators located the proximal locus of dyslexia at the word recognition level (e.g., Gough & Tunmer, 1986; Morrison, 1984, 1987; Perfetti, 1985; Siegel, 1985, 1988; Siegel & Faux, 1989; Stanovich, 1986b, 1988b; Vellutino, 1979) and continue searching for the locus of the flaw in the word recognition module. Research in the last 10 years focuses intensively on phonological processing abilities. It is now well established that dyslexic children display deficits in various aspects of phonological processing. They have difficulty making explicit reports about sound segments at the phoneme level; they display naming difficulties; their utilization of phonological codes in short-term memory is inefficient, and their categorical perception of certain phonemes may be other than normal (Cossu, Shankweiler, Liberman, Katz, & Tola, 1988; Kamhi & Catts, 1989; Liberman & Shankweiler, 1985; Lieberman, Meskill, Chatillon, & Schupack, 1985; Mann, 1986; Pennington, 1986; Pratt & Brady, 1988; Snowling, Goulandris, Bowlby, & Howell, 1986; Taylor, Lean, & Schwartz, 1989; Wagner & Torgesen, 1987; Werker & Tees, 1987; Williams, 1984, 1986; Wolf & Goodglass, 1986). Importantly, there is increasing evidence that the linkage from phonological processing ability to reading skill is a causal one (Bradley & Bryant, 1985; Bryant, Bradley, Maclean, & Crossland, 1989; Liberman & Shankweiler, 1985; Lundberg, Frost, & Peterson, 1988; Maclean, Bryant, & Bradley, 1987; Stanovich, 1986b, 1988b; Wagner, 1988; Wagner & Torgesen, 1987). Presumably, lack of phonological sensitivity makes the learning of grapheme-to-phoneme correspondences very difficult.

In short, there is now voluminous evidence indicating that phonological deficits are the basis of the dyslexic performance pattern. This is an oversimplification, because it ignores, at least temporarily, the possibility of core deficits in the realm of orthographic processing. I believe that there is growing evidence for the utility of distinguishing a group of dyslexics who have severe problems in accessing the lexicon on a visual/orthographic basis (see Stanovich, in press; Stanovich & West, 1989). Suggestive evidence comes from the work on acquired reading disability that reveals the existence of surface dyslexia (Patterson, Marshall, & Coltheart, 1985) and from multivariate investigations indicating that efficient phonological processing is a necessary but not sufficient condition for attaining advanced levels of word recognition skill (Juel, Griffith, & Gough, 1986; Tunmer & Nesdale, 1985).

However, as regards orthographic processing deficits, two crucial caveats are in order. First, there is a very large body of evidence indicating that this group of children must be numerically quite smaller than the group with phonological difficulties (Aaron, 1989; Freebody & Byrne, 1988; Gough & Hillinger, 1980; Liberman, 1982; Liberman & Shankweiler, 1985; Pennington, 1986; Perfetti, 1985; Rayner & Pollatsek, 1989; Vellutino, 1979). Logically, they must be very small in number, because they have not obscured the identification of phonological problems in samples that were not preselected for subtypes. Secondly, I believe that the problem encountered by these children is not similar to the "visual perception" problems popular in the early history of the study of dyslexia but now widely recognized to have been overstated (Aman & Singh, 1983; Morrison, Giordani, & Nagy, 1977; Stanovich, 1986a; Vellutino, 1979). The actual problems in orthographic processing must be much more subtle and localized than these older views suggested. Nevertheless, any smaller group of dyslexics with orthographic-core deficits would mirror the phonological-core group in all of the other processing characteristics of the model. What are those characteristics?

The critical processing differences that distinguish dyslexic readers from garden-variety poor readers can be considered within the context of the two critical designs that were introduced earlier. First, the baseline performance of garden-variety poor readers has become considerably clarified in recent years. It appears that, on a wide variety of reading-related cognitive tasks, the performance of garden-variety poor readers mirrors that of younger, skilled children who are reading at the same level (Stanovich, Nathan, & Vala-Rossi, 1986; Stanovich, Nathan, & Zolman, 1988). Compared to chronological-age peers, they do display deficits in phonological processing, but these are in addition to other processing and cognitive deficits (see Aaron, 1989).

In contrast, discrepancy-defined dyslexic readers are actually inferior in phonological processing compared to reading-level controls, but they do not have as many other cognitive deficits when compared with chronological-age controls (Baddeley et al., 1982; Bradley & Bryant, 1978; Holligan & Johnston, 1988; Kochnower et al., 1983; Olson et al., 1985, 1989; Siegel & Faux, 1989; Siegel & Ryan, 1988; Snowling, 1980, 1981; Snowling et al., 1986). Although there are some exceptions to this pattern (Beech & Harding, 1984; Treiman & Hirsh-Pasek, 1985),

Olson, Wise, Conners, and Rack (in press) recently reported a meta-analysis of this literature that explains most of the extant discrepancies.

Compared to the data attained from the reading-level design, it has been more difficult to differentiate dyslexic subjects empirically in garden-variety designs. Although some garden-variety comparisons support the idea of qualitative difference (Horn & O'Donnell, 1984; Jorm et al., 1986; Rutter & Yule, 1975; Silva et al., 1985), other investigations demonstrate that it can often be surprisingly difficult to differentiate discrepancy-defined dyslexic readers from garden-variety poor readers (Fredman & Stevenson, 1988; Siegel, 1988; Taylor et al., 1979). Nevertheless, some research provides indications of the compensatory processing pattern when dyslexics and garden-variety poor readers are matched on reading comprehension: poorer word recognition but superior "horizontal faculties" on the part of the dyslexics (Aaron, 1989; Bloom, Wagner, Reskin, & Bergmann, 1980; Fredman & Stevenson, 1988; Seidenberg et al., 1985). Similarly, dyslexics matched with garden-variety CA controls on word recognition skill have displayed superior reading comprehension and horizontal faculties (see Bloom et al., 1980; Ellis & Large, 1987; Jorm et al., 1986; Silva et al., 1985).

In the phonological-core variable-difference model, the term *variable differences* refers to the key performance contrasts between the garden-variety and the dyslexic, poor reader. As outlined before, the cognitive status of the garden-variety poor reader is well described by a developmental lag model (Stanovich et al., 1988). Cognitively, they are remarkably similar to younger children reading at the same level. A logical corollary of this pattern is that the garden-variety reader will have a wide variety of cognitive deficits when compared to CA controls who are reading at normal levels. However, it is important to understand that the garden-variety poor reader does share the phonological problems of the dyslexic reader, and these deficits appear also to be a causal factor in their poor reading (Perfetti, 1985; Stanovich, 1986b). However, for the garden-variety reader, the deficits, relative to CA controls, extend into a variety of domains (see Ellis & Large, 1987), and some of these (e.g., vocabulary, language comprehension) may also be causally linked to reading comprehension. Such a pattern does not characterize the dyslexic who has a deficit localized in the phonological core.

The phonological-core variable-difference model assumes multidimensional continuity for reading ability in general and for all its related cognitive subskills. That is, it conceives of all of the relevant distributions of reading-related cognitive skills as being continuously arrayed in a multidimensional space and not distributed in clusters. There is considerable evidence from a variety of different sources supporting such a continuity assumption (Ellis, 1985; Jorm, 1983; Olson et al., 1985; Scarborough, 1984; Seidenberg et al., 1985; Share et al., 1987; Silva et al., 1985; Vogler, Baker, Decker, DeFries, & Huizenga, 1989). However, the fact that the distribution is a graded continuum does not render the concept of dyslexia scientifically useless, as many critics would like to argue. As mentioned previously, Ellis (1985) drew the analogy with obesity. No one doubts that it is a very real health problem, despite the fact that it is operationally defined in a somewhat arbitrary way by choosing a criterion in a continuous distribution.

The framework of the phonological-core variable-difference model meshes nicely with the multidimensional continuum notion. Consider the following characterization. As we move in multidimensional space from the dyslexic to the garden-variety poor reader, we will move from a processing deficit localized in the phonological core to the global deficits of the developmentally lagging, garden-variety poor reader. Thus, the actual cognitive differences that are displayed will be variable depending on the type of poor reader who is the focus of the investigation. The differences on one end of the continuum will consist of deficits located only in the phonological core (i.e., the dyslexic) and will increase in number as we run through the intermediate cases that are less and less likely to pass strict psychometric criteria for dyslexia. Eventually we will reach the part of the multidimensional space containing relatively "pure" garden-variety poor readers who clearly will not qualify for the label dyslexic by either regression or exclusionary criteria, will have a host of cognitive deficits, and will have the cognitively immature profile of a developmentally lagging individual.

This framework provides an explanation for why almost all processing investigations of reading disability have uncovered phonological deficits but also why some investigations have found deficits in *other* areas as well (see Stanovich, 1988b). This outcome is predictable from the fact that the phonological-core variable-difference model posits that virtually all poor readers have a phonological deficit but that other processing deficits emerge as one drifts in the multidimensional space from "pure" dyslexics toward garden-variety poor readers. Thus, the model's straightforward prediction is that the studies that revealed a more isolated deficit will be those that had more psychometrically select dyslexic readers. In short, the reading/IQ discrepancy of the subject populations should be significantly greater in those studies displaying more specific deficits. Presumably, studies finding deficits extending beyond the phonological domain are in the "fuzzy" area of the multidimensional space and are identifying the increasing number of processing differences that extend beyond the phonological domain as one moves toward the garden-variety area of the space.

Definitional Issues Within the Phonological-Core Variable-Difference Framework

The research literature on the comparative cognitive characteristics of dyslexic children does provide some support for the construct validity of the concept of developmental reading disorder. However, several critical caveats are in order. First, we must always be cognizant that continuity characterizes the distribution of reading ability and that of various reading-related subskills. We should not let connotations of discreteness creep into our thinking about developmental reading disorder merely because we have imposed a discrete classification on an underlying continuum of ability.

A second related point is that the continuity inherent in the distribution of reading skill will make it difficult to differentiate dyslexic children if liberal classification criteria are used. Indeed, the failure of dyslexic subjects to separate from garden-variety controls in many investigations (e.g., Siegel, 1988; Taylor et al., 1979) still provides justification for the argument that IQ discrepancies play no useful role in discussions of reading difficulty (Seidenberg et al., 1986; Siegel, 1988, 1989). Although it has proven easier to demonstrate the dyslexic children are differentiable from younger controls in reading-level designs (see Olson et al., 1990), it still appears that such a differentiation is only achieved when the dyslexic subjects are psychometrically very select; that is, when they display extremely severe discrepancies (see Stanovich et al., 1988). As Stanovich previously argued, "Only by exercising some restraint and specificity in our application of the label will we arrive at a concept that has some scientific and educational utility (1986a, p. 109), and "It is only by isolating the true outliers that researchers can hope to obtain the evidence for specificity that the dyslexia concept requires if it is to be of scientific and practical utility. . . . Groups who have pushed for ever-more-inclusive definitions of dyslexia . . . are indirectly undermining the concept. The wider the net that is cast, the greater will be the difficulties in distinguishing dyslexia from other educational designations (e.g., borderline retardation, EMR). Lack of restraint in applying the label is in part responsible for the failure of researchers to demonstrate consistently that the performance profiles of disabled subjects differ reliably from those of other poor readers" (1986b, p. 387).

In this context, the DSM-III-R prevalence statement (APA, 1987) that "estimates of the prevalence of the disorder in school-age children have ranged from 2% to 8%" (p. 44) is probably appropriate, and it is worth emphasizing the following: (a) Such estimates are to some extent arbitrary (Reynolds, 1985), because they simply result from the adoption of a criterion in a continuous distribution (as does the situation of classifying someone as "obese"; see Ellis, 1985); and (b) the adoption of estimates closer to the 2% end of the prevalence range will be more likely to result in a scientifically justifiable subgroup. It is noteworthy that the seminal Rutter and Yule (1975) epidemiological investigation, one of the more successful demonstrations of differences in a garden-variety control design, adopted a criterion that resulted in approximately 3.7% of their sample being classified as specifically reading retarded.

Remaining Issues of Construct Validity

Although the review of research on the cognitive differentiability of dyslexic children just concluded does seem to support the construct validity of the developmental reading disorder classification, there is still inadequate data on other foundational assumptions. For example, outside of the pioneering work of Lyon (1985), there are insufficient data on differential response to treatment. There are, for instance, no good data indicating that discrepancy-defined dyslexics respond differently to various educational treatments than do garden-variety readers of the same age or than younger nondyslexic children reading at the same level (Pressley & Levin, 1987; van der Wissel, 1987). Although more data exists on the issue of differential developmental growth curves for reading, the data are contradictory. Rutter and Yule (1975) found a differential prognosis for specifically disabled and garden-variety poor readers. The garden-variety poor readers displayed greater growth in reading but less growth in arithmetic ability than the specifically disabled children. However, this finding of differential reading growth rates fails to replicate in some other studies (Labuda & DeFries, 1989; McKinney, 1987; Share et al., 1987; van der Wissel & Zegers, 1985).

Until convincing data on these two issues are provided, the utility of the concept of developmental reading disorder will continue to be challenged, because the reading disabilities field will have no rebuttal to assertions that it is more educationally and clinically relevant to define reading disability without reference to IQ discrepancy (Seidenberg et al., 1986; Siegel, 1988, 1989). For example, Share, McGee, and Silva (1989) still saw fit to argue that "it may be timely to formulate a concept of reading disability which is independent of any consideration of IQ. Unless it can be shown to have some predictive value for the nature of treatment or treatment outcome, considerations of IQ should be discarded in discussions of reading difficulties" (p. 100). No amount of clinical evidence, case studies, or anecdotal reports will substitute for the large-scale experimental demonstrations that, compared to groups of garden-variety poor readers, discrepancy-defined poor readers show differential response to treatment and differential prognosis. Similarly, no amount of clinical evidence, case studies, or anecdotal reports will substitute for further evidence that the reading-related cognitive profiles of these two groups are reliably different.

Miscellaneous Issues Concerning the DSM-III-R Definition

Under "Associated Features," DSM-III-R states that "deficits in expressive language and speech discrimination are usually present, and may be severe enough to warrant the additional diagnosis of Developmental Expressive or Receptive Language Disorder" (APA, 1987, p. 43). This is quite consistent with much research that localizes the most critical subskills of the reading process in the speech/language domain (Liberman & Shankweiler, 1985; Mann, 1986; Pennington, 1986) and is also congruent with many researchers who argued that reading disability is best viewed as a type of developmental language disorder (Gathercole & Baddeley, 1987; Kahmi & Catts, 1989). Kamhi and Catts (1989) reviewed a considerable amount of research consistent with this position and argued the strong view that "dyslexia is more than a reading failure; it is a language problem that begins early in life and continues throughout childhood, adolescence, and into adulthood" (p. 41). Furthermore, they wrote, "it is not possible to have a reading disability without a deficit in some aspect of phonological, syntactic, semantic, or discourse processing" (p. 59). Gathercole and Baddeley (1987) endorsed a position that views reading disability as a point on a continuum of language disorder:

> Although language problems are typically detected prior to the children receiving reading instruction . . . it is possible that the alphabetic literacy skills required in reading may be more sensitive to the adequacy of speech analytic skills than other aspects of normal linguistic development, such that a mild deficit may only be detectable in reading performance. More severe subjects may result in the more generalized symptom complex associated with developmental language disorder. . . . This is also clearly consistent with the notion that the two populations may quantitatively differ rather than qualitatively. (p. 464)

Although the close relationship between developmental reading and developmental language disorders is acknowledged, DSM-III-R warns that "visual perceptual deficits are seen in only about 10% of cases"

(APA, 1987, p. 43). Whereas virtually all researchers are in agreement that visual deficits are the primary cause of dyslexia in only a very few instances and that problems with phonological processing are the primary cause of the disability (Aaron, 1989; Liberman & Shankweiler, 1985; Olson et al., 1985; Olson et al., 1989; Read & Ruyter, 1985; Vellutino, 1979; Vellutino & Scanlon, 1987), whether or not pure visual deficits, in the absence of phonological problems, can independently cause the disability is still a matter of dispute (Hulme, 1988). Furthermore, even among researchers who support the possibility of visual deficits in some cases of dyslexia, there would be little agreement on the prevalence of such cases. The DSM-III-R figure of 10% is little more than a guess and probably serves no useful purpose. It might be preferable to replace the 10% with the phrase "very few." Additionally, it would be useful to include a warning that verbal deficits can easily be mistaken for visual processing problems, if the proper tasks and controls for verbal processing are not utilized (Calfee, 1977; Vellutino, 1979; Stanovich, 1978; Stanovich & Purcell, 1981). For example, any task employing stimuli that can be verbally labeled simply cannot be interpreted as a primary measure of visual processing.

In discussing age of onset, DSM-III-R states that "the disorder is usually apparent by age 7 (second grade). In severe cases, evidence of reading difficulty may be apparent as early as age 6 (first grade)." This statement rightly assumes that a statistical method for defining discrepancy more complicated than the outmoded "2 years below grade level (or expected grade level)" is being employed. However, some cautions regarding the developmental changes in the reliabilities of various aptitude and achievement tests might be included. These reliability differences render a discrepancy classification less reliable at younger ages. Such early classification should not, of course, be discouraged, but a tentative stance toward discrepancy classifications made at younger ages should be advocated. Encouragingly, in a longitudinal investigation, Share and Silva (1986) found considerable stability in regression-based discrepancy classifications in a large groups of children at age 7 and age 11.

Regarding age of onset, the tendency to view dyslexia as a type of developmental language disorder (Kamhi & Catts, 1989) with a primary causal locus in deficient phonological awareness and phonological sensitivity (Bradley & Bryant, 1985; Liberman & Shankweiler, 1985; Wagner & Torgesen, 1987) has very positive implications for future developments in early diagnosis. Some of the most diagnostic of the phonological processing tasks that are strongly predictive of reading failure can be administered to children at very young ages, before they attend school (Berninger, Thalberg, DeBruyn, & Smith, 1987;

Blachman, 1989; Bradley & Bryant, 1985; Fox & Routh, 1975; Maclean et al., 1987; Stanovich, Cunningham, & Cramer, 1984; Torgesen & Bryant, 1989; Williams, 1984). As further work refines these tasks, and they are developed as psychometric instruments, clinicians will have greater ability to diagnose developmental reading disorder at younger ages than is now possible. Early prevention efforts based on remediating phonological processes have generally met with success (Blachman, 1989; Bradley, 1987; Bradley & Bryant, 1985; Lundberg et al., 1988).

Finally, other aspects of the DSM-III-R definition reflect fairly uncontroversial and established conclusions; for example, "many of the signs of the disorder remain for life" (Feagans, 1983; Finucci, 1986; Horn, O'Donnell, & Vitulano, 1983; Johnson, 1988; Scarborough, 1984; Spreen, 1982) and that there is a familial pattern to dyslexia (Childs & Finucci, 1983; Decker & Bender, 1988; Pennington, 1986; Scarborough, 1989).

DIAGNOSTIC AND ASSESSMENT PROCEDURES

As previously mentioned, in being fairly prototypic of current legal, professional, and research definitions of dyslexia, DSM-III-R shares the data base that has been built on these definitions. It also, quite naturally, shares some of the weaknesses of the traditional definitions. For example, the emphasis on exclusionary criteria has long been the subject of criticism (e.g., Rutter, 1978). However, there are fundamental conceptional problems with the current classification procedures that go beyond the exclusionary problem. Increasingly, researchers are raising questions not merely about the statistical procedures necessary to define severe discrepancy reliably (e.g., Reynolds, 1985) but with the foundational benchmark from which discrepancy is measured: intelligence. Indeed, it is surprising that, for so long, the concept of intelligence received so little discussion in the learning disabilities literature. Researchers and practitioners in this field seem not to have realized that it is a foundational concept for the study of reading disability. As currently defined, IQ is a superordinate construct for the classification of a child as reading disabled. Without a clear conception of the construct of intelligence, the notion of a reading disability, as currently defined, dissolves into incoherence.

The problem is that one would be hard-pressed to find a concept more controversial than intelligence in all of psychology. It has been the subject of dispute for decades, and this shows no sign of abating. Current work on individual differences in intelligent functioning continues to produce exciting findings and interesting theories (Baron,

1985; Ceci, 1990; Ceci & Liker, 1986; Sternberg, 1985, 1988) but no consensual view of the intelligence concept (Sternberg & Detterman, 1986). Although much progress has been made in both empirical and theoretical domains, quite fundamental disputes remain. For example, some investigators recently emphasized more contextualized approaches to the study of intelligence (Ceci, 1990; Ceci & Liker, 1986; Sternberg, 1985, 1988; Sternberg & Wagner, 1986), whereas others have been advocating more decontextualized biological approaches (Vernon, 1987).

Yet despite the controversy surrounding intelligence in the cognitive, developmental, and psychometric literature, it was adopted as a foundational construct for the definition of dyslexia. The choice of IQ test performance as the baseline from which to measure achievement discrepancies was accepted by teachers, clinicians, professional organizations, and government agencies in the absence of much critical discussion or research evidence. Until quite recently—although see Reed (1970) for an early discussion of the complications involved in the discrepancy definition—the field seems never to have grappled very seriously with the question of why the benchmark should have been IQ.

One reason why professional assent to the use of IQ test scores in the discrepancy definition was given so readily derived from the belief that IQ scores were valid measures of intellectual potential. One major problem, however, was that most psychometricians, developmental psychologists, and educational psychologists long ago relinquished the belief that IQ test scores measured potential in any valid sense. Indeed, standard texts in educational measurement and assessment routinely warn against interpreting IQ scores as measures of intellectual potential (Anastasi, 1988; Cronbach, 1984; Thorndike, 1963). As Eisenberg (1978) argued, "there is no better reason for assuming that the measured IQ represents intellectual potential than that reading level reflects reading potential" (p. 37).

At their best, IQ test scores are gross measures of current cognitive functioning (Detterman, 1982). Indeed, many theorists would dispute even this characterization. Siegel (1988, 1989), for example, attacked the representativeness of several of the typical tasks that are on IQ tests and gave the outlines of the objections of many theorists. Lyon (1987) noted:

> Tacitly ingrained within this assumption is the notion that learning-disabled children can attain levels of achievement commensurate with their IQ if they were assessed and taught properly. However, both the assumption of discrepancy and the implicit message that it connotes stand on shaky conceptual and logical grounds . . . we do not know whether psychometric discrepancies tell us anything of value about how children respond to instruction. (pp. 78–79)

However, even if we were to concede that the idea of measuring aptitude/achievement discrepancies is a valid educational procedure, there are other, more educationally relevant choices for a benchmark than IQ. As the work of several investigators suggests (Aaron, 1989; Carver, 1981; Gillet & Temple, 1986; Hood & Dubert, 1983; Royer, Kulhavy, Lee, & Peterson, 1986; Spring & French, 1990; Sticht & James, 1984), measuring the discrepancy between reading comprehension and listening comprehension would seem to have been a more logical choice in the first place. Certainly, a discrepancy calculated in this way seems to have more face validity and educational relevance than the traditional procedure (Aaron, 1989; Durrell & Hayes, 1969; Spache, 1981). Children who understand written material less well than they would understand the same material if it were read to them appear to be in need of educational intervention. Presumably, their listening comprehension exceeds their reading comprehension, because word recognition processes are inefficient and are a "bottleneck" that impedes comprehension (Perfetti, 1985; Perfetti & Lesgold, 1977).

There are, of course, several obstacles to implementing procedures of measuring reading disability by reference to discrepancies from listening comprehension. For example, whereas several individual measures of listening comprehension ability have been published (CTB/McGraw-Hill, 1981; Durrell & Hayes, 1969; Spache, 1981), it may be the case that none have been standardized across the range of ages, nor attained the psychometric properties, to serve as an adequate measure from which to assess discrepancy (Johnson, 1988; but see Aaron, 1989). Nevertheless, it may be worth the research effort to develop such instruments further, because assessing discrepancies from listening comprehension could help to mitigate against many of the paradoxes that plague discrepancy measures based on the more nebulous concept of intelligence. For example, it has often been pointed out that changes in the characteristics of the IQ test being used will result in somewhat different subgroups of children being identified as discrepant and also alter the types of processing deficits that they will display in comparison studies (e.g., Bowers, Steffy, & Tate, 1988; Lindgren, DeRenzi, & Richman, 1985; Reed, 1970; Shankweiler, Crain, Brady, & Macaruso, in press; Siegel & Heaven, 1986; Stanley, Smith, & Powys, 1982; Torgesen, 1985; Vellutino, 1978).

Further, despite many recommendations that performance IQ measures provide "fairer" measures of the reading potential of disabled children (e.g., Siegel & Heaven, 1986; Stanovich, 1986a), it is not at all clear that the spatial abilities, fluid intelligence, and problem-solving abilities tapped by most performance tests provide the best measures of the potential to comprehend verbal material. To the contrary, it would appear that verbally loaded measures and/or listening comprehension would provide the best estimates of how much a dyslexic child could comprehend from written text, if their deficit decoding skills were to be remediated. As Hessler (1987) noted:

> There are different types of intelligence, and they predict academic achievement differently. . . . In fact, the performance score accounts for so little academic achievement that there is reason to question its relevance for use as an ability measure to predict academic achievement. It is therefore a mistake to use any test of intelligence as an ability measure for predicting academic achievement in a severe discrepancy analysis simply because it is called a test of intelligence, cognitive ability, scholastic aptitude, or whatever, without demonstrating some ability to predict academic achievement. (p. 45)

Consistent with this interpretation, van der Wissel (1987) demonstrated via data simulation that the extent to which an IQ subtest separates dyslexic from garden-variety children is *inversely* related to how highly the subtest correlates with reading achievement. It is a paradoxical situation indeed when the indicators that best determine this subgroup discrimination are those that do not relate to the criterion performance that drew professional attention in the first place: reading failure.

Of course, the common recommendation against the use of verbally loaded tests stems from the either tacit or explicit assumption that the reading difficulties themselves may lead to depressed performance on such measures. Some recognition of these "Matthew effects" involved in reading (i.e., rich-get-richer and poor-get-poorer effects, see Stanovich, 1986b) is thus implicit in this recommendation. However, such a suggestion tacitly undermines the whole notion of discrepancy measurement by weakening the distinction between aptitude and achievement; moreover, it serves to remind us that, whereas the logic of the learning disabilities field has implicitly given all the causal power to IQ (i.e., it is reading that is considered discrepant from IQ rather than IQ that is discrepant from reading), this is a vast oversimplification, because there are potent effects running in both directions. As Senf (1986) noted, "logically, we are imputing more validity to the predictor (IQ) than we are to the resulting behavior (achievement)" (p. 39). Doehring, Trites, Patel, and Fiedorowicz (1981) argued similarly:

> The usual definitions of developmental dyslexia do not recognize the interactive development of reading, language, and cognition. . . . Cognitive and language abilities essential to reading could be excluded from the measures used to define normal intelligence, but we will not be entirely sure what these abilities are until there is a comprehensive theo-

> ry of human ability that provides a firmer basis for defining intelligence. Even then we will have to recognize the probability that the difficulty in learning to read has, itself, impaired the development of important aspects of intelligence such as vocabulary and comprehension. (p. 6)

Much evidence has now accumulated to indicate that reading itself is a moderately powerful determinant of vocabulary growth, verbal intelligence, and general comprehension ability (Juel, 1988; Share & Silva, 1987; Share et al., 1989; Stanovich, 1986b; Stanovich & West, 1989). These Matthews effects highlight a further problematic aspect of discrepancy-based classification. Do we really want to withhold certain types of educational treatments from children whose poor reading is accompanied by equally subpar IQs (or listening comprehension) when we know that the poor reading may be, at least in part, a direct cause of the low IQs and listening comprehension ability? The possibility of a Matthew effect precludes us from assuming that the poor listening comprehension or verbal intelligence could not be enhanced by better reading.

Thus, the fact that some verbal subtests tap expressive language skills and vocabulary are not criticisms of the use of such tests for the purposes of discrepancy measurement, if verbal facility is part of our conception of intelligence or of our conception of the potential to understand written language. Similarly, if one's conception of intelligence emphasizes acquired skills and knowledge rather than a relatively nonmalleable type of "mental power" (Spearman & Wynn-Jones, 1950), then the fact that some subtests tap acquired knowledge is again not a criticism. If we wish to employ the construct of intelligence, then we must live with the implications of the conceptualization that we choose. If our conception of intelligence does have a verbal component—and we construct an IQ test accordingly—and if reading disabled children have poor expressive language skills and vocabulary, then they are simply less intelligent! Hessler (1987) argued:

> Using a nonverbal test of intelligence because an individual has better nonverbal cognitive abilities than verbal cognitive abilities does not, of course, remove the importance of verbal processing and knowledge structures in academic achievement; it only obscures their importance and perhaps provides unrealistic expectations for an individual's academic achievement. (p. 46)

Lyon (1987) pressed a similar point:

> Many samples of learning-disabled children are selected for study on a basis of a discrepancy between the Performance IQ derived from the

> WISC–R and measures of academic achievement. Unfortunately, the Performance IQ, and even the Full Scale IQ, bear little psychometric relationship to measures of academic achievement, particularly in the oral language, reading, and written language domains, thus negating the predictive power attributed to discrepancies between measures of nonverbal intelligence and school learning. (pp. 78–79)

Similarly, if vocabulary size is a component of a knowledge structure that supports intelligent behavior and if vocabulary acquisition is partially determined by text exposure differences that are themselves a product of reading ability, as the concept of "Matthew effects" (Stanovich, 1986b) proposed, then these effects should be viewed as actually depressing intelligence. Again we must live with the logical implications of the concept of intelligence we posit. The belief that acquired vocabulary is somehow an inappropriate measure of intelligence, simply because it is partially a product of differential environmental opportunities, seems to be a remnant of the view that intelligence is a fixed quantity uninfluenced by the environment. If our view of intelligence is instead that environmental opportunities provided by reading directly affect cognitive functioning, then, in constructing a test consistent with such a conception, we would actually *want* indicators—such as vocabulary items—that are influenced by environmental opportunity. If we do not like some of the logical implications of such arguments, then it becomes even more useful to move toward measuring discrepancies from listening comprehension and to move away from implicating conceptions of intelligence in our definitions of dyslexia.

SUMMARY AND DIRECTIONS

The definition of developmental reading disorder in DSM-III-R is fairly typical in that it mirrors in many respects longstanding research, professional, and legal definitions. As a prototypic definition, the developmental reading disorder classification of DSM-III-R shares many of the strengths and weaknesses of other definitions that have been based on aptitude/achievement discrepancies and exclusionary qualifications. The definition is bolstered by recent, long-overdue research indicating that it is possible to differentiate dyslexia from garden-variety poor reading. However, this trend is stronger in studies employing reading-level matches with younger children than it is in studies employing garden-variety controls of the same CA as the dyslexics. Thus, further work on the cognitive differentiability of discrepancy-defined poor

Table 6.2
Suggestions for Changes in the DSM-III-R Definition and Description of Developmental Reading Disorder

1. Consider using discrepancies from listening comprehension as the key defining feature of the disorder, thus avoiding some of the psychometric, conceptual, and logical paradoxes connected with the use of intelligence measures.

2. Connect developmental reading disorder even more closely with developmental language disorder, emphasizing the possibility that reading disabilities represent the mild end of the continuum of language disorder and that the processing deficit in the case of developmental language disorder is likely to be more specifically localized in the domain of phonological processing.

3. Identification of an early delay in phonological sensitivity as the genesis of reading disorder in later life suggests an emphasis on early diagnosis and remediation. The DSM-III-R description should emphasize the positive outcomes of early remediation efforts focused on phonological processing and that early diagnostic instruments for phonological sensitivity exist and continue to be developed.

4. The integrity of the concept developmental reading disorder will best be maintained if the DSM-III-R stresses that only extreme discrepancies will result in the identification of reading disabled children who differ in any appreciable extent from children who are poor readers but who are not given the developmental reading disorder classification.

5. The definition might consider a candid admission that we lack adequate data that children with developmental reading disorder differ from ordinary poor readers in prognosis or response to treatment.

6. Rather than simply guessing that visual processing deficits characterize 10% of children with developmental reading disorder, the definition should simply state that the number of such children with a visual processing deficit in the absence of phonological problems is thought to be extremely small.

readers is still needed. It is additionally unfortunate that we still have no good evidence that dyslexics differ either in their prognosis or their response to treatment from poor readers without severe discrepancies. A further problem is that there are a variety of conceptual and psychometric paradoxes surrounding the construct of intelligence that by logical necessity transfer to reading disability definitions that are based on such a construct.

Table 6.2 lists some of the suggestions for revision in the DSM-III-R definition and description of developmental reading disorder that follow from this review of the literature. The most important and substantive suggestion is that researchers explore the possibility of using listening comprehension as a baseline from which to measure reading discrepancies. Listening comprehension has the advantages of conceptual coherence, ecologically validity to educators, and a less contentious history in educational psychology. Some recent research on the processing of oral versus written language will aid in our explorations of the potential of such an alternative baseline (Aaron, 1989; Carlisle, 1989; Carver, 1981; Daneman & Blennerhassett, 1984; Danks & End, 1987; Hood & Dubert, 1983; Hoover & Gough, 1990; Royer et al., 1986; Spring & French, 1990; Sticht & James, 1984).

The increasing tendency to view reading disability as a type of developmental language disorder (see Kamhi & Catts, 1989) has potential implications for assessment. Processing problems in the phonological domain have been increasingly singled out as the most salient cause of reading disabilities. Such a localization for the disability raises the possibility of assessing, before school entry, the critical phonological skills that are necessary prerequisites to the development of adequate reading skill. Successful remediation based on such early assessment is already an empirically demonstrated fact (Blachman, 1989; Bradley, 1987; Bradley & Bryant, 1985; Lundberg et al., 1988). A standardized assessment device for young children has recently been developed by Torgesen and Bryant (1989).

REFERENCES

Aaron, P. G. (1989). *Dyslexia and hyperlexia*. Dordrecht: Kluwer Academic.

Aman, M., & Singh, N. (1983). Specific reading disorders: Concepts of etiology reconsidered. In K. Gadow & I. Bialer (Eds.), *Advances in learning and behavioral disabilities* (Vol. 2, pp. 1–47). Greenwich, CT: JAI.

American Psychiatric Association. (1987). *Diagnostic and statistical manual of mental disorders* (3rd ed., rev.). Washington, DC: Author.

Anastasi, A. (1988). *Psychological testing* (6th ed.). New York: Macmillan.

Applebee, A. N. (1971). Research in reading retardation: Two critical problems. *Journal of Child Psychology & Psychiatry, 12*, 91–113.

Baddeley, A. D., Ellis, N. C., Miles, T. R., & Lewis, V. J. (1982). Developmental and acquired dyslexia: A comparison. *Cognition, 11*, 185–199.

Baddeley, A. D., Logie, R., & Ellis, N. C. (1988). Characteristics of developmental dyslexia. *Cognition, 30*, 197–227.

Baron, J. (1985). *Rationality and intelligence*. Cambridge: Cambridge University Press.

Bateman, B. (1979). Teaching reading to learning disabled and other hard-to-teach children. In L. Resnick & P. Weaver (Eds.), *Theory and practice of early reading* (Vol. 1, pp. 227–259). Hillsdale, NJ: Lawrence Erlbaum Associates.

Beech, J., & Harding, L. (1984). Phonemic processing and the poor reader from a developmental lag viewpoint. *Reading Research Quarterly, 19*, 357–366.

Berninger, V. W., Thalberg, S. P., DeBruyn, I., & Smith, R. (1987). Preventing reading disabilities by assessing and remediating phonemic skills. *School Psychology Review, 16*, 554–565.

Blachman, B. A. (1989). Phonological awareness and word recognition: Assessment and intervention. In A. G. Kamhi & H. W. Catts (Eds.), *Reading disabilities* (pp. 133–158). Boston, MA: College-Hill.

Bloom, A., Wagner, M., Reskin, L., & Bergman, A. (1980). A comparison of intellectually delayed and primary reading disabled children on measures of intelligence and achievement. *Journal of Clinical Psychology, 36*, 788–790.

Board of Trustees of the Council for Learning Disabilities. (1987). The CLD position statements. *Journal of Learning Disabilities, 20*, 349–350.

Bowers, P., Steffy, R., & Tate, E. (1988). Comparison of the effects of IQ control methods on memory and naming speed predictors of reading disability. *Reading Research Quarterly, 23*, 304–319.

Bradley, L. (1987, December). *Categorising sounds, early intervention and learning to read: A follow-up study.* Paper presented at the meeting of the British Psychological Society, London.

Bradley, L., & Bryant, P. E. (1978). Difficulties in auditory organization as a possible cause of reading backwardness. *Nature, 271*, 746–747.

Bradley, L., & Bryant, P. E. (1985). *Rhyme and reason in reading and spelling.* Ann Arbor: University of Michigan Press.

Bryant, P. E., Bradley, L., Maclean, M., & Crossland, D. (1989). Nursery rhymes, phonological skills and reading. *Journal of Child Language, 16*, 407–428.

Bryant, P. E., & Goswami, U. (1986). Strengths and weaknesses of the reading level design: A comment on Backman, Mamen, and Ferguson. *Psychological Bulletin, 100*, 101–103.

Calfee, R. C. (1977). Assessment of independent reading skills: Basic research and practical applications. In A. Reber & D. Scarborough (Eds.), *Toward a psychology of reading* (pp. 289–323). Hillsdale, NJ: Lawrence Erlbaum Associates.

Carlisle, J. F. (1989). The use of the sentence verification technique in diagnostic assessment of listening and reading comprehension. *Learning Disabilities Research, 5*, 33–44.

Carver, R. P. (1981). *Reading comprehension and rauding theory.* Springfield, IL: Thomas.

Ceci, S. J. (1986). *Handbook of cognitive, social, and neuropsychological aspects of learning disabilities* (Vol. 1). Hillsdale, NJ: Lawrence Erlbaum Associates.

Ceci, S. J. (1990). *On intelligence more or less: A bio-ecological treatise on intellectual development.* Englewood Cliffs, NJ: Prentice-Hall.

Ceci, S. J., & Liker, J. K. (1986). A day at the races: A study of IQ, expertise, and cognitive complexity. *Journal of Experimental Psychology: General, 115*, 255–266.

Childs, B., & Finucci, J. M. (1983). Genetics, epidemiology, and specific reading disability. In M. Rutter (Ed.), *Development neuropsychology* (pp. 507–519). New York: Guilford.

Coles, G. S. (1978). The learning-disabilities test battery: Empirical and social issues. *Harvard Educational Review, 48*, 313–340.

Coles, G. S. (1987). *The learning mystique.* New York: Pantheon.

Cone, T. E., & Wilson, L. R. (1981). Quantifying a severe discrepancy: A critical analysis. *Learning Disability Quarterly, 4*, 359–371.

Cossu, G., & Marshall, J. C. (1986). Theoretical implications of the hyperlexia syndrome: Two new Italian cases. *Cortex, 22*, 579–589.

Cossu, G., Shankweiler, D., Liberman, I. Y., Katz, L., & Tola, G. (1988). Awareness of phonological segments and reading ability in Italian children. *Applied Psycholinguistics, 9*, 1–16.

Critchley, M. (1970). *The dyslexic child.* London: Heinemann Medical Books.

Cronbach, L. J. (1984). *Essentials of psychological testing* (4th ed.). New York: Harper & Row.

CTB/McGraw-Hill. (1981). *Listening test.* Monterey, CA: Author.

Daneman, M., & Blennerhassett, A. (1984). How to assess the listening comprehension skills of prereaders. *Journal of Educational Psychology, 76*, 1372–1381.

Danks, J. H., & End, L. J. (1987). Processing strategies for reading and listening. In R. Horowitz & S. J. Samuels (Eds.), *Comprehending oral and written language* (pp. 271–294). San Diego: Academic Press.

Decker, S. N., & Bender, B. G. (1988). Converging evidence for multiple genetic forms of reading disability. *Brain and Language, 33*, 197–215.

Detterman, D. (1982). Does "g" exist? *Intelligence, 6*, 99–108.

Doehring, D. G. (1978). The tangled web of behavioral research on developmental dyslexia. In A. L. Benton & D. Pearl (Eds.), *Dyslexia* (pp. 123–135). New York: Oxford University Press.

Doehring, D. G., Trites, R. L., Patel, P. G., & Fiedorowicz, C. A. M. (1981). *Reading disabilities: The interaction of reading, language, and neuropsychological deficits.* New York: Academic Press.

Duane, D. (1985). Written language underachievement: An overview of the theoretical and practical issues. In F. Duffy & N. Geschwind (Eds.), *Dyslexia: A neuroscientific approach to clinical evaluation* (pp. 3–32). Boston: Little, Brown.

Durrell, D., & Hayes, M. (1969). *Durrell listening-reading series.* New York: Psychological Corporation.

Eisenberg, L. (1978). Definitions of dyslexia: Their consequences for research and policy. In A. L. Benton & D. Pearl (Eds.), *Dyslexia* (pp. 29–42). New York: Oxford University Press.

Eisenberg, L. (1979). Reading disorders: Strategies for recognition and management. *Bulletin of the Orton Society, 29*, 39–55.

Ellis, A. W. (1985). The cognitive neuropsychology of developmental (and acquired) dyslexia: A critical survey. *Cognitive Neuropsychology, 2*, 169–205.

Education for All Handicapped Children Act (PL 94–142). (1975).

Ellis, N., & Large, B. (1987). The development of reading: As you seek so shall you find. *British Journal of Psychology, 78*, 1–28.

Feagans, L. (1983). Medical progress: A current view of learning disabilities. *The Journal of Pediatrics, 102*, 487–493.

Finucci, J. M. (1986). Follow-up studies of developmental dyslexia and other learning disabilities. In S. D. Smith (Ed.), *Genetics and learning disabilities* (pp. 97–121). San Diego: College-Hill.

Fodor, J. (1983). *Modularity of mind.* Cambridge: MIT Press.

Fox, B., & Routh, D. K. (1975). Analyzing spoken language into words, syllables, and phonemes: A developmental study. *Journal of Psycholinguistic Research, 4*, 331–342.

Frankenberger, W., & Harper, J. (1987). States' criteria and procedures for identifying learning disabled children: A comparison of 1981/82 and 1985/86 guidelines. *Journal of Learning Disabilities, 20*, 118–121.

Fredman, G., & Stevenson, J. (1988). Reading processes in specific reading retarded and reading backward 13-year-olds. *British Journal of Developmental Psychology, 6*, 97–108.

Freebody, P., & Byrne, B. (1988). Word-reading strategies in elementary school children: Relations to comprehension, reading time, and phonemic awareness. *Reading Research Quarterly, 23*, 441–453.

Gardner, H. (1983). *Frames of mind.* New York: Basic.

Gaskins, I. (1982). Let's end the reading disabilities/learning disabilities debate. *Journal of Learning Disabilities, 15*, 81–83.

Gathercole, S. E., & Baddeley, A. D. (1987). The processes underlying segmental analysis. *European Bulletin of Cognitive Psychology, 7*, 462–464.

Gillet, J. W., & Temple, C. (1986). *Understanding reading problems: Assessment and instruction* (2nd ed.). Boston: Little, Brown.

Goswami, U., & Bryant, P. E. (1989). The interpretation of studies using the reading level design. *Journal of Reading Behavior, 21*, 413–424.

Gough, P. B., & Hillinger, M. L. (1980). Learning to read: An unnatural act. *Bulletin of the Orton Society, 30*, 171–176.

Gough, P. B., & Tunmer, W. E. (1986). Decoding, reading, and reading disability. *Remedial and Special Education, 7*, 6–10.

Hall, J., & Humphreys, M. (1982). Research on specific learning disabilities: Deficits and remediation. *Topics in Learning and Learning Disabilities, 2*, 68–78.

Hammill, D., Leigh, J., McNutt, G., & Larsen, S. (1981). A new definition of learning disabilities. *Learning Disability Quarterly, 4*, 336–342.

Hessler, G. L. (1987). Educational issues surrounding severe discrepancy. *Learning Disabilities Research, 3*, 43–49.

Hinshelwood, J. (1895). Word-blindness and visual memory. *Lancet, 2*, 1564–1570.

Hinshelwood, J. (1917). *Congenital word-blindness*. London: Lewis.

Holligan, C., & Johnston, R. S. (1988). The use of phonological information by good and poor readers in memory and reading tasks. *Memory & Cognition, 16*, 522–532.

Hood, J., & Dubert, L. A. (1983). Decoding as a component of reading comprehension among secondary students. *Journal of Reading Behavior, 15*, 51–61.

Hooper, S. R., & Willis, W. (1989). *Learning disability subtyping*. New York: Springer-Verlag.

Hoover, W. A., & Gough, P. B. (1990). The simple view of reading. *Reading and Writing: An Interdisciplinary Journal, 2*, 127–160.

Horn, W. F., & O'Donnell, J. (1984). Early identification of learning disabilities: A comparison of two methods. *Journal of Educational Psychology, 76*, 1106–1118.

Horn, W. F., O'Donnell, J. P., & Vitulano, L. (1983). Long-term follow-up studies of learning-disabled persons. *Journal of Learning Disabilities, 16*, 542–555.

Hulme, C. (1988). The implausibility of low-level visual deficits as a cause of children's reading difficulties. *Cognitive Neuropsychology, 5*, 369–374.

Hynd, G., & Cohen, M. (1983). *Dyslexia: Neuropsychological theory, research, and clinical differentiation*. New York: Grune & Stratton.

Hynd, G., & Hynd, C. (1984). Dyslexia: Neuroanatomical/neurolinguistic perspectives. *Reading Research Quarterly, 19*, 482–498.

Hynd, G., & Semrud-Clikeman, M. (1989). Dyslexia and neurodevelopmental pathology: Relationships to cognition, intelligence, and reading skill acquisition. *Journal of Learning Disabilities, 22*, 204–220.

Jackson, N. E., & Butterfield, E. C. (1989). Reading-level match designs: Myths and realities. *Journal of Reading Behavior, 21*, 387–412

Johnson, D. J. (1988). Review of research on specific reading, writing, and mathematics disorders. In J. F. Kavanaugh, & T. J. Truss (Eds.), *Learning disabilities: Proceedings of the national conference* (pp. 79–163). Parkston, MD: York.

Jorm, A. (1983). Specific reading retardation and working memory: A review. *British Journal of Psychology, 74*, 311–342.

Jorm, A., Share, D., Maclean, R., & Matthews, R. (1986). Cognitive factors at school entry predictive of specific reading retardation and general reading backwardness: A research note. *Journal of Child Psychology and Psychiatry, 27*, 45–54.

Juel, C. (1988). Learning to read and write: A longitudinal study of 54 children from first through fourth grade. *Journal of Educational Psychology, 80*, 437–447.

Juel, C., Griffith, P. L., & Gough, P. B. (1986). Acquisition of literacy: A longitudinal study of children in first and second grade. *Journal of Educational Psychology, 78*, 243–255.

Kamhi, A., & Catts, H. (1989). *Reading disabilities: A Developmental language perspective*. Boston: College-Hill.

Kavale, K. A. (1987). Theoretical issues surrounding severe discrepancy. *Learning Disabilities Research, 3,* 12–20.

Kavale, K. A., & Nye, C. (1981). Identification criteria for learning disabilities: A survey of the research literature. *Learning Disability Quarterly, 4,* 363–388.

Kavanagh, J. F., & Truss, T. J. (Eds.). (1988). *Learning disabilities: Proceedings of the national conference.* Parkston, MD: York.

Kirk, S. (1963). *Behavioral diagnosis and remediation of learning disabilities.* Paper presented at the Conference on the Exploration into the Problems of the Perceptually Handicapped Child. Evanston, IL: Fund for the Perceptually Handicapped Child.

Kirk, S., & Elkins, J. (1975). Characteristics of children enrolled in the child service demonstration centers. *Journal of Learning Disabilities, 8,* 630–637.

Kochnower, J., Richardson, E., & DiBenedetto, B. (1983). A comparison of the phonic decoding ability of normal and learning disabled children. *Journal of Learning Disabilities, 16,* 348–351.

Labuda, M., & DeFries, J. C. (1989). Differential prognosis of reading-disabled children as a function of gender, socioeconomic status, IQ, and severity: A longitudinal study. *Reading and Writing: An Interdisciplinary Journal, 1,* 25–36.

Lerner, J. (1975). Remedial reading and learning disabilities: Are they the same or different? *Journal of Special Education, 9,* 119–131.

Lerner, J. (1985). *Learning disabilities* (4th ed.). Boston, MA: Houghton Mifflin.

Liberman, I. (1982). A language-oriented view of reading and its disabilities. In H. Mykelbust (Ed.), *Progress in learning disabilities* (Vol. 5, pp. 81–101). New York: Grune & Stratton.

Liberman, I. Y., & Shankweiler, D. (1985). Phonology and the problems of learning to read and write. *Remedial and Special Education, 6,* 8–17.

Lieberman, P., Meskill, R. H., Chatillon, M., & Schupack, H. (1985). Phonetic speech perception deficits in dyslexia. *Journal of Speech and Hearing Research, 28,* 480–486.

Lindgren, S. D., De Renzi, E., & Richman, L. C. (1985). Cross-national comparisons of developmental dyslexia in Italy and the United States. *Child Development, 56,* 1404–1417.

Lundberg, I., Frost, J., & Peterson, O. (1988). Effects of an extensive program for stimulating phonological awareness in preschool children. *Reading Research Quarterly, 23,* 263–284.

Lyon, G. R. (1985). Educational validation studies of learning disability subtypes. In B. P. Rourke (Ed.), *Neuropsychology of learning disabilities* (pp. 228–253). New York: Guilford.

Lyon, G. R. (1987). Learning disabilities research: False starts and broken promises. In S. Vaughn & C. S. Bos (Eds.), *Research in learning disabilities* (pp. 69–85). Boston, MA: College-Hill.

Maclean, M., Bryant, P., & Bradley, L. (1987). Rhymes, nursery rhymes, and reading in early childhood. *Merrill–Palmer Quarterly, 33,* 255–281.

Mann, V. (1986). Why some children encounter reading problems. In J. Torgesen & B. Wong (Eds.), *Psychological and educational perspectives on learning disabilities* (pp. 133–159). New York: Academic Press.

McKinney, J. D. (1987). Research on the identification of learning-disabled children: Perspectives on changes in educational policy. In S. Vaughn & C. Bos (Eds.), *Research in learning disabilities* (pp. 215–233). Boston: College-Hill.

Morgan, W. P. (1896). A case of congenital word-blindness. *British Medical Journal, 2,* 1378.

Morrison, F. (1984). Word decoding and rule-learning in normal and disabled readers. *Remedial and Special Education, 5,* 20–27.

Morrison, F. J. (1987). The nature of reading disability: Toward an integrative framework. In S. Ceci (Ed.), *Handbook of cognitive, social, and neuropsychological aspects of learning disabilities* (pp. 33–62). Hillsdale, NJ: Lawrence Erlbaum Associates.

Morrison, F., & Giordani, B., & Nagy, J. (1977). Reading disability: An information processing analysis. *Science, 196,* 77–79.

Olson, R. K., Kliegl, R., Davidson, B., & Foltz, G. (1985). Individual and developmental differences in reading disability. In T. Waller (Ed.), *Reading research: Advances in theory and practice* (Vol. 4, pp. 1–64). London: Academic Press.

Olson, R. K., Wise, B., Conners, F., & Rack, J. (1990). Organization, heritability, and remediation of component word recognition and language skills in disabled readers. In T. Carr & B. A. Levy (Eds.), *Reading and its development: Component skills approaches* (pp. 261–322). New York: Academic Press.

Olson, R. K., Wise, B., Conners, F., Rack, J., & Fulker, D. (1989). Specific deficits in component reading and language skills: Genetic and environmental influences. *Journal of Learning Disabilities, 22,* 339–348.

Orton, S. T. (1925). "Word-blindness" in school children. *Archives of Neurology and Psychiatry, 14,* 581–615.

Orton, S. T. (1928). Specific reading disability—strephosymbolia. *Journal of the American Medical Association, 90,* 1095–1099.

Orton, S. T. (1937). *Reading, writing, and speech problems in children.* New York: Norton.

Patterson, K., Marshall, J., & Coltheart, M. (1985). *Surface dyslexia.* London: Lawrence Erlbaum Associates.

Pennington, B. F. (1986). Issues in the diagnosis and phenotype analysis of dyslexia: Implications for family studies. In S. D. Smith (Ed.), *Genetics and learning disabilities* (pp. 69–96). San Diego: College-Hill.

Perfetti, C. A. (1985). *Reading ability.* New York: Oxford University Press.

Perfetti, C. A., & Lesgold, A. M. (1977). Discourse comprehension and sources of individual differences. In M. Just & P. Carpenter (Eds.), *Cognitive processes in comprehension* (pp. 141–183). Hillsdale, NJ: Lawrence Erlbaum Associates.

Pratt, A. C., & Brady, S. (1988). Relation of phonological awareness to reading disability in children and adults. *Journal of Educational Psychology, 80,* 319–323.

Pressley, M., & Levin, J. R. (1987). Elaborative learning strategies for the inefficient learner. In S. J. Ceci (Ed.), *Handbook of cognitive, social, and neuropsychological aspects of learning disabilities* (Vol. 2, pp. 175–212). Hillsdale, NJ: Lawrence Erlbaum Associates.

Rayner, K., & Pollatsek, A. (1989). *The psychology of reading.* Englewood Cliffs, NJ: Prentice-Hall.

Read, C., & Ruyter, L. (1985). Reading and spelling skills in adults of low literacy. *Remedial and Special Education, 6,* 43–52.

Reed, J. C. (1970). The deficits of retarded readers—Fact or artifact? *The Reading Teacher, 23,* 347–357.

Reynolds, C. R. (1981). The fallacy of "two years below grade level for age" as a diagnostic criterion for reading disorders. *Journal of School Psychology, 19,* 350–358.

Reynolds, C. R. (1984–1985). Critical measurement issues in learning disabilities. *Journal of Special Education, 18,* 451–475.

Reynolds, C. R. (1985). Measuring the aptitude–achievement discrepancy in learning disability diagnosis. *Remedial and Special Education, 6,* 37–55.

Reynolds, C. R., & Stowe, M. (1985). *Severe discrepancy analysis.* Philadelphia: TRAIN.

Royer, J., Kulhavy, R., Lee, S., & Peterson, S. (1986). The relationship between reading and listening comprehension. *Educational and Psychological Research, 6,* 299–314.

Rutter, M. (1978). Prevalence and types of dyslexia. In A. Benton & D. Pearl (Eds.),

Dyslexia: An appraisal of current knowledge (pp. 5–28). New York: Oxford University Press.

Rutter, M., & Yule, W. (1975). The concept of specific reading retardation. *Journal of Child Psychology and Psychiatry, 16,* 181–197.

Scarborough, H. S. (1984). Continuity between childhood dyslexia and adult reading. *British Journal of Psychology, 75,* 329–348.

Scarborough, H. S. (1989). Prediction of reading disability from familial and individual differences. *Journal of Educational Psychology, 81,* 101–108.

Seidenberg, M. S., Bruck, M., Fornarolo, G., & Backman, J. (1985). Word recognition processes of poor and disabled readers? Do they necessarily differ? *Applied Psycholinguists, 6,* 161–180.

Seidenberg, M. S., Bruck, M., & Fornarolo, G., & Backman, J. (1986). Who is dyslexic? Reply to Wolf. *Applied Psycholinguistics, 7,* 77–84.

Senf, G. F. (1986). LD Research in sociological and scientific perspective. In J. K. Torgeson & B. Y. L. Wong (Eds.), *Psychological and educational perspectives on learning disabilities* (pp. 27–53). Orlando, FL: Academic Press.

Shankweiler, D., Crain, S., Brady, S., & Macaruso, P. (in press). Identifying the causes of reading disability. In P. B. Gough, L. Ehri, & R. Treiman (Eds.), *Reading acquisition.* Hillsdale, NJ: Lawrence Erlbaum Associates.

Share, D. L., McGee, R., McKenzie, D., Williams, S., & Silva, P. A. (1987). Further evidence relating to the distinction between specific reading retardation and general reading backwardness. *British Journal of Developmental Psychology, 5,* 35–44.

Share, D. L., McGee, R., & Silva, P. (1989). IQ and reading progress: A test of the capacity notion of IQ. *Journal of the American Academy of Child and Adolescent Psychiatry, 28,* 97–100.

Share, D. L., & Silva, P. A. (1986). The stability and classification of specific reading retardation: A longitudinal study from age 7 to 11. *British Journal of Educational Psychology, 56,* 32–39.

Share, D. L., & Silva, P. A. (1987). Language deficits and specific reading retardation: Cause or effect? *British Journal of Disorders of Communication, 22,* 219–226.

Shepard, L. (1980). An evaluation of the regression discrepancy method for identifying children with learning disabilities. *Journal of Special Education, 14,* 79–91.

Siegel, L. S. (1985). Psycholinguistic aspects of reading disabilities. In L. Siegel & F. Morrison (Eds.), *Cognitive development in atypical children* (pp. 45–65). New York: Springer-Verlag.

Siegel, L. S. (1988). Evidence that IQ scores are irrelevant to the definition and analysis of reading disability. *Canadian Journal of Psychology, 42,* 201–215.

Siegel, L. S. (1989). IQ is irrelevant to the definition of learning disabilities. *Journal of Learning Disabilities, 22,* 469–479.

Siegel, L. S., & Faux, D. (1989). Acquisition of certain grapheme–phoneme correspondences in normally achieving and disabled readers. *Reading and Writing: An Interdisciplinary Journal, 1,* 37–52.

Siegel, L. S., & Ryan, E. B. (1988). Development of grammatical-sensitivity, phonological, and short-term memory skills in normally achieving and learning disabled children. *Developmental Psychology, 24,* 28–37.

Siegel, L. S., & Heaven, R. K. (1986). Categorization of learning disabilities. In S. J. Ceci (Ed.), *Handbook of cognitive, social, and neuropsychological aspects of learning disabilities* (Vol. 1, pp. 95–121). Hillsdale, NJ: Lawrence Erlbaum Associates.

Silva, P. A., McGee, R., & Williams, S. (1985). Some characteristics of 9-year-old boys with general reading backwardness or specific reading retardation. *Journal of Child Psychology and Psychiatry, 26,* 407–421.

Snowling, M. (1980). The development of grapheme–phoneme correspondence in normal and dyslexic readers. *Journal of Experimental Child Psychology, 29,* 294–305.
Snowling, M. (1981). Phonemic deficits in developmental dyslexia. *Psychological Research, 43,* 219–234.
Snowling, M., Goulandris, N., Bowlby, M., & Howell, P. (1986). Segmentation and speech perception in relation to reading skill: A developmental analysis. *Journal of Experimental Child Psychology, 41,* 489–507.
Snowling, M., Stackhouse, J., & Rack, J. (1986). Phonological dyslexia and dysgraphia—a developmental analysis. *Cognitive Neuropsychology, 3,* 309–339.
Spache, G. D. (1981). *Diagnostic reading scales.* Monterey, CA: CTB/McGraw-Hill.
Spearman, C., & Wynn-Jones, L. (1950). *Human ability.* London: Macmillan.
Spreen, O. (1982). Adult outcomes of reading disorders. In R. Malatesha & P. Aaron (Eds.), *Reading disorders: Varieties and treatments* (pp. 473–498). New York: Academic Press.
Spring, C., & French, L. (1990). Identifying reading-disabled children from listening and reading discrepancy scores. *Journal of Learning Disabilities, 23,* 53–58.
Stanley, G., Smith, G., & Powys, A. (1982). Selecting intelligence tests for studies of dyslexic children. *Psychological Reports, 50,* 787–792.
Stanovich, K. E. (1978). Information processing in mentally retarded individuals. In N. R. Ellis (Ed.), *International review of research in mental retardation* (Vol. 9, pp. 29–60). New York: Academic Press.
Stanovich, K. E. (1986a). Cognitive processes and the reading problems of learning disabled children: Evaluating the assumption of specificity. In J. Torgesen & B. Wong (Eds.), *Psychological and educational perspectives on learning disabilities* (pp. 87–131). New York: Academic Press.
Stanovich, K. E. (1986b). Matthew effects in reading: Some consequences of individual differences in the acquisition of literacy. *Reading Research Quarterly, 21,* 360–407.
Stanovich, K. E. (1988a). Explaining the differences between the dyslexic and the garden-variety poor reader: The phonological-core variable-difference model. *Journal of Learning Disabilities, 21,* 590–612.
Stanovich, K. E. (1988b). The right and wrong places to look for the cognitive locus of reading disability. *Annals of Dyslexia, 38,* 154–177.
Stanovich, K. E. (1989). Learning disabilities in broader context. *Journal of Learning Disabilities, 22,* 287–297.
Stanovich, K. E. (in press). Speculations on the causes and consequences of individual differences in early reading acquisition. In P. Gough, L. Ehri, & R. Treiman (Eds.), *Reading acquisition.* Hillsdale, NJ: Lawrence Erlbaum Associates.
Stanovich, K. E., Cunningham, A. E., & Cramer, B. (1984). Assessing phonological awareness in kindergarten children: Issues of task comparability. *Journal of Experimental Child Psychology, 38,* 175–190.
Stanovich, K. E., Nathan, R., & Vala-Rossi, M. (1986). Developmental changes in the cognitive correlates of reading ability and the developmental lag hypothesis. *Reading Research Quarterly, 21,* 267–283.
Stanovich, K. E., Nathan, R. G., & Zolman, J. E. (1988). The developmental lag hypothesis in reading: Longitudinal and matched reading-level comparisons. *Child Development, 59,* 71–86.
Stanovich, K. E., & Purcell, D. G. (1981). Comments on "Input Capability and Speed of Processing in Mental Retardation." *Journal of Abnormal Psychology, 90,* 168–171.
Stanovich, K. E., & West, R. F. (1989). Exposure to print and orthographic processing. *Reading Research Quarterly, 24,* 402–433.

Sternberg, R. (1985). *Beyond IQ: A triarchic theory of human intelligence*. Cambridge: Cambridge University Press.

Sternberg, R. J. (1988). *The triarchic mind*. New York: Viking.

Sternberg, R. J., & Detterman, D. K. (1986). *What is intelligence?* Norwood, NJ: Ablex.

Sternberg, R. J., & Wagner, R. K. (1986). *Practical intelligence*. Cambridge: Cambridge University Press.

Sticht, T. G., & James, J. H. (1984). Listening and reading. In P. D. Pearson (Ed.), *Handbook of reading research* (pp. 293–317). New York: Longmen.

Taylor, H. G., Lean, D., & Schwartz, S. (1989). Pseudoword repetition ability in learning-disabled children. *Applied Psycholinguistics, 10*, 203–219.

Taylor, H. J., Satz, P., & Friel, J. (1979). Developmental dyslexia in relation to other childhood reading disorders: Significance and clinical utility. *Reading Research Quarterly, 15*, 84–101.

Thorndike, R. L. (1963). *The concepts of over- and under-achievement*. New York: Teachers College, Columbia University.

Tindal, G., & Marston, D. (1986). Approaches to assessment. In J. K. Torgeson & B. Y. L. Wong (Eds.), *Psychological and educational perspectives on learning disabilities* (pp. 55–84). Orlando: FL: Academic Press.

Torgesen, J. K., & Bryant, B. R. (1989). *Screening Test of Phonological Awareness*. Austin, TX: Pro-Ed.

Torgesen, J. K. (1985). Memory processes in reading disabled children. *Journal of Learning Disabilities, 18*, 350–357.

Treiman, R., & Hirsh-Pasek, K. (1985). Are there qualitative differences in reading behavior between dyslexics and normal readers? *Memory and Cognition, 13*, 357–364.

Tunmer, W. E., & Nesdale, A. R. (1985). Phonemic segmentation skill and beginning reading. *Journal of Educational Psychology, 77*, 417–427.

van der Wissel, A. (1987). IQ profiles of learning disabled and mildly mentally retarded children: A psychometric selection effect. *British Journal of Developmental Psychology, 5*, 45–51.

van der Wissel, A., & Zegers, F. E. (1985). Reading retardation revisited. *British Journal of Developmental Psychology, 3*, 3–9.

Vaughn, S., & Bos, C. S. (1987). *Research in learning disabilities: Issues and future directions*. Boston: College-Hill.

Vellutino, F. (1978). Toward an understanding of dyslexia: Psychological factors in specific reading disability. In A. L. Benton & D. Pearl (Eds.), *Dyslexia* (pp. 59–111). New York: Oxford University Press.

Vellutino, F. (1979). *Dyslexia: Theory and research*. Cambridge, MA: MIT Press.

Vellutino, F., & Scanlon, D. (1987). Phonological coding, phonological awareness, and reading ability: Evidence from a longitudinal and experimental study. *Merrill–Palmer Quarterly, 33*, 321–363.

Vernon, P. A. (Ed.). (1987). *Speed of information-processing and intelligence*. Norwood, NJ: Ablex.

Vogler, G., Baker, L. A., Decker, S. N., DeFries, J. C., & Huizinga, D. (1989). Cluster analytic classification of reading disability subtypes. *Reading and Writing: An Interdisciplinary Journal, 1*, 163–177.

Wagner, R. K. (1988). Causal relations between the development of phonological processing abilities and the acquisition of reading skills: A meta-analysis. *Merrill–Palmer Quarterly, 34*, 261–279.

Wagner, R. K., & Torgesen, J. K. (1987). The nature of phonological processing and its causal role in the acquisition of reading skills. *Psychological Bulletin, 101*, 192–212.

Werker, J. F., & Tees, R. C. (1987). Speech perception in severely disabled and average reading children. *Canadian Journal of Psychology, 41,* 48–61.

Williams, J. (1984). Phonemic analysis and how it relates to reading. *Journal of Learning Disabilities, 17,* 240–245.

Williams, J. P. (1986). The role of phonemic analysis in reading. In J. Torgesen & B. Wong (Eds.), *Psychological and educational perspectives on learning disabilities* (pp. 399–416). New York: Academic Press.

Wilson, L. R., & Cone, T. (1984). The regression equation method of determining academic discrepancy. *Journal of School Psychology, 22,* 95–110.

Wolf, M., & Goodglass, H. (1986). Dyslexia, dysnomia, and lexical retrieval: A longitudinal investigation. *Brain and Language, 28,* 154–168.

Yule, W. (1984). The operationalizing of "underachievement"—doubts dispelled. *British Journal of Clinical Psychology, 23,* 233–234.

CHAPTER 7

Language and Speech Disorders

Rhea Paul
Portland State University

HISTORICAL BACKGROUND

Descriptions of the syndrome of disorders of language learning in children date back at least to the early 19th century. Gall (1825) was perhaps the first to describe children with poor understanding and use of speech and to differentiate them from the mentally retarded. The syndrome was referred to by early writers as "childhood aphasia" because of the supposed associations between deficits seen in children's language development and those observed in adults with specific neurological lesions. These writers distinguished between what they called *congenital aphasia*, a developmental difficulty in the acquisition of language, and *acquired aphasia* in childhood, a syndrome similar to that seen in adults with neurological traumas. Broadbent (1872) emphasized that acquired childhood aphasias were usually transient. This transience, which is noted by a variety of writers (summarized by Lennenberg, 1967), serves as a primary distinction between acquired aphasias, which have known points of onset and verifiable neurological correlates, and developmental disorders of language learning, which are recognized when the child fails to learn to speak normally but have no obvious point of onset or identifiable neurological basis.

The concept of acquired aphasia in childhood has remained quite stable over the century and a half in which it has been studied, and, generally, aphasias acquired in childhood can be classified using the same terms that are used for the adult pathologies, with no separate diagnostic system necessary. The only exception is the acquired childhood aphasia that appears to develop spontaneously after some language learning has occurred and is accompanied by seizures. This type is sometimes called *Landau* or *Landau–Kleffner syndrome*. It is seen exclusively in children and has no adult correlate.

In contrast to the acquired aphasias, the notions involved with the congenital type of child language disorder have undergone great changes. One way in which this is evident is in terminology. Nineteenth-century neurologists referred to disorders of language learning as "aphasias," analogous to the disorders seen in adults. McCall (1911) introduced the term *word deafness* to describe the child with normal hearing but inability to associate words with their meaning. Karlin (1954) suggested the term *congenital verbal auditory agnosia*. Until recently, these labels were used interchangeably to refer to children with developmental language learning disorders.

Myklebust (1954) proposed the term *language pathology*, which is distinct from the other terms because it does not make any assumptions about neurological origins of the disorder. Although *childhood aphasia* and related terms continued to be used throughout the 1950s and 1960s, by the late 1970s many writers (Bloom & Lahey, 1978; Johnston, 1982) were uncomfortable with the implicit analogy to specific CNS pathology inherent in the *aphasia* label. The reason for this discomfort grew out of the failure of a variety of studies to establish reliable neurological correlates for childhood language pathology. Whereas children with poor language acquisition have always been assumed to be victims of some form of brain damage, often referred to as "minimal brain dysfunction," intensive efforts to establish the site of lesion in this population have been unable to demonstrate expected physical evidence of central nervous system pathology.

Electroencephalographic studies, for example, have shown that, whereas 40% of language-impaired children produced abnormal EEG records (Forest, Eisenson, & Stark, 1966; Goldstein, Landau, & Kleffner, 1960), the majority of subjects did not. Further, because, the incidence of abnormal EEG in normal children ranges from 17% to 30% (Satterfield, 1973), it is difficult to argue that EEG findings alone can identify children with language disorders. Classic neurological examination has shown a higher incidence of motor abnormalities in language-impaired than in deaf children, but 32% of the language-disordered group showed no neurological abnormality at all (Goldstein et al., 1960). Computed tomographic (CT) studies (Harcherik et al., 1985) show no differences in findings for this population than for other neuropsychiatrically impaired youngsters, nor was any particular type of structural abnormality associated with the language impairment. In an earlier study (Caparulo, Cohen, & Rothman, 1981), 56% of the children with various kinds of neuropsychiatric disorders, including language impairment, were found to have normal CT scans. Studies aimed at establishing cerebral lateralization for language function, using dichotic listening techniques (Springer & Eisenson, 1977), show no strong evidence of neural pathology in terms of ear preference for verbal material. In summary, although these group studies of language-impaired children have supported the assumption of CNS etiology for the language disorder by showing high proportions of abnormal findings, many individuals in the language-impaired group do not conform to the abnormal patterns.

Although early conceptions of childhood aphasias focused on neurological phenomena, interest in the 1960s and 1970s began to move away from the neurological level to interest in more psychological aspects of the syndrome, particularly those involved with auditory perception and discrimination. Rosenthal (1972) showed that language-impaired children required longer interstimulus intervals (ISI), of over 300 msec as opposed to 36 msec for normals, to judge the order in which two acoustic stimuli were presented. Tallal and her colleagues (Tallal, 1988; Tallal & Piercy, 1975) also showed that language-impaired children are significantly poorer than normal peers at judging the order in which they heard two brief acoustic stimuli presented with ISIs of 20 to 60 msec. When the duration of the auditory signals was increased, however, language-impaired subjects were able to make these order judgments. Tallal and Piercy (1975) concluded that it is the short duration of auditory stimuli, so typical of the stream of ordinary speech, that causes difficulty for the language-impaired population. Thus, they attributed to this group "an impaired rate of processing auditory information." Eisenson (1972) also reported deficits in a variety of perceptual abilities, with auditory perceptual problems being the most pronounced.

In summary, the concept of a relatively specific disorder of language acquisition has been present in the research literature for over 100 years. The most dramatic change in our conceptualization of this syndrome has been a move away from descriptions that imply analogies to acquired aphasias in adulthood, principally because many attempts to identify pathognomonic features have been unsuccessful. Current attempts to characterize SDLDs focus on their neuropsychological aspects.

DEFINITIONAL ISSUES

The modern definition of developmental disorders of language in children was developed by Benton (1964) and is structured by exclusion. A child is said to evidence a language disorder if the language difficulty cannot be accounted for by hearing impairment, mental retardation, neuromotor dysfunction, environmental deprivation, autism, or emotional disturbance. Specific language disorders are frequently present in families with a history of language or learning problems in other family members, and are about four times more prevalent in boys than in girls.

This rather dry, vague definition masks the profound and painful consequences of this syndrome. Children with language disorders, who have the normal social motivation to communicate and interact with others and the cognitive abilities to perceive their differences from peers, suffer constant frustration and anxiety from their inability to communicate (Caparulo & Cohen, 1977). Some children develop self-abusive behavior, such as head-banging, in response to this frustration. Others show withdrawal or depressive behavior as they try to get their ideas across and find no viable avenue (Paul & Cohen, 1982). Unlike retarded children, who are often unaware of their difficulties and are apparently content, these children frequently present as unhappy and painfully aware of their limitations (Caparulo & Cohen, 1977).

Equally poignant is the effect of their affliction on their families. Having bonded to what they considered to be a normal baby, parents of these children often go through long, agonizing periods of searching for the cause of the child's failure to speak, during which they receive conflicting information from professionals, some of whom counsel them to take the child home, love him (or her), and simply wait for the child to grow out of the delay. Other professionals say or imply that the child's delay is due to parenting practices, either telling the parents that they are "doing too much" for the child and not "forcing" him or her to communicate, or imputing the lack of speech to negativism or symbiotic processes induced by cold, demanding parents or overprotective ones.

Even when an appropriate diagnosis is finally achieved, the long-term, poor prognosis for those with severe disorders is slow to dawn on families and becomes another adjustment they need to make after so many earlier disappointments. In addition, the daily difficulties of dealing with a frustrated, hyperactive, uncommunicative child, who may be unable to follow simple instructions or to communicate basic needs and desires, weighs heavily on parents and siblings. Although many parents may initially welcome the diagnosis of specific language disorder as preferable to retardation or autism, which they may have suspected, the disorder is often not much easier for families to address than the other apparently "worse" classifications.

Diagnostic Labels

Although CNS pathology is still widely assumed to underlie the disorders of language development seen in children, a movement away from terminology that implies neuropathological analogs to adult aphasia has occurred. Terminology in the spirit of that proposed by Myklebust (1971) is now used more frequently by practitioners. This terminology avoids unsubstantiated assumptions about neuropathology but is unfortunately not free of its own problems. There are still a multitude of terms that are used to describe the syndrome, and some are not so well defined as they might be. Several terms, including *language impaired, language disordered, language deviant,* and *language delayed,* are often used interchangeably by clinicians practicing in the field.

Language deviant and *language delayed* are the most problematic terms, because they make claims about the type of problem the patient presents. That is, the former asserts that the course of language growth differs from the normal sequence, whereas the latter implies that growth follows the normal progression, but at a slowed rate, and will eventually approach a normal outcome. Research comparing grammatical development in normal and disordered children generally shows that the grammars developed by the two groups are very similar (Morehead & Ingram, 1973), which would call into question the implication of the term *deviant.* Although most researchers in language pathology would agree that children with language disorders generally show a sequence of acquisition that mirrors the normal process but is slower, the outcome for children with language disorders is, in any individual case, quite uncertain. Leonard (1972) argued against the use of *language delay,* because children with language disorders so often plateau in their development and never achieve the normal endpoint. For these reasons, it is advisable to avoid these two, more assumption-laden terms and make use of labels that provide as objective a description of the syndrome as possible. *Language impaired* and *language disordered* are the terms used by the majority of language pathologists to refer to the syndrome. The term *developmental* is often affixed to specify that the syndrome being discussed occurred before language had been acquired and has no known point of onset or specific etiology. Thus, DSM-III (American Psychiatric Association [APA], 1980) and

Table 7.1
DSM-III-R Criteria for Specific Developmental Language Disorders

Expressive Disorders

A. The score obtained from a standardized measure of expressive language is substantially below that obtained from a standardized measure of nonverbal intellectual capacity (as determined by an individually administered IQ test).

B. The disturbance in A significantly interferes with academic achievement or activities of daily living requiring the expression of verbal (or sign) language. This may be evidenced in severe cases by use of a markedly limited vocabulary, by speaking only in simple sentences, or by speaking only in the present tense. In less severe cases, there may be hesitations or errors on recalling certain words, or errors in the production of long or complex sentences.

C. Not due to pervasive developmental disorder, defect in hearing acuity, or a neurological disorder (aphasia).

Receptive Disorders

A. The score obtained from a standardized measure of receptive language is substantially below that obtained from a standardized measure of nonverbal intellectual capacity (as determined by an individually administered IQ test).

B. The disturbance in A significantly interferes with academic achievement or activities of daily living requiring the comprehension of verbal (or sign) language. This may be manifested in more severe cases by an inability to understand simple words or sentences. In less severe cases, there may be difficulty in understanding only certain types of words, such as spatial terms, or an inability to comprehend longer or more complex statements.

C. Not due to a pervasive developmental disorder, defect in hearing acuity, or a neurological disorder (aphasia).

Note: From *Diagnostic and Statistical Manual of Mental Disorders* (3rd ed., rev.) by the American Psychiatric Association, 1987, Washington, DC: Author. Reprinted by permission.

DSM-III-R (APA, 1987) use the term *specific developmental language disorder* (SDLD) to describe the syndrome under discussion. The DSM-III-R criteria for specific developmental language disorders are listed in Table 7.1.

As these criteria make clear, DSM-III and DSM-III-R define two types of specific developmental language disorders: expressive (315.31) and receptive (315.31). Expressive disorders are defined as those that affect only the production of language—that is, talking. Receptive disorders, on the other hand, are defined as those that affect only the comprehension or understanding of language. This classification system mirrors that used for adult aphasias and grew out of the acquired aphasia tradition. However, in children who never learned to speak, the possibility of a circumscribed disorder of receptive language, analogous to a Wernicke's aphasia, simply does not exist. Any disorder affecting receptive language, which interferes with the child's ability to extract meaning from the linguistic input and deduce linguistic rules, must also affect the child's ability to produce spoken language. Cohen, Paul, and Volkmar (1986) argued for the need to discard the deceptive label of "specific developmental receptive language disorder," because it leads the clinician to expect to find a type of disorder that may not exist (i.e., one in which a child with a developmental deficit shows intact expression in the presence of poor comprehension). Instead, a term such as "specific developmental language disorder–global type" would be a more accurate label. It would make clear to the clinician that, in the case in which receptive language is affected in a child with no history of normal language development, expressive skills will inevitably be affected as well.

Differential Diagnosis

The term *specific* refers to the fact that, in order to qualify for this diagnosis, the patient must present a deficit in language learning over and above what could be predicted on the basis of hearing impairment, mental retardation, neuromotor dysfunction, emotional disturbance, environmental deprivation, or pervasive developmental disorder. This qualification is important, because some degree of language impairment is expected in each of these conditions.

Hearing-impaired children show slow oral language growth and may require language habilitation or an alternative communication modality, such as sign language, in order to communicate effectively. Mentally retarded individuals would be expected to achieve language development levels commensurate with their overall mental age and would, therefore, not show language use appropriate for their chronological level. Children with neuromotor dysfunction have normal receptive language but may be precluded from producing speech because of motoric difficulties. Children with emotional disturbance may show elective mutism or schizophrenic language usage. Children with pervasive developmental disorders may be mute or delayed in the acquisition of language or may show idiosyncratic language patterns and poor motivation to communicate. Neglected or abused children may have communication difficulties. Thus, language problems are associated with all of these syndromes; but *specific* developmental language disorders are defined as those that cannot be attributed to any of the preceding conditions or that exist over and above the difficulty the syndrome would predict.

For example, a hearing-impaired child with a 50% loss in both ears would be expected to have difficulty learning to understand and speak but to produce some oral language and to improve with hearing aids. If

Table 7.2
A Sample of Diagnostic Instruments in Communication Disorders

Nonverbal Intelligence Tests:	
Test	*Age Range*
Leiter International Performance Scale (Arthur, 1952)	2–18
Hiskey–Nebraska Test of Learning Aptitude (Hiskey, 1966)	3–16
Test of Nonverbal Intelligence (Brown, Sherbenou, & Dollar, 1982)	5–adult
Wechsler Preschool and Primary Scale of Intelligence–Revised (Wechsler, 1971)	3–7
Wechsler Intelligence Scale for Children–Revised (Wechsler, 1974)	6–17

a child with this level of hearing impairment were unable, even with special education and amplification, to produce or understand any spoken language, a specific developmental language disorder could be diagnosed in addition to a hearing impairment. Similarly, a mentally retarded 8-year-old with an IQ of 50 would be expected to have language skills approximately commensurate with a normal 4-year-old. If this child produced only one- to two-word sentences and were unable to follow simple commands, a specific developmental language disorder in addition to the retardation could also be diagnosed. However, on the other hand, a mentally retarded 6-year-old with an IQ of 50, whose language skills resembled those of a normal 3-year-old would be considered to be performing at the level of language competence expected for his or her mental age and would not be diagnosed as having a language disorder.

Differential diagnosis of SDLD from these disorders that closely resemble it is a crucial aspect of its evaluation. Differentiating SDLD from mental retardation requires the use of a nonverbal assessment of cognitive functioning so as not to penalize the patient for the verbal disability (see Table 7.2). Any child with poor language development who scored above the mentally retarded range (above 70) on a nonverbal standardized IQ measure could be considered to have a language disorder. So could a retarded child whose nonverbal mental age was higher than would be predicted on the basis of his or her language performance. Differentiating SDLD from hearing impairment would be done using a similar logic, largely based on the findings of audiometric assessment.

Differentiating SDLD from elective mutism requires the determination of the prevasiveness of the child's failure to communicate. In order to be considered SDLD, it must be shown that the child does not talk or understand better in any environment; that communication with all interlocutors, including parents, is equally depressed. Elective mutism, on the other hand, can only be diagnosed when the child shows normal communication skills in some environment(s) and a marked lack of the same skill in others.

Distinguishing SDLD from autism or pervasive developmental disorder is the distinction that has received the most attention in the literature. De Hirsch (1967) and Wing and Attwood (1987) discussed the difficulty of making this decision. There are many similarities between the two groups of patients, including poor auditory discrimination, echolalia, and limited verbal production. The primary differences include the short auditory memory of the "aphasic" youngster compared to the autistic child who may show superior mnestic capacity. The autistic disturbance of prosodic features (i.e., pitch, stress, and intonation), the autistic's idiosyncratic use of words, and general lack of motivation to communicate in the autistic child also distinguish the two groups. Caparulo and Cohen (1977) pointed to the spontaneous development of gestural language systems by children with SDLD as a compensatory communicative mechanism and to their history of normal behavior in the preverbal period as characteristic features. Autistic children, on the other hand, show little desire to communicate, even in infancy. This failure of social communication extends beyond language and can be seen in the autistic child's deviant use of eye contact, lack of nonverbal communicative functions, and history of disinterest in preverbal communication in "baby games" such as peek-a-boo (Volkmar, 1987).

Differentiating SDLD from environmental deprivation must be done on the basis of the rate of recovery upon the advent of treatment. That is, once placed in a therapeutic language-learning environment, a child with normal language potential should show substantial gains in language development rather quickly (Culp, Heide, & Richardson, 1987). It should be noted that, except in the most extreme circumstances of pathological abuse and neglect, inadequate environmental stimulation is almost never the primary cause of language disorder. This conclusion arises from findings that the amount of language a child needs to hear in order to acquire normal competence has been shown to be quite small (Schiff, 1979; Skuse, 1984). Although parents from "underprivileged backgrounds" may speak differently to children and use language for different purposes than parents from middle-class homes, children with normal language potential will learn to speak and understand regardless of these differences (Naremore, 1980). Although environmental factors may be maintaining factors in a language disorder, they are rarely the initiating factor.

Associated Deficits

How specific is a "specific developmental language disorder?" That is, can a child show language deficits without an effect on other areas of development? Numerous studies of correlates of language disorders suggest that a primary disorder of language learning has broad implications for the child's development in a variety of areas. For example, Paul, Spangle-Looney, and Dahm (in press) reported that 90% of toddlers with expressive language delays showed deficits in socialization as assessed on the *Vineland Adaptive Behavior Scale* (Sparrow, Balla, & Chiccetti, 1984). Learning disorders have also been frequently associated with language deficits, particularly in the areas of reading, writing, and spelling (Catts & Kahmi, 1986; Maxwell & Wallach, 1984). Cantwell, Baker, and Mattison (1980) and Baker, Cantwell, and Mattison (1980) reported higher incidences of psychiatric diagnoses in children diagnosed as language disordered than those seen in the general population. Attention deficit disorder, particularly with hyperactivity, is the diagnosis given most frequently in the language-disordered sample. Tallal (1988) questioned whether the "psychiatric" problems in these studies are not more properly thought of as neurological, because the items that differentiate the language-impaired from the normal children include problems with clumsiness, concentration, and speech rather than more purely emotional difficulties such as mood, depression, anxiety, or conduct. However, empirical studies do show that language-impaired children show a high incidence of hyperactivity and/or attention deficit. For example, Paul, Cohen, and Caparulo (1983) reported that 60% of their sample of language disordered youngsters could be diagnosed as having ADD with hyperactivity.

It is unclear how these disorders interface with the primary language deficit. They could conceivably be the cause of the failure to acquire language normally by interfering with optimal processing of language input. On the other hand, it is possible to see them as sequelae, resulting from the child's inability to operate in normal social interactions mediated by language and to use internal language to organize thought and modulate action. Finally, it could be that social and behavioral problems accompany language disorders, because they arise from the same underlying neurological diatheses. Whatever the relationship, specific developmental language disorders appear to have far-reaching effects on the child's social development.

Intellectual development too is affected in these disorders. Although children diagnosed as language impaired in the preschool period are often found to have IQs in the normal range, cross-sectional studies (Eisenson, 1972) show that IQ tends to move down into the mildly retarded range by early school age. Paul and Cohen (1984) reported, in a longitudinal study, that 72% of their sample of adolescents diagnosed in early childhood as language disordered tested in the mentally retarded range on nonverbal IQ measures at a 10-year follow-up. This profile differs from that seen in children with primary mental retardation, who show subnormal IQs from the early preschool period, with further declines noted with advancing age. Again it is difficult to specify the direction of causality underlying this observation. Language disorders may be the cause of the gradual decrement in intellectual development, as symbolic function mediated by language becomes the primary avenue of thought by early school age. It is also possible, however, that language disorders are caused by subtle deficits in cognitive ability that are not tapped by tests of early intellectual functioning. Kahmi (1981) suggested that circumscribed cognitive deficits are present even in language disordered children who score within the normal range on IQ measures. Others (Morehead & Ingram, 1973; de Ajuriaguerra et al., 1976) have hypothesized that language-impaired children do not have a selective linguistic deficit but rather a general difficulty in any type of mental representation. Tallal (1988) disputed this conclusion and contended that the deficits are more perceptual than conceptual, arising from an inability to process incoming stimuli rapidly, regardless of the mode of presentation. Whatever the resolution of this issue, Paul and Cohen (1984) and de Ajuriaguerra et al. (1976) showed that nonverbal IQ is a good predictor of language outcome in this population. Valid assessment of nonverbal IQ is, therefore, an important aspect of assessment in this population.

School achievement, particularly in reading, writing, and spelling, is very frequently affected by developmental language disorders. Orton (1937) emphasized the relationship between disorders of language learning and school academic difficulties. This continuity of disorders, with speech and language difficulties in the preschool period often developing into learning disabilities during school age, has continued to be observed. Some investigators (Myklebust, 1971; Tallal, 1988) argued that learning disabilities are in fact the manifestation of language disorders at school age, particularly in children who have a milder or more subtle form of disorder that may become evident only when the child must make the complex intersensory connections between auditory and visual forms of language. Other researchers, such as Johnson (1988), have concluded that there are several subtypes of learning disability, with the language-based type being perhaps the most prevalent one. Whatever the true relationship between language and literacy development, there do appear to be strong connections between disorders in both realms, and, in some cases, language disorders may in fact develop into learning disabilities. That is, learning disabilities may be simply the maturation of early SDLDs, and children with learning disability are known frequently to have histories of slow language growth during the preschool period (Catts & Kahmi, 1986). It may be the case that, whereas all children with language disabilities are so labeled in the preschool period, those with milder forms of deficit that persist into the school years and affect the acquisition of literacy skills are labeled learning disabled. Conversely, those with more severe disorders that preclude the acquisition of literacy maintain the label of SDLD. In any case, language disorders that persist beyond age 4 are very frequently associated with less than average school achievement (Schery, 1985).

OTHER DISORDERS OF COMMUNICATIVE FUNCTION

DMS-III and DSM-III-R mention several other disorders associated with communicative functioning in addition to SDLD. These include the developmental articulation disorder (315.39), cluttering, and stuttering (307.00). Articulation disorders affect the production of speech sounds and can be diagnosed independently of language disorders, although the two types of disorders frequently cooccur. DSM-III-R criteria for developmental articulation disorder are presented in Table 7.3.

Cluttering and stuttering, like the other speech and language disorders, primarily affect communicative function, although they are not classified with the others in specific developmental disorders in DSM-III-R. Instead they appear in a separate section under "Speech Disorders Not Otherwise Classified." DSM-III-R diagnostic criteria for these disorders are presented in Tables 7.4 and 7.5.

TABLE 7.3
Developmental Articulation Disorder

A. Consistent failure to use developmentally expected speech sounds. For example, in a 3-year-old, failure to articulate *p*, *b*, and *t*, and in a 6-year-old, failure to articulate *r*, *sh*, *th*, *f*, *z*, and *l*.
B. Not due to a pervasive developmental disorder, mental retardation, defect in hearing acuity, disorders of the oral speech mechanism, or a neurologic disorder.

TABLE 7.4
Stuttering

Frequent repetitions or prolongations of sounds or syllables that markedly impair the fluency of speech.

In summary, modern definitions of SDLD are structured by exclusion, that is, by describing what SDLD is not. These definitions often fail to capture the severity of the disorder. The multitude of terms used to refer to SDLD in the literature and the difficulty of distinguishing it from other disorders affecting communication, such as autism, have caused problems for both clinicians and researchers. Despite being called specific, SDLDs are very frequently associated with a variety of deficits apart from communication including poor social skills, hyperactivity, and school learning problems. These associated difficulties are not fully addressed in DSM-III-R. Similarly, the relations among SDLDs and other communicative disorders, such as articulation disorders, stuttering, and cluttering are not well specified.

ALTERNATIVE DIAGNOSTIC SCHEMES

As should be evident from the foregoing discussion, the DSM-III-R diagnostic criteria for specific developmental language disorders, although an improvement over the "pre-Benton" system of treating them as analogous to adult aphasias, has not taken advantage of all the information that has come to bear on diagnostic issues.

First, and most obviously, DSM-III-R appears to make a spurious distinction between SDLDs and "other speech disorders," such as cluttering and stuttering. Stuttering could be called a *speech* disorder, affecting only the output of linguistic communication, not the entire process, which is referred to as *language*. However, DSM-III-R places articulation disorders, which also affect only speech and not language within the SDLD section. Cluttering, stuttering, articulation, and lan-

TABLE 7.5
Cluttering

A disorder of speech fluency involving both the rate and the rhythm of speech and resulting in impaired speech intelligibility. Speech is erratic and dysrhythmic, consisting of rapid and jerky spurts that usually involve faulty phrasing patterns (e.g., alternating pauses and bursts of speech that produce groups of words unrelated to the grammatical structure of the sentence).

guage disorders are most frequently diagnosed and treated by speech-language pathologists; thus, it makes sense to include these under one rubric—"Disorders Affecting Communicative Function." Such a change would allow for uniform coding of these disorders that are usually managed by one group of practitioners, speech-language pathologists, and would stress their common path of expression.

Second, the continuity of SDLDs with learning disabilities has been well established in the literature (Catts & Kahmi, 1986; Johnson, 1988; Maxwell & Wallach, 1984; Tallal, 1988; Weiner, 1985). This fact should be reflected in DSM criteria. One way to approach this problem would be to have the academic disorders (i.e., arithmetic, reading, and writing) follow, rather than precede, the Disorders Affecting Communicative Function and to emphasize under "Associated Features" of SDLDs that they often develop into learning disabilities rather than merely stating that LDs can be concurrent diagnoses, which implies they have independent etiologies. Not to do so could impede the progress in research on both types of disorders, if they are treated as disjunctive when in fact they are intimately related. Table 7.6 provides a listing of proposed criteria addressing these concerns.

Tallal (1988) argued for further modification within the SDLD category, based not on the exclusionary criteria developed by Benton but rather on inclusionary criteria derived from current research. In her 1988 extensive summary of her recent longitudinal research, she contended that neuropsychological deficits, such as the inability of language impaired children to discriminate, sequence, and remember rapidly presented stimuli, form the basis for the classically recognized subgroups of language disorders: expressive and receptive types. This research also suggests that other perceptual and motor functions, including crossmodal integration, tactile discrimination, visual integration, and double spontaneous tactile stimulation accurately distinguish language impaired from normal children without recourse to higher-level linguistic functions or use of standardized tests. Further, she argued that reading disabled children show the same types of perceptual and motor deficits, whereas articulation disordered youngsters do not. She asserted that these perceptual deficits are not restricted to the auditory mode but impact on the processing of any rapidly presented stimuli, particularly those involving spoken and written language, and crossmodal integration of information. Tallal contended that her research indicates that exclusionary diagnostic systems such as that used by DSM-III-R should be replaced by inclusionary systems that focus on the neuropsychological features she claimed are known to characterize language impaired children and that stress the continuities between language and learning disability.

TABLE 7.6
Proposed Descriptions and Criteria for Disorders of Communicative Function (to precede "Academic Skills Disorders")

Disorders Affecting Speech Alone

Stuttering
Definitions and criteria to remain the same as in *DSM-III-R*.
Articulation Disorder
Definitions and criteria to remain the same as in *DSM-III-R*.

Disorders Affecting Speech and Language

Cluttering
Definitions and criteria to remain the same as in *DSM-III-R* except the initial definition and "Diagnostic Criteria," which would be amended as follows: "Cluttering is a disorder of language fluency involving a disability in formulating sentences resulting in impairment in the rate, rhythm, and intelligibility of speech. Speech is erratic and dysrhythmic, consisting of rapid and jerky spurts that usually involve faulty phrasing patterns (e.g., alternating pauses and bursts of speech that produce groups of words unrelated to the grammatical structure of the sentence), giving the impression of hurried and confused production and a difficulty in expressing ideas in words. Reading and writing are almost always affected."

Developmental Language Disorder
Expressive Type:
Associated Features:
Developmental articulation disorder is often present. When DLD-E persists to school age, academic learning difficulties, particularly reading and writing disorders, almost always result. When the DLD–E at school age is subtle, reading disorder may become the primary diagnosis. A history of delay in reaching some motor milestones, developmental coordination disorder, and attention deficit disorder with Hyperactivity are not uncommon. Emotional problems, social withdrawal or immaturity, and behavioral difficulties may be present.
Age at Onset:
May first be recognized at 18–24 months when speech fails to appear. Moderate to severe forms usually recognized by age 3. Less severe forms may not be recognized until school age. When this is the case, a diagnosis of reading disorder often accompanies the DLD–E diagnosis or, if the DLD–E is subtle enough, replaces it.
Course, Prevalence, Familial Pattern, Differential Diagnosis:
Same as *DSM-III-R*.

Diagnostic Criteria:
A. The child's "expressive language quotient"—the average or median of several age-equivalent scores obtained from a series of well-standardized measures of expressive language (ELA) divided by the child's mental age (MA) as measured by a well-standardized, individually administered nonverbal intelligence test (ELA/MA)—is less than 70.
B. Same as *DSM-III-R*.
C. Same as *DSM-III-R*.

(*continued*)

TABLE 7.6
(Continued)

Disorders Affecting Speech and Language

Global Type:

This disorder affects the development of receptive language, which in turn affects expressive ability, because a child in the language-acquisition period cannot learn to talk without having learned to understand language. Thus, this diagnosis subsumes developmental language disorder–expressive type. The disorder is not explainable by mental retardation or inadequate schooling and is not due to a pervasive developmental disorder, hearing impairment, or neurological disorder. The diagnosis is made only if the impairment significantly intereferes with academic achievement or with activities of daily living that require the use of verbal (or sign) language.

The comprehension deficit varies depending on the severity of the disorder and the age of the child. In mild cases, there may be only difficulties in understanding particular types of words (e.g., spatial terms) or statements (e.g., complex "if–then" sentences). In more severe cases, there may be multiple disabilities, including an inability to understand basic vocabulary or simple sentences, and deficits in various areas of auditory processing (e.g., discrimination of sounds, association of sounds and symbols, storage, recall, and sequencing). Severity of the expressive deficit associated with this disorder may also vary.

Associated Features:

Developmental articulation disorder and attention deficit disorder with hyperactivity are very often present. When DLD–G at school age is sublte, reading disorder may become the primary diagnosis. Perceptual disorders affecting auditory, visual, and tactile channels, as well as difficulties in crossmodal integration and performance of certain motor tasks, are very often detected when specialized testing is undertaken. Functional enuresis, developmental coordination disorder, EEG abnormalities, and other social and behavioral problems are less commonly present.

Age at Onset:

The disorder is typically recognized by age 4. Severe forms are apparent by age 2; mild forms of the disorder, however, may not be evident until the child is in school, when language demands become more complex. When this is the case, a diagnosis of reading disorder often accompanies the DLD–G diagnosis or, if the DLD–G is subtle enough, replaces it.

Course, Prevalence, Familial Pattern, Differential Diagnosis:

Same as *DSM-III-R*.

Diagnostic Criteria:

A. The child's "expressive and receptive language quotients"—the average or median of several age-equivalent scores obtained from a series of well-standardized measures of expressive (ELA) and receptive language (RLA), each divided by the child's mental age (MA) as measured by a well-standardized, individually administered nonverbal intelligence test (ELA/MA; RLA/MA)—are each less than 70.

B. Same as *DSM-III-R*.

C. Same as *DSM-III-R*.

Other investigators (Aram, 1988; Johnson, 1988; Rapin, 1988) advocated a more cautious approach to changing diagnostic approaches. First, Rapin (1988) pointed to a failure to replicate some of Tallal's findings on perceptual disorders in language impaired children and questioned whether the disorders that Tallal explicated are universal in language impaired children. Rapin (1988) also pointed out that language disorders have multiple etiologies, some genetic and some having to do with prenatal and perinatal trauma such as prematurity and anoxia. Given the multiple pathways to the disorder, it is possible that the pathophysiology is also diverse. Second, Aram (1988) pointed out the difficulty of making an inclusion-based definition because of the lack of psychometrically adequate instruments for assessing children in the areas Tallal advocated. Third, the lack of adequate natural history information, validating different outcomes for children with different specific symptom profiles, makes it difficult to predict which patients will or will not cross diagnostic boundaries in the course of development. Although the current exclusionary type of definition is unsatisfactory in many ways, it would appear that it is premature to depart from it entirely until further empirical evidence on inclusionary criteria has been accumulated. However, *DSM* criteria should take advantage of progress in research by listing the frequently reported perceptual, cognitive, and neurological correlates of these disorders under the appropriate Associated Features categories.

One further area in which *DSM* criteria could be improved is in the specification of severity necessary to obtain the diagnosis. DSM-III-R offers no guidance on this point, and, like other points, it is a matter of controversy in the field. The earliest studies of language development (Brown, 1973) showed that there is great individual variation in the rate of acquisition of language; therefore, it is often quite difficult to determine if a preschooler falls outside the normal range or is simply at the lower end of it. Fey (1986) discussed the difficulties of establishing what constitutes a "significant delay" in language development in children. Scores of more than two standard deviations below the norm or those below the tenth percentile on a standardized language measure are often used to determine significant deficit. However, the lack of properly standardized assessment instruments in the field, as well as the fact that children may perform differently on different tests or have deficits in one aspect of language and not in another, limits the utility of such a scheme. Stark and Tallal (1981) proposed developing a Language Quotient based on averaging age-equivalent scores from several norm-referenced instruments and dividing the average language age by chronological age to derive a quotient similar to IQ. Miller (1981) argued that this ratio comparison should not include chronological age but rather mental age as tested by a nonverbal, intellectual assessment instrument. In this way, it is possible to restrict the diagnosis of SDLD to only those children who have language deficits over and above any that would be predicted by subnormal cognitive function. This quotient could be derived separately for expressive and receptive areas, and a diagnosis of SDLD–expressive type given, if the quotient falls below 70 on the expressive portion only. Similarly, a diagnosis of SDLD–global type would be awarded if the quotient for both expressive and receptive areas were below 70. Another alternative would be to use the median age-equivalent score from the verbal and nonverbal measures, because this would obviate some of the problems in applying ratio level statistics, such as the mean, to ordinal level data, such as age equivalents. Both these schemes would take into account the differences in rates of acquisition at various age levels. For example, it would not make sense to call a 7-year-old with a 6-month delay SDLD, whereas it would make sense to diagnose a 14-month-old who was 6 months behind.

In summary, it has been suggested that DSM-III-R, although an improvement over older classification systems, retains some weakness. These include a spurious distinction between various types of speech and language disorders, a failure to take into account the frequent continuity between language and academic learning disorders, and a lack of current data on deficits often associated with SDLDs. An alternative diagnostic scheme has been proposed that attempts to address some of these weaknesses.

ASSESSMENT PROCEDURES

Table 7.7 presents a sample of some of the assessment instruments used most frequently in diagnosing SDLDs and related syndromes. In evaluating a patient for disorders of communicative function, five aspects of language are generally considered.

Phonology refers to the production of speech sounds and the rules for combining them in words. Traditionally, phonology has been conceived in terms of articulation skills and has been assessed by asking children to name pictures of words chosen for their inclusion of a range of target sounds in various word positions (e.g., Goldman & Fristoe, 1969). More contemporary methods look at the rules the child uses to simplify difficult sounds in words and sometimes employ analysis of conversational speech (Shriberg & Kwiatkowski, 1981). Phonology is usually evaluated only in the expressive modality, although tests of auditory perception and discrimination of speech sounds are some-

TABLE 7.7
Language Tests

		Area Assessed:						
		Expressive:			*Receptive:*			
Test	*Age Range*	*Phon.*	*Sem.*	*Syn.*	*Phon.*	*Sem.*	*Syn.*	*Pragmatics*
Arizona Articulation Proficiency Scale (Fudula, 1974)	3–11	•						
Assessment of Children's Language Comprehension (Foster, Giddan, & Stark, 1973)	3–7					•	•	
Boehm Test of Basic Concepts (Boehm, 1971)	5–7					•		
Carrow Elicited Language Inventory (Carrow, 1974)	3–8			•				
Clinical Evaluation of Language Functions (Semel & Wiig, 1980)	5–18	•	•	•	•	•	•	
Deep Test of Articulation (McDonald, 1964)	all	•						
Developmental Articulation Test (Hejna, 1968)	all	•						
Developmental Sentence Structures (Lee, 1974)	2–7			•				
Environmental Language Inventory (McDonald, 1978)	1–3		•	•				

(Continued)

TABLE 7.7
(Continued)

		Area Assessed:						
		Expressive:			Receptive:			
Test	*Age Range*	*Phon.*	*Sem.*	*Syn.*	*Phon.*	*Sem.*	*Syn.*	*Pragmatics*
Evaluating Communicative Competence (Simon, 1984)	5–18		•	•		•	•	
Expressive One-Word Picture Vocabulary Test (Gardner,1979)	3–12		•					
Fisher–Logemann Test of Articulation Competence (Fisher & Logemann, 1971)	all	•						
Fullerton Language Test for Adolescents (Thorum, 1980)	11–18		•	•		•	•	•
Goldman–Fristoe Test of Articulation (Goldman & Fristoe, 1969)	6–18	•						
Illinois Test of Psycholinguistic Abilities (Kirk, McCarthy, & Kirk, 1968)	2–10		•	•	•	•	•	
Let's Talk Inventory for Adolescents (Wiig, 1982)	9–adult							•
Northwestern Screening Syntax Test (Lee, 1971)	3–8			•			•	

Oral Language Sentence Imitation Test (Zachman, Husingh, Jorgensen, & Barrett, 1977)	3–adult			•			
Photo Articulation Test (Pendergast, Dickey, Selmar, & Sodor, 1969)	3–12	•					
Preschool Language Assessment Instrument (Blank, Rose, & Berlin, 1978)	3–12		•	•		•	•
Preschool Language Scale (Zimmerman, Steiner, & Evatt, 1969)	1–7		•	•		•	•
Receptive-Expressive Emergent Language Scale (Bzoch & League, 1970)	0–3		•	•		•	•
Sequenced Inventory of Communicative Development (Hedrick, Prather, & Tobin, 1975)	0–4		•	•		•	•
Templin–Darley Tests of Articulation (Templin & Darley, 1969)	3–8	•					
Test of Auditory Comprehension of Language–Revised (Carrow-Woolfolk, 1985)	3–7					•	•

(Continued)

TABLE 7.7
(Continued)

		Area Assessed:						
		Expressive:			*Receptive:*			
Test	*Age Range*	*Phon.*	*Sem.*	*Syn.*	*Phon.*	*Sem.*	*Syn.*	*Pragmatics*
Test of Adolescent Language (Hammill, Brown, Larsen, &Wiederholt, 1980)	11–18		•	•		•	•	
Test of Early Language Development (Hresko, Reid, & Hammill, 1981)	3–8		•	•		•	•	
Test of Language Development–Primary (Newcomer & Hammill, 1988)	4–8	•	•	•	•	•	•	
Test of Language Development–Intermediate (Newcomer & Hammill, 1984)	8–13		•	•		•	•	
Token Test for Children (DiSimoni, 1978)	3–12					•	•	
Vocabulary Comprehension Scale (Bangs, 1975)	2–6					•		
Peabody Picture Vocabulary Test (Dunn & Dunn, 1981)	2–adult					•		

times given. The predictive validity of these tests has been questioned, however.

Semantics is the aspect of language that has to do with the meanings encoded in words and sentences. It is most commonly evaluated by means of vocabulary tests. These can examine both expressive (naming) and receptive (picture identification) skills. Other areas of semantics, such as word definitions, recognition of absurdities, analogies, and relationships among words, are frequently included in instruments that examine a variety of linguistic skills through a battery of subtests.

Syntax refers to the ability to construct and understand grammatical sentences. Receptive syntax is often tested in picture-pointing formats by reading a sentence and asking the child to point to which of two or three pictures correctly depict its content. Expressive syntax, the ability to formulate grammatical sentences, has traditionally been evaluated by asking children to repeat or imitate sentences read to them. Recent research calls this technique into question, however. For example, it has been shown that the ability to imitate a sentence is not the same as the ability to formulate it in conversational speech and that children make different mistakes in imitation than they do when they produce their own sentences spontaneously (Hood & Lightbown, 1978; Prutting, Gallagher, & Mulac, 1975). Contemporary practice (Miller, 1981; Lahey, 1988) advocates collecting samples of spontaneous speech from children in naturalistic settings and analyzing them, using one of the procedures available in the literature (Tyack & Gotsleban, 1974; Lee, 1974; Miller & Chapman, 1985), for sentence length and syntactic maturity. Speech samples collected in this way can also be used to look at the other aspects of language development and provide the most ecologically valid sample of the child's language performance.

Morphology includes the rules for the formation of words by the addition of units of meaning, such as inflections. The ability to use plurals, possessives, past tenses, and so on relies on morphological knowledge. This skill is generally tested expressively by asking children to add morphemes to nonsense words ("This is a gleep. Now there are two of them. There are two ______.") or to real ones in sentence completion tasks ("Martha had a dress. Now she has two of them. She has two ______."). Testing of receptive morphological knowledge is frequently included within receptive syntax tests.

Pragmatics is the use of language in the social context to achieve interpersonal goals. Sentences in real conversations are not generated simply to use syntactic rules, pronounce sounds, or add morphemes but to express an intention on the part of the speaker or to have an effect on the listener. This ability to get a message across can be evaluated independent of the form of the message. Some children with relatively poor vocabulary and sentence structures can nevertheless be quite efficient in communicating their meaning using prosodic features and gestures, for example, in combination with their limited verbal repertoires. On the other hand, some children with comparatively mild disorders of linguistic skill use the language at their disposal very inefficiently or simply produce little spontaneous verbal output. Thus, children may have different profiles of pragmatic skill that have little to do with their other levels of linguistic functioning. There are few standardized measures of expressive or receptive pragmatic ability. Again, observation of language use in a social context is generally the best way to evaluate this domain. Bedrosian (1985), Chapman (1981), Lund and Duchan (1983), Paul (1987), and Prutting and Kirchner (1983) presented methods for performing these evaluations.

Collateral areas. In addition to examining language performance itself, the assessment of children suspected of showing SDLDs must include evaluations that will rule out other specifiable causes for the disorder. Hearing must be assessed by a trained audiologist in order to preclude hearing impairment as a primary cause of the disorder. Nonverbal intelligence should be established by a psychologist, using a procedure that allows a performance score to be calculated to avoid penalizing the child for his language difficulties. Only children whose language levels fall significantly below their nonverbal developmental level can receive the diagnosis of SDLD. Autism and emotional disturbance as well as elective mutism should be investigated as possible diagnoses by a child psychiatrist or other experienced clinician, using appropriate interview schedules and observations.

In addition, assessment of SDLD children should examine areas in which these patients frequently show associated deficits. Assessment for ADD with hyperactivity should be part of the diagnostic workup. Other adaptive functions as well as maladaptive behaviors can be examined by interviewing the primary caregiver with the *Vineland Adaptive Behavior Scales* (Sparrow et al., 1984). The child should also be examined for auditory perceptual disorders, crossmodal integration deficits, and the other neuropsychological difficulties Tallal (1988) suggested. Dyslexia and other learning difficulties must be evaluated in language impaired children of school age using standardized batteries designed for this purpose.

In summary, the assessment of children with SDLD requires the participation of a multidisciplinary team both to establish that the primary deficit is in selected language functions and that it is not secondary to some other disorder. All associated areas of deficit should also be identified. Although language disorders in children are called specific, which could imply limited impact, their actual effect on the child's development in a variety of areas can be profound.

A typical assessment battery for a 4-year-old child suspected of having SDLD might include the following:

1. Assessment of receptive vocabulary, such as the *Peabody Picture Vocabulary Test* (Dunn & Dunn, 1981).
2. Assessment of receptive syntax and morphology, such as the *Test of Auditory Comprehension of Language* (Carrow-Woolfolk, 1985).
3. Assessment of articulation, such as the *Goldman–Fristoe Test of Articulation* (Goldman & Fristoe, 1969), or phonology, such as the *Natural Process Analysis* (Shriberg & Kwiatkowski, 1981).
4. Assessment of language production in free speech, such as the *Developmental Sentence Score* (Lee, 1974), or a computerized analysis, such as the *System for Analyzing Language Transcripts* (Miller & Chapman, 1985).
5. Observational rating of pragmatic skills, such as that given by Prutting and Kirchner (1983).
6. Assessment of nonverbal intelligence, such as the *Leiter International Performance Scale* (Arthur, 1952).
7. Assessment of social skill, such as the *Vineland Adaptive Behavior Scale* (Sparrow, Balla, & Chiccetti, 1984).
8. Assessment of auditory, visual, and tactile deficits following Tallal (1988).

CONCLUSIONS

Specific developmental language disorders have been recognized for over 100 years as a serious disability of childhood. Although classification schemes have changed over the years, the difficulty of clearly defining the disorder and distinguishing it from similar syndromes affecting the developing child persists. DSM-III-R has made some progress in diagnostic issues, but problems such as the relation among the various types of speech and language disorders, the role of associated deficits, and the relation to school learning problems remain to be addressed in a nosological scheme.

The assessment of the child with a language impairment requires an intensive multidisciplinary evaluation in order to identify all aspects of deficits and identify channels of strength to be used for remediation. Only comprehensive and careful diagnostic procedures will lead to fuller understanding of these serious disorders and eventually, it is hoped, to a brighter prognosis than they currently enjoy.

REFERENCES

American Psychiatric Association. (1980). *Diagnostic and statistical manual of mental disorders* (3rd ed.). Washington, DC: Author.

American Psychiatric Association. (1987). *Diagnostic and statistical manual of mental disorders* (3rd ed., rev.). Washington, DC: Author.

Aram, D. M. (1988). Discussion. In J. F. Kavanagh & T. J. Truss, Jr. (Eds.), *Learning disabilities: Proceedings of the national conference* (pp. 285–289). Parkton, Md: York.

Arthur, G. (1952). *Arthur Adaptation of the Leiter International Performance Scale*. Washington, DC: Psychological Services Center.

Baker, L., Cantwell, D. P., Mattison, R. E. (1980). Behavior problems in children with pure speech disorders and in children with combined speech and language disorder. *Journal of Abnormal Psychology, 8*, 245–256.

Bangs, T. E. (1975). *Vocabulary Comprehension Scale*. Boston: Teaching Resources.

Bedrosian, J. L. (1985). An approach to developing conversational competence. In D. N. Ripich & F. M. Spinelli (Eds.), *School discourse problems* (pp. 231–255). San Diego, CA: College-Hill.

Benton, A. L. (1964). Developmental aphasia and brain damage. *Cortex, 1*, 40–52.

Blank, M., Rose, S. A., & Berlin, L. J. (1978). *Preschool Language Assessment Instrument: The language of learning in practice*. New York: Grune & Stratton.

Bloom, L., & Lahey, M. (1978). *Language development and language disorders*. New York: Wiley.

Boehm, A. (1971). *Boehm Test of Basic Concepts*. New York: Psychological Corporation.

Broadbent, W. (1872). Cerebral mechanisms of speech and thought. *Medico-Chirurgical Transactions, 55*, 145.

Brown, R. (1973). *A first language*. Cambridge, MA: Harvard University Press.

Brown, L., Sherbenou R. J., & Dollar, S. J. (1982). *Test of Nonverbal Intelligence*. Austin, TX: Pro-ed.

Bzoch, D. R., & League, R. L. (1970). *Receptive-Expressive Emergent Language Scale: For the measurement of language skills in infancy*. Baltimore, MD: University Park Press.

Cantwell, D. P., Baker, L., & Mattison, R. E. (1980). Factors associated with the development of psychiatric disorder in children with speech and language retardation. *Archives of General Psychiatry, 37*, 423–426.

Caparulo, B. K., & Cohen, D. J. (1977). Cognitive structures, language, and emerging social competence in autistic and aphasic children. *Journal of the American Academy of Child Psychiatry, 16*, 620–645.

Caparulo, B. K., Cohen, D. J., & Rothman, S. (1981). Computed tomographic brain scanning in children with developmental neuropsychiatric disorders. *Journal of the American Academy of Child Psychiatry, 20*, 338–357.

Carrow, E. (1974). *Carrow Elicited Language Inventory*. Austin, TX: Learning Concepts.

Carrow-Woolfolk, E. (1985). *Test for Auditory Comprehension of Language*. Allen, TX: Teaching Resources.

Catts, H., & Kahmi, A. (1986). The linguistic basis of reading disorders: Implications for the Speech-Language Pathologist. *Language, Speech and Hearing Services in Schools, 17*, 329–341.

Chapman, R. S. (1981). Exploring children's communicative intents. In J. Miller (Ed.), *Assessing language production in children: Experimental procedures* (pp. 111–136). Baltimore, MD: University Park Press.

Cohen, D. J., Paul, R., & Volkmar, F. R. (1986). Issues in the classification of pervasive and other developmental disorders: toward DSM-IV. *Journal of the American Academy of Child Psychiatry, 25*, 213–220.

Culp, R. G., Heide, J., & Richardson, M. T. (1987). Maltreated children's developmental scores: treatment versus nontreatment. *Child Abuse and Neglect, 2,* 29–34.

de Ajuriaguerra, J., Jaeggi, A., Guignard, F., Kocher, F., Maquard, M., Roth, S., & Schmid, E. (1976). The development and prognosis of dysphasia in children. In D. Morehead & A. Morehead (Eds.), *Normal and deficient child language* (pp. 345–386). Baltimore, MD: University Park Press.

De Hirsch, K. (1967). Differential diagnosis between aphasic and schizophrenic language in children. *Journal of Speech and Hearing Disorders, 32,* 3–10.

DiSimoni, R. (1978). *The Token Test for Children.* Boston: Teaching Resources Corporation.

Dunn, L., & Dunn, L. (1981). *Peabody Picture Vocabulary Test–Revised.* Circle Pines, MN: American Guidance Service.

Eisenson, J. (1972). *Aphasia in children.* New York: Harper & Row.

Fey, M. (1986). *Language intervention with young children.* San Diego: CA: College-Hill.

Fisher, H., & Logemann, J. (1971). *The Fisher–Logemann Test of Articulation Competence.* Boston: Houghton Mifflin.

Forest, T., Eisenson, J., & Stark, J. (1966). EEG findings in 113 non-verbal children. Abstract. *Electroencephalogy and Clinical Neurophysiology, 22,* 291.

Foster, R., Giddan, J. J., & Stark, J. (1973). *Assessment of children's language comprehension.* Palo Alto, CA: Counseling Psychologists Press.

Fudala, J. B. (1974). *Arizona Articulation Proficiency Scale.* Los Angeles, CA: Western Psychological Services.

Gall, F. (1825). On the function of the brain and each of its parts (Vols. 1–6). *Phrenological Library.* Boston: March, Capen, & Lyon.

Gardner, M. F. (1979). *Expressive One-Word Picture Vocabulary Test.* Novato, CA: Academic Therapy.

Goldman, R., & Fristoe, M. (1969). *Goldman–Fristoe Test of Articulation.* Circle Pines, MN: American Guidance Service.

Goldstein, R., Landau, W., & Kleffner, F. (1960). Neurologic observations on a population of deaf and aphasic children. *Annals of Otology, Rhinology and Laryngology, 69,* 756–767.

Hammill, D. D., Brown, V. L., Larsen, S. C., & Wiederholt, J. L. (1980). *Test of Adolescent Language.* Austin, TX: Pro-ed.

Harcherik, D. F., Cohen, D. J., Ort, S., Paul, R., Shaywitz, B. A., Volkmar, F. R., Rothman, S. L. G., & Leckman, J. F. (1985). Computed tomographic brain scanning in four neuro-psychiatric disorders of childhood. *American Journal of Psychiatry, 142,* 731–734.

Hedrick, D. L., Prather, E. M., and Tobin, A. R. (1975). *Sequenced Inventory of Communication Development.* Seattle, WA: University of Washington Press.

Hejna, R. F. (1968). *Developmental Articulation Test.* Ann Arbor, MI: Speech Materials.

Hiskey, M. S. (1966). *Hiskey–Nebraska Test of Learning Aptitude.* Lincoln, NE: Union College Press.

Hresko, W. P., Reid, D. K., & Hammill, D. D. (1981). *Test of Early Language Development.* Austin, TX: Pro-ed.

Hood, L., & Lightbown, P. (1978). What do children do when asked to "Say what I say": Does elicited imitation measure linguistic knowledge? *Allied Health, 1,* 195–220.

Johnson, D. J. (1988). Review of research on specific reading, writing, and mathematics disorders. In J. F. Kavanagh & T. J. Truss, Jr. (Eds.), *Learning disabilities: Proceedings of the National Conference* (pp. 79–163). Parkton, MD: York Press.

Johnston, J. R. (1982). The language disordered child. In N. J. Lass, L. V. McReynolds, J. L. Northern, & D. E. Yoder (Eds.), *Speech, language, and hearing volume II: Pathologies of speech, language, and hearing* (pp. 780–801). Philadelphia, PA: Saunders.

Kahmi, A. G. (1981). Nonlinguistic symbolic and conceptual abilities of language-impaired and normally developing children. *Journal of Speech and Hearing Research, 24,* 446–453.

Karlin, I. (1954). Aphasia in children. *American Journal Disorders of Childhood, 87,* 752–767.

Kirk, S. A., McCarthy, J. J., & Kirk, W. D. (1968). *The Illinois Test Of Psycholinguistic Abilities* (rev. ed.). Urbana: University Of Illinois Press.

Lahey, M. (1988). *Language Disorders and Language Development.* New York: Macmillan.

Lee, L. (1971). *Northwestern Syntax Screening Test.* Evanston, IL: Northwestern University Press.

Lee, L. (1974). *Developmental Sentence Analysis.* Evanston, IL: Northwestern University Press.

Lennenberg, E. (1967). *The biological foundations of language.* New York: Wiley.

Leonard, L. B. (1972). What is deviant language? *Journal of Speech and Hearing Disorders, 37,* 427–446.

Lund, N. J., & Duchan, J. F. (1983). *Assessing children's language in naturalistic contexts.* Englewood Cliffs, NJ: Prentice-Hall.

Maxwell, W., & Wallach, G. (1984). The language learning disabilities connection: Symptoms of early language disability change over time. In G. Wallach & K. Butler (Eds.), *Language and learning disabilities in school-aged children* (pp. 15–33). Baltimore, MD: Williams & Wilkins.

McCall, E. (1911). Two cases of congenital aphasia in children. *British Medical Journal, 1105,* 1407.

McDonald, E. E. (1964). *A Deep Test of Articulation.* Pittsburgh: Stanwix House.

McDonald, J. D. (1978). *Environmental Language Inventory.* Columbus, OH: Merrill.

Miller, J. F. (1981). *Assessing Language Production in Children.* Baltimore: University Park Press.

Miller, J. F., & Chapman, R. S. (1985). *System for Analyzing Language Transcripts.* Madison, WI: University of Wisconsin.

Morehead, D., & Ingram, D. (1973). The development of base syntax in normal and linguistically deviant children. *Journal of Speech and Hearing Research, 16,* 330–352.

Myklebust, H. (1954). *Auditory disorders in children.* New York: Grune & Stratton.

Myklebust, H. (1971). Childhood aphasia: An evolving concept. In L. Travis (Ed.), *Handbook of speech pathology and audiology* (pp. 1181–1202). Englewood Cliffs, NJ: Grune & Stratton.

Naremore, R. (1980). Language variation in a multi-cultural society. In T. J. Hixon, L. D. Schriberg, & J. H. Saxman (Eds.), *Introduction to communication disorders* (pp. 177–216). Englewood Cliffs, NJ: Prentice-Hall.

Newcomer, P. L., & Hammill, D. D. (1982). *Test of Language Development–Intermediate.* Austin, TX: Pro-ed.

Newcomer, P. L., & Hammill, D. D. (1988). *Test of Language Development-Primary.* Austin, TX: Pro-ed.

Orton, S. (1937). *Reading, writing and speech problems in children.* New York: Norton.

Paul, R. (1987). A model for the assessment of communication disorders in infants and toddlers. *NSSLHA Journal, 15,* 88–105.

Paul, R., & Cohen, D. J. (1982). Communication development and its disorders. *Schizophrenia Bulletin, 3,* 279–294.

Paul, R., & Cohen, D. J. (1984). Outcomes of severe disorders of language acquisition. *Journal of Autism and Developmental Disorders, 14,* 405–421.

Paul, R., Cohen, D. J., & Caparulo, B. K. (1983). A longitudinal study of patients with severe developmental disorders of language learning. *Journal of the American Academy of Child Psychiatry, 22,* 525–534.

Paul, R., Spangle-Looney, S., & Dahm, P. S. (in press). Communication and socialization in "late talking" toddlers. *Journal of Speech and Hearing Research.*

Pendergast, K., Dickey, S., Selmar, J., & Soder, A. (1969). *Photo Articulation Test.* Danville, IL: Interstate Press.

Prutting, D. A., & Kirchner, D. M. (1983). Applied pragmatics. In T. Gallagher & C. Prutting (Eds.), *Pragmatic assessment and intervention issues in language* (pp. 29–64). San Diego: College-Hill.

Prutting, D., Gallagher, T., & Mulac, A. (1975). The expressive portion of the N.S.S.T. compared to a spontaneous language sample. *Journal of Speech and Hearing Disorders, 40,* 40–49.

Rapin, I. (1988). Discussion. In J. F. Kavanagh & T. J. Truss, Jr. (Eds.), *Learning disabilities: Proceedings of the National Conference* (pp. 273–280). Parkton, MD: York.

Rosenthal, W. (1972). Auditory and linguistic interaction in developmental aphasia: Evidence from two studies of auditory processing. *Papers and Reports in Child Language Development, 4.* Palo Alto, CA: Stanford University.

Satterfield, J. (1973). EEG issues in children with minimal brain dysfunction. *Seminars in Psychiatry, 5,* 35–46.

Schery, T. (1985). Correlates of language development in language disordered children. *Journal of Speech and Hearing Disorders, 50,* 73–83.

Schiff, N. B. (1979). The influence of deviant maternal input on the development of language during the preschool years. *Journal of Speech and Hearing Research, 22,* 581–603.

Semel, E., & Wiig, E. (1980). *Clinical Evaluation of Language Functions.* Columbus, OH: Merrill.

Skuse, D. (1984). Extreme deprivation in early childhood-II: Theoretical issues and a comparative review. *Journal of Child Psychology and Psychiatry, 25,* 543–572.

Shriberg, L. D., & Kwiatkowski, J. (1981). *Natural Process Analyses of Continuous Speech Samples.* New York: Wiley.

Simon, C. (1984). *Evaluating communicative competence.* Tucson, AZ: Communication Skill Builders.

Sparrow, S., Balla, D. A., & Chiccetti, D. V. (1984). *Vineland Adaptive Behavior Scales.* Circle Pines, MN: American Guidance Service.

Springer, S., & Eisenson, J. (1977). Hemispheric specialization for speech in language-disordered children. *Neuropsychologia, 15,* 287–293.

Stark, R. E., & Tallal, P. (1981). Selection of children with specific language deficits. *Journal of Speech and Hearing Disorders, 46,* 114–122.

Tallal, P. (1988). Developmental language disorders. In J. F. Kavanagh & T. J. Truss, Jr. (Eds.), *Learning disabilities: Proceedings of the National Conference* (pp. 181–272). Parkton, MD: York.

Tallal, P., & Piercy, M. (1975). Developmental aphasia: The perception of brief vowels and extended stop consonants. *Neuropsychologia, 13,* 69–74.

Templin, M., & Darley, F. (1969). *The Templin–Darley Tests of Articulation.* Iowa City: Bureau of Educational Research and Services, University of Iowa.

Thorum, A. R. (1980). *The Fullerton Language Test for Adolescents.* Palo Alto, CA: Consulting Psychologists Press.

Tyack, D., & Gotsleban, R. (1974). *Language sampling, analysis and training.* Palo Alto, CA: Consulting Psychologists Press.

Volkmar, F. R. (1987). Social development. In D. S. Cohen, A. M. Donnellan, & R. Paul (Eds.), *Handbook of autism and pervasive developmental disorders* (pp. 41–60). New York: Wiley.

Wechsler, D. (1971). *Wechsler Pre-school and Primary Scale of Intelligence–Revised.* New York: Psychological Corporation.

Wechsler, D. (1974). *Wechsler Intelligence Scale for Children–Revised.* New York: Psychological Corporation.

Weiner, P. (1985). The value of follow-up studies. *Topics in Language Disorders, 5,* 78–92.

Wiig, E. (1982). *Let's talk inventory for adolescents.* Columbus, OH: Merrill.

Wing, L., & Attwood, A. (1987). Syndromes of autism and atypical development. In D. J. Cohen, A. M. Donnellan, & R. Paul (Eds.), *Handbook of autism and pervasive development disorders* (pp. 3–19). New York: Wiley.

Zachman, L., Huisingh, R., Jorgensen, C., & Barrett, M. (1977). *The Oral Language Sentence Imitation Test.* Moline, IL: Linguistic Systems.

Zimmerman, I. L., Steiner, V. G., & Evatt, R. (1969). *Preschool Language Scale.* Columbus, OH: Merrill.

CHAPTER 8

Motor Skill Disorders

Ruthmary K. Deuel
Washington University School of Medicine

As early as Orton's (1925) description of the "motor incoordinate" type of child with normal intelligence, it was recognized that abnormalities of motor performance commonly accompanied what would now be called specific reading and other academic skill disorders. It has since been frequently documented that not only children with specific learning disabilities, but also those with Attention Deficit-Hyperactivity Disorder (ADHD) and more pervasive developmental disorders, such as mental retardation and autism, may have difficulties with motor learning and motor execution (Bakwin, 1968; Barlow, 1974; Bender, 1970; Nichols & Chen, 1981), and some individuals suffer from motor incoordination without any other symptoms (Ford, 1960). The fact that these groups of children differ from their "normal" age mates on various measures of motor performance has been solidly established (Clinton & Boyce, 1975; Denckla & Rudel, 1978; Deuel, Feeley & Bonskowski, 1984; Hertzig, 1981; Keogh, 1986; Paine, Werry, & Quay, 1968; Peters, Romine, & Dykman, 1975; Strauss & Lehtinen, 1947; Wolff & Hurwitz, 1966), but whether there are specific patterns of motor learning and execution deficits that accompany specific developmental neuropsychiatric disorders and thus have significance for neuropsychiatric syndrome diagnosis remains somewhat controversial (Deuel & Robinson, 1987; Yule & Taylor, 1987). Other current, related issues are: (a) whether mild and moderate motor performance deficits contribute to the total burden of handicap in a given child and (b) when should the clinician recommend specific remediation of motor performance deficits. Future progress on answers to these clinical questions would be aided by consensus on a comprehensive categorization of the motor deficits. Such consensus would also allow results of different research studies to be compared in detail.

In addition, studies to date strongly suggest a need for new theoretical approaches to motor performance deficits. The deficits clearly do not form evidence for static focal brain damage (Rutter, 1981; Denckla, LeMay, & Chapman, 1985) but are of high incidence in neuropsychiatric disorders. This seeming paradox may be resolved by considering old and new conceptual models of brain processing as regards behaviorally measurable voluntary motor acts. Previously, motor function was thought to be the result of serial hierarchical processing within fixed neuroanatomical pathways. Each structure along the pathway was assigned an invariate "function." If the structure was destroyed, motor behavior would be invariably altered by subtraction of that structure's function. Recently, properties that suggest another type of neural processing have been documented for some higher mental functions such as laterally directed spatial attention and motor behavior (Deeke, Kornhuber, Lang, Lang, & Schreiber, 1985; Gevins et al., 1989; Wise & Desimone, 1988). In this type of processing, different networks of neuroanatomic entities are configured by different functional demands, and a given neuroanatomic entity may contribute different influences to different functions (Deuel & Collins, 1984; Grimm, 1983; Gur et al., 1983). Surprisingly clear and concrete examples of such processing are found in simple nervous systems in which a single neuron at one moment functions as a pacemaker and at another as an inhibitory "gate keeper" (Getting, 1989). If control of higher order motor behavior is conceptualized as occurring in functionally determined networks (Mesulam, 1981), then deficits in performance could result from interactions among neural systems. Such deficits would appear only during specific processing stages and/or under conditions of simultaneous heavy demand on several participating neural networks (Kinsbourne & Cook, 1971). Such a concept of dynamic interactive neural deficits is amenable to experimental test and may provide the missing link between observed motor performance deficits in children with developmental neuropsychiatric disorders but without fixed brain damage. The great majority of children with developmental motor performance deficits are indeed without fixed or relevant brain damage (Denckla et al., 1985; Gomez, 1967; Rutter, 1981, 1982, 1983).

HISTORICAL PERSPECTIVE

After Orton's (1925, 1937) description of motor performance difficulties in learning disabled children, little mention was made of such difficulties for a long time, although Annell (1949) wrote a sensitive description of a child who played hookey on gym days because of embarrassment over his motor performance. His motor performance was so poor that he could not change his clothes in time to take the gym class or change back in time to be dressed for the next academic class.

In the early 1960s, Walton (1961), Walton, Ellis, and Court (1962), and later Gubbay, Ellis, Walton and Court (1965) carefully described motor performance deficits in learning disabled children. They pointed out that these motor deficits were clearly a primary handicapping factor within the spectrum of neuropsychiatric dysfunctions of their subjects. These investigators defined a higher-order cognitive (i.e., apraxic and agnostic) component of the motor deficits they described, thus linking cognitive and motor disablities. However, separation of the apraxic and agnostic type of motor performance deficit from more elementary motoric abnormalities was not attempted at that juncture. In Ford's (1960) *Textbook of Pediatric Neurology*, a unitary entity, "congenital maladroitness," was recognized that was similar to the current DSM-III-R diagnostic category, Developmental Coordination Disorder (American Psychiatric Association, 1987), and the ICD-9-CM (World Health Organization, 1989) category, Specific Motor Retardation (315.4), although the latter now allows for a "Clumsiness" versus a "Dyspraxia" syndrome.

Throughout the 1960s and 1970s, a large number of test batteries for evaluating the motor performance deficits of childhood was developed, some of which are cited here in chronological order (Kennard, 1960; Gubbay et al., 1965; Roach & Kephart, 1966; Spreen and Gaddes, 1969; Gubbay, 1973; Hart, Rennick, & Klinge, 1974; Peters et al., 1975; Adams, Kocsis, & Estes, 1979). In part through their use, it has become obvious that motor performance deficits do not carry simple diagnostic significance of the sort conceptualized in the 1960s. At that time, using the serial hierarchical processing model, it had been considered that any motor functional deficit must be evidence of "damage" to a fixed neuroanatomical pathway. Therefore, motor "soft signs" were valued as diagnostic evidence for "brain damage or brain dysfunction" (Gomez, 1967). Prior to the 1960s, many of the developmental neuropsychiatric syndromes had been considered the result of abnormal parenting, with treatment solely with extended psychotherapy often being attempted.

Although discovery of soft signs, used as indicators of organic brain dysfunction, served to relieve the parents of the guilt of poor management and the children of ineffectual extended psychotherapy (Peters, 1987), it seems that the actual data used to support brain dysfunction is tenuous (Shaffer et al., 1985; Shafer, Shaffer, O'Connor, & Stokman, 1983). In fact, work by Rutter shows that, whereas severe brain damage is much more likely than mild brain damage to be accompanied by a variety of neuropsychiatric disturbances, the only specific neuropsychiatric symptom of severe brain damage that can be isolated as a true "sequela" (i.e., as related to the damage per se) is social disinhibition (Rutter, 1981, 1982, 1983). Therefore, the conclusion that, even if the motor performance deficits were to denote brain damage or fixed brain dysfunction, they would not have discriminating diagnostic significance for neuropsychiatric disorders is hard to avoid. Currently, many authorities would not assign any significance to neurological findings of motor performance deficits in syndrome diagnosis (Bigler, 1988), and some would even dispense with the neurological examination altogether in certain instances (Peters, 1987).

PERSPECTIVES ON DEVELOPMENTAL TRAJECTORIES IN MOTOR PERFORMANCE

The use of the term *developmental* for a deficit often is intended to indicate that what is normal at an early age becomes abnormal at a later age, or, in other words, shows a slow or arrested development of normal activity patterns. Thus, not all types of motor performance deficits to be considered are clearly developmental. For example, chorea, tics, tremor, and some forms of dyspraxia seem to be abnormal at any developmental stage. It is important to note that the term *developmental* does not necessarily imply that the child will grow out of a motor deficit eventually or is just taking longer to develop a certain capacity, although this may prove to be the case in a given individual (Knuckey & Gubbay, 1983). Sometimes the term is simply used when there is no catastrophic or even any discernable etiology for the condition (i.e., the condition is idiopathic). Sometimes *developmental* is used to describe an abnormality that appears only as lack of developmental attainment. With this variety of uses, the term *developmental,* when applied to an abnormality, boils down to one used when it appears in childhood and carries little more precision than that. To disambiguate these meanings as applied to motor performance deficits would involve a lot of work. The ideal study would include intensive longitudinal follow-up of a large number of subjects from birth through high school graduation to determine the normal sequence of development in boys and girls from a range of racial and socioeconomic conditions. The range of physical growth stages or pubertal stages at which various types of motor learning and performances start, become skilled, or even regress or become superceded would also be evaluated. This ideal data base would allow construction of empirically derived motor performance developmental trajectory charts. When more precise description and understanding of neural processes during the learning and consolidation of various types of motor performance becomes available, the term *developmental* as applied to motor learning and performance deficits will be amenable to more accurate use than at present. The perinatal project (Nichols & Chen, 1981; Nichols, 1987) is the closest extant approximation of the ideal study just outlined, but, because it includes such a broad range of developmental issues, the details of motor performance are not all covered. In the meantime, there are many cross-sectional studies that provide useful information on specific performance items (Adams et al., 1979; Annett, 1973; Denckla, 1974; Goodenough, 1935; Gubbay, 1975; Gubbay et al., 1965; Hart, et al., 1974; Hulme, Biggerstaff, & Morgan, 1982; Kennard, 1960; Kornse, Manni, Rubenstein, & Gratziani, 1981; Peters et al., 1975; Rapin, Tourke, & Costa, 1960; Roach & Kepart, 1966; Rudel, Healey, & Denckla, 1984; Spreen & Gaddes, 1969; Wolff, Gunnoe, & Cohen, 1983, to include a few).

In defining motor performance deficits, the normal developmental trajectory of motor performance is most important, but the ontogeny of motor development is just beginning to be studied in precise detail. Thus, at the present time, the basis for expectations are poorly understood; that is, should motor performance follow physical or mental development? There is evidence for both (Cratty, 1986; Godfrey & Kephart, 1969). In physical development, is size, bone maturity, or pubertal stage to be the index (Cratty, Cratty, & Cornell, 1986)? Of course, the type of motor performance followed is also quite germane to this question. One of the most primitive types of movement that begins in early fetal human life and also in early fetal life of other vertebrates (Hamburger, 1971) are cyclic spontaneous movement patterns (Almli, personal communication, 1989). Such movements, which at the time of birth involve mouth, head, limbs, and trunk of the human infant, seem to be generated solely by the central nervous system, as no sensory stimulation precedes them and no externally obvious goal is served by the periodic muscle contractions. Evidence from animal research suggests that at early fetal stages these movements are initiated by spinal neural activity, whereas at later fetal stages, as encephalization proceeds, the telencephalon is the controlling neural substrate (Almli, Miller, Orup, & Galiano, 1988). Almli studied these movements in rats and preterm infants and found that they are stable for the first 3 postnatal weeks in normal human preterms, occurring about every 90 seconds and with the same periodicity in all body segments. These phenomena seem to be independent of external stimulation (e.g., heart rate accelerations and decelerations that result from light touch or external pressure stimulation) (Almli & Lawter, 1984) but may instead be altered by brain damage or medication (DeVries, Visser, & Prechtl, 1985; Parmelee & Sigmon, 1976). Generally, these movements appear preparatory to interactive nervous system function, but, whereas such movements are not recognized as pathological, their *sui generis* appearance is similar to some types of adventitious movements, namely tics and chorea.

Superimposed on the substrate of spontaneous (internally generated) neural motor activation are responsive motor behaviors of the automatic and voluntary musculature, visual, somatosensory, and auditory stimuli (Almli, personal communication, 1989; Peiper, 1963). Some newborn motor behaviors, however, fall into a third category, that is, neither simply reflexive nor completely unrelated to environmental events: namely, movements directed by neonatal thought. Eye movements, for instance, clearly can be directed by mental operations, such as memory (e.g., visual recognition of a novel stimulus) (Fantz & Fagen, 1975; Parmelee & Sigmon, 1976). It is true that reflex limb withdrawal to noxious stimuli continues, and postural patterns are integrated into more reflex patterns (Peiper, 1963) that have a clearly established developmental trajectory in the first weeks and months of life (Farmer, 1975). The important point is that, from birth, the full-term newborn is able to use cognitive mechanisms to direct motor behaviors, and, the older the infant, the more clear it is that motor behaviors are purposeful (i.e., initiated with a behaviorally determined goal). As soon as mental operations begin to direct motor output programs, the potential for a type of dyspraxia exists, and indeed dyspraxia of gaze is clinically recognizable in infants (Cogan, 1966).

In the first year of life, innervation patterns develop that override segmental and reflex mechanisms and allow differential activation of even individual fingers. Thus, the forefinger and thumb may cooperate in a "pincer grasp," whereas the other digits are differentially extended. Clumsiness, as further defined a bit later, may result from failure of such late infantile innervatory patterns. There is much random movement of the upper extremities that may appear random to be seen in 4- to 10-month-olds. This has been described as "arm waving" by Cratty (1986) and Thielen (1979). These movements subside once effective reaching, grasping, and transferring are achieved. Similar types of behaviors also are noted before actual crawling, walking (Cratty, 1986), and talking (e.g., where long strings of unrecognizable syllables may precede actual meaningful words). Many investigators and observant parents argue that these are anticipatory "practice" movements occurring purposefully before a clear, goal-directed activity can take place. These manual behaviors, when retained, may be the basis of some of the adventitious hand movements of retarded and autistic children and those with Rett syndrome (Bord & Gascon, 1988).

Once the infant has successfully grasped and transferred objects, and has successfully removed its whole body from one location to the other, it becomes a toddler and is on the way to self-regulated motor learning. In the words of Berta Bobath (1971), "a child's emotional and social development is affected by his ability to move, because movement makes him independent from his mother's constant attention." Some roots of the interactions between the motor, perceptual, and social and emotional aspects of childhood neuropsychiatric disorders are likely to be discovered during these periods of development when the baby is teaching itself to carry out goal-directed activities that result in the gross motor milestones (Forssberg & Nashner, 1982).

Clear motor hand preference is detectable between 1 and 2 years (Deuel & Moran, 1980; Ramsey, Campos, & Fenson, 1979; Rice, Plomin, & DeFries, 1984). From about 1 year of age, the hands also may carry a considerable communicative burden. Gestures, such as waving goodbye and clapping, often precede verbal communications. In some children, actions are used in place of words, and handling objects may enhance memory of their names (Uzqiris, 1967). In deaf people, indeed the entire linguistic structure of communication may be managed by the hands. In fact, there is even evidence from adult signers who have had strokes that the very same motor acts may be performed readily in one context (e.g., to express a syntactical relationship) but be absent in another (e.g., to describe a spatial array in a right hemispheric lesioned subject) (Bellugi, Klima, & Poizner, 1988). Further, many fine-motor skills, such as drawing and writing, are like language (Bates & Marchman, 1988) and must be learned. Material specific dyspraxias (see Table 8.6) may be recognized in the inability to acquire these skills (Kelso & Tuller, 1981).

DEFINITIONAL ISSUES

The old term *soft neurological signs* has been associated with the "minimal brain dysfunction" diagnostic category (Clements, 1966) and has been used commonly to denote any number of undefined types of motor deficit as well as anatomical anomalies (Thompson & O'Quinn,

1979) and sensory neurological abnormalities (Taft & Barowsky, 1989). In its abbreviated form, *soft signs*, the term frequently covers entirely nonneurological anatomical anomalies in addition to sensory and motor deficits. Therefore, such terms are best avoided, not only because of different meanings in different contexts but also because they connote elusiveness, inconsistency, and difficulty of ascertainment of the abnormalities (Herzig & Shapiro, 1987). One source of this reputed difficulty is a failure to recognize that many of the "soft motor signs" are *performance* deficits. They are neither elusive nor subtle, but they cannot be ascertained without observing the patient during motor performance. A neurological exam is not an invariant procedure but includes general assessments of several domains with more-detailed assessments of one or more of these domains. Choice of detailed assessments is dependent on the index of suspicion of abnormalities held by the examiner. Thus, if the examiner does not suspect that the right hand will be moving while the left hand is voluntarily attempting successive finger on thumb tapping, and so omits the right hand from his field of observation, even the most severe mirror synkinesis will prove very elusive on this examiner's neurological exam. Many of the specific motor deficits discussed in this chapter might better be named "dynamic" than "soft" signs, as they can be observed only during some sort of activity. Thus, as soon as a performance is carefully observed—even so rudimentary a one as tapping a finger repetitively against the thumb (Denckla, 1974; Spreen & Gaddes, 1969)—the deficit looses its elusive or inconsistent features.

A quantitative standard for assessment is helpful, and of course developmentally appropriate normative values are important. Current lack of normative data hampers precise accuracy in developmental assessment for many types of motor performance, and the situation is even worse for quantitative evaluation of motor learning. Nevertheless, if no performance is required or observed, not only will a developmental level be misassigned and severity of deficit underestimated or overestimated, but the most obvious deficit may be missed entirely.

There are in fact useful developmental normative values for finger tapping and peg moving (as detailed in a later section of this chapter): A normal 3-year-old taps slower than a normal 8-year-old, as has been apparent for over 50 years (Goodenough, 1935). Beyond these behaviors, there is variable and sometimes poorly reliable developmental normative data. As noted frequently (David et al., 1981; Yule & Taylor, 1987; Taft & Barowsky, 1989), the lack of a commonly recognized typology, nosology, or categorization for motor disturbances is a major block to consensus on what type of motor disturbances actually do occur in neuropsychiatric disorders of childhood. This lack of a com-

TABLE 8.1
DSM–III–R Diagnostic Criteria for Developmental Coordination Disorder (315.40)

A. The person's performance in daily activities requiring motor coordination is markedly below the expected level, given the person's chronological age and intellectual capacity. This may be manifested by marked delays in achieving motor milestones (walking, crawling, sitting), dropping things, "clumsiness," poor performance in sports, or poor handwriting.
B. The disturbance in A significantly interferes with academic achievement or activities of daily living.
C. Not due to a known physical disorder, such as cerebral palsy, hemiplegia, or muscular dystrophy.

Note: From *Diagnostic and Statistical Manual of Mental Disorders* (3rd ed., rev.) (p. 49) by the American Psychiatric Association, 1987, Washington, DC: Author.

monly recognized comprehensive categorization does not obviate consensus about any of the clearly handicapping, easily recognizable primary motoric disturbances such as spasticity, weakness, severe choreoathetosis, or ataxia. For example, spasticity and ataxia are abnormal at any age, are recognized as the result of anatomic disruption of specific elements of hierarchical serial neural pathways, are found with cerebral palsy, muscle disease, acute infantile hemiplegia or after major head injury, are usually revealed by history, and are easily confirmed by even a rudimentary physical examination. The lack of a generally accepted and comprehensive categorization thus applies to those very motor performance disturbances that are most commonly found in developmental neuropsychiatric disorders without frank neurological damage (Fletcher & Taylor, 1982; Denckla et al., 1985) and are called *elusive* and *soft*.

The DSM-III-R diagnosis of Developmental Coordination Disorder (APA, 1987, p. 49) was developed to set the entity of motor performance deficit apart from other specific developmental disorders. This broad diagnostic category requires that "the person's performance in daily activities requiring motor coordination is markedly below the expected level, given the person's chronological age and intellectual capacity" and goes on to say that up to 6% of school children may be affected. Developmental coordination disorders should, therefore, allow for proper diagnosis and management of handicapping motor performance deficits in up to 6% of the primary school population. However, as the category currently is described in DSM-III-R, there is insufficient specificity to make it optimally useful for either diagnosis or prognosis. This seems to flow in part from its inclusion of all types of motor performance deficit into one broad category. In contrast, the recent literature indicates that there are specific subtypes of motor performance deficit that are clinically and neurologically separable (Cermak, 1985; Deuel, et al., 1984; Jones & Prior, 1985; Lazarus & Todor, 1987; Taft & Borowsky, 1989). These subtypes each require a separate differential diagnosis and may indicate different treatments. It would be diagnostically and prognostically more useful to acknowledge their discriminability in the same way that developmental language disorders are divided at least into expressive and receptive problems. ICD-9-CM (WHO, 1989) does divide the category into clumsiness and dyspraxia. Inclusion of delays in gross motor milestones (e.g., walking, crawling, sitting) broadens the DSM-III-R motor skills disorder category very greatly and makes it require a very large differential diagnosis. On the other hand, the perinatal project data of Nichols (1987) does suggest that gross motor abnormalities (e.g., poor hopping) occur frequently and are significantly more common in children with fine-motor performance deficits than in those without. Further investigation of the relationship of gross motor delay of known and unknown etiology to motor performance deficits (of unknown etiology) is clearly in order.

In addition to the unidimensional nature of the DSM-III-R category, its severity requirement is problematic. If only very severely and obviously handicapping motor performances are considered abnormal, the important question of whether less-severe motor performance deviations are subtly but pervasively handicapping, or handicapping one in one, limited area, cannot be pursued. Further, the question of whether such deficits should predicate specific and helpful management techniques may be overlooked (Deuel, 1981, 1988; Peters et al., 1975). Finally, on theoretical grounds, it is important to identify motor performance deficits of mild and moderate degrees. For one thing, it is necessary to quantitatively describe each individual performance item in order to construct its normal developmental trajectory so that abnormal may be reliably stated. For another, if one considers the spectrum of developmental neuropsychiatric disorders including motor performance deficits to be a functionally determined set of neural processing deficits, then again reliable identification and quantification of even mild deficits is necessary to allow an estimation of how the members of the set interact with each other.

In practical, clinical terms, determining the presence, let alone the severity, of motor performance deficits is difficult, principally because they do not often form a complaint in the clinical history (Frei, 1986; Taft & Barowsky, 1989). A presenting history typical of conduct disorder is often given for children with moderate to severe clumsiness and apraxia (Gubbay, 1975). The disability that underlies the child's refusals may go undetected, because the secondary behavioral symptoms are more recognizable and annoying to adults and are interpreted as proceeding from an emotional disturbance. Thus, the severity requirement of the DSM-III-R category may actually operate to prevent ascertainment of motor skill disorders. In any event, demanding evidence of significant interference with academic achievement or daily living as a diagnostic criterion leaves the determination of motor skills disorder totally up to the personal judgment of the clinician, as there are few objective criteria that help determine what is "significant."

In sum, it would be better to make identification (diagnosis) into an objective, criterion-based process for each individual performance deficit and then to apply a second set of rules for quantification (severity) of each identified deficit. Once deficits have been ascertained and severity determined, the third step does remain a judgment of the role the deficits may play in that particular patient's profile of cognitive and emotional strengths and weaknesses. For some of the 3% to 6% of primary school children who display them, motor performance deficits may be a primary cause of a cascade of adverse conduct and attentional problems; in others, they may play little or no role in the overall handicap displayed by the child. Placing the motor performance deficits in the perspective of the child's individual strengths and weaknesses requires a broad range of information concerning the child, his environment, and his personal interactions that cannot be derived from the directed motor performance exam alone. Thus, the motor performance exam is necessary for identification but not sufficient to carry through with determination of significance and appropriate management.

Lack of motor performance demands in "the standard neurological exam" seems to have led to a large number of extensive special performance batteries (Adams et al., 1979; Gubbay, 1973; Levine, 1987; Peters et al., 1975; Spreen & Gaddes, 1969; Voeller, 1981) as well as single function tests (Rapin, Tourke, & Costa, 1960; Rudel et al., 1984; Wolff et al., 1983). Probably because there is no one universally accepted categorization for motor performance deficits, nor even a universally accepted concept of what motor activities are to be examined in a comprehensive motor performance battery, functional definitions for individual disorders abound. Almost every research study that investigates such matters uses a different set of tests and/or a different conceptual basis of abnormality, with abnormality being defined in terms of the unique test or concept (Taylor, Powell, Cherland, & Vaughan, 1988).

In conclusion, there presently seems to be a major problem in coverage and specificity in the diagnostic terms used for motor performance deficits as well as controversy about how much and what kind of specialized examination is required to reveal them. Given the customary lack of appearance of motor performance deficits in complaints and

TABLE 8.2
Proposed Diagnostic Criteria for Motor Skill Disorders (Developmental Motor Performance Deficits)

A. Motor skill disorders (developmental motor performance deficits), including clumsiness, dyspraxia, and/or adventitious movements, are recognizable when the child's tested motor performance is below expectation for age and/or verbal intelligence.
B. Not due to other recognizable cerebral, brainstem, cerebellar, spinal cord, nerve, or muscle disease.
C. The degree to which motor performance deficits disrupt daily living, academic achievement, and self-esteem must be determined for each individual within the context of the total cognitive and emotional disability.

Note: Adapted from *Proposed Nosology of Disorders of Higher Cerebral Function in Children* (Task Force on Nosology of Disorders of Higher Cerebral Function in Children) by R. David, R. Deuel, P. Ferry, G. Gascon, G. Golden, I. Rapin, P. Rosenberger, and B. Shaywitz, 1981, Minneapolis, MN: The Child Neurology Society.

history volunteered by child and parent, and the uncertainty on the part of most examiners as to whether motor deficits could be causing a handicap or not, motor problems are often ignored. However, if one bears in mind the conceptual possibilities, the deficits may all be elicited and directly observed without lengthy or highly stereotyped batteries of multiple, specific tests.

A comprehensive but specific categorization of the spectrum of motor performance deficits commonly encountered in developmental neuropsychiatric disorders is needed, at least provisionally. One such categorization scheme, which David et al. (1981) first proposed, is based on motor performance abnormalities that are observable in most clinical settings and sometimes reported in the patient's history. The outline is shown in Table 8.2.

It provides enough categories to allow coverage of all the higher-order motor execution deficits to be encountered. For example, it separates dyspraxia from adventitious movements and clumsiness. It does not purport to cover primary motoric abnormalities such as paresis, spasticity, and ataxia.

PROPOSED DIAGNOSTIC CATEGORIES, ASSESSMENT PROCEDURES, AND DIFFERENTIAL DIAGNOSIS

In an attempt to provide a comprehensive conceptual categorization of motor performance deficits, David et al. (1981) proposed a general typology of disorders of motor execution. This typology divides the motor abnormalities into simple categories that are readily distinguished by screening with a directed clinical history and neurological examination and, when combined with observations of hand performance and preference (Oldfield, 1971), includes not only the entire spectrum of motor performance deficits but information about laterality as well. Use of an inclusive categorization would promote a common vocabulary and perspective on the entire range of disorders of motor execution, at least in school-aged children, and also allow enough specificity of diagnosis to provide educational, psychological, and medical guidelines for management. The general categories advocated by these investigators include clumsiness, adventitious movements, and dyspraxia. Within each of these categories, independent subtypes exist. Clinically, a given child may display more than one general category of deficit and more than one subtype within each category.

For diagnostic assessment, there are four principles:

1. Motor performance and learning disorders are often "silent" and must be sought by appropriate specific questions in the history and specific observation during the appropriate motor performance in the physical examination.
2. Clumsiness, dyspraxia, and adventitious movements must be separately considered, as the physical examination and laboratory tests for the one will not ascertain the others.
3. Motor performance, per se, is quantifiable. Although specific normative data is not available for all motor performance deficits, many published tests with normative data are available now, and several projects are underway to provide more such data.
4. Not all measurable deficits in motor performance are significantly handicapping in a given child.

In operational terms, at least the essentials shown in Table 8.3, I (History) and II (Physical Examination), should be carried out. As has been noted, although a directed history should be attempted and may prove very informative, motor performance milestones may be forgotten; caretakers may be unaware of current motor performance deficits, and so an actual measure of motor performance in the three major areas should be applied. Another approach, if a physical examination is not available, is to use formal tests and batteries (Item III of Table 8.3). If only formalized tests are to be used, then a good screening test, such as the Gubbay (1973), should be followed by specific assessments for clumsiness, such as the Denckla (1974) finger-tapping test; and apraxia, such as the Lincoln–Oseretsky tests (Sloan, 1948). The most efficient way to ascertain the range of adventitious movements remains examination by a trained neurological clinician, although the Fog test (Fog & Fog, 1963) is a formal means of evaluation specifically for synkinesis.

TABLE 8.3
Outline of Diagnostic Assessment for Motor Skill Disorders

I. History—General pediatric and neurologic history with emphasis of:
   a. Gross motor milestone
   b. Fine motor milestones (tie shoes, button buttons, manipulate safety pin, snap together toys and models, sew, work with tools)
   c. Current fine and gross motor activities

II. Physical Examination—General pediatric and neurologic examination with emphasis of:
   a. Observation for adventitious movements
   b. Motor performance items
      1. Finger tapping, hand turning, toe tapping
      2. Sequential motor acts on imitation, pantomime on command, and with objects
      3. Writing and drawing

III. Formal Tests and Test Batteries
   a. Screening tests with normative data
   b. Specific item tests
      1. Clumsiness
      2. Apraxia
      3. Adventitious movements
      4. Handedness assessments

In regard to quantitation of motor performance deficit severity, in formal tests, there is generally chronological age-based normative data, and the number of standard deviations from the age mean may be used as a measure of severity. However, as most complex motor activities are more or less efficiently performed according to overall intelligence, how is one to make the designation of dyspraxia in a person with a low or low normal IQ? In general, it is legitimate to use the verbal IQ as a standard. Thus, if the verbal IQ is 85, but the motor performance is below the 10th percentile for age, then the discrepancy between IQ and motor performance may be assigned, if no other cause is obvious, to a specific motor performance deficit. Use of verbal ability as a standard for motor performance is not of course helpful in children with language disorders. It is of interest, however, that, in a single study of children with large verbal versus performance WISC–R discrepancies, subjects had evidence of dyspraxia whether the verbal was much better than the performance or vice versa (Cermak, Coster & Drake, 1983).

Some treatable or genetically informative diseases exist in which motor performance deficits and accompanying behavioral and attentional abnormalities also may lead to recognition or high suspicion of the disease being present. These diseases, therefore, form a differential diagnosis not only of the motor performance deficits but also of the neuropsychiatric disorders. In sum, when a child presents with certain motor performance deficits in the context of a neuropsychiatric syndrome, the diagnostician should be alert to the possible presence of certain diseases, etiological considerations, and recognizable conditions that affect the central nervous system.

Clumsiness

Clumsiness may be defined as slowness and inefficiency in performing all elemental fine-motor movements (e.g., single-joint movements). In neurological terms, this definition of clumsiness implies abnormality in the primary interactions of different muscle groups with each other or a deficit in alteration of joint angles by effector structures, irrespective of cause. Although most often it results from faulty speed and direction specified by cerebral innervation or from faulty control of agonist and antagonist muscles, it could also result from bony malformations. This use of the term *clumsiness* differs from that of English investigators (Dare & Gordon, 1970; Gubbay, 1975; Gubbay et al., 1965; Walton, 1961; Walton et al., 1962) who use the term very generally to include all measurable performance deficits (e.g., slow primary-joint movements, adventitious movements, motor planning, and sequencing disorders). The commonly employed Gubbay (1973, 1975) test for apraxia, although it enjoys developmentally sound normative values, still has an outcome metric that does not allow for separation of clumsiness from adventitious movements or dyspraxia. A disadvantage of conceptually lumping these performance deficits together, or testing for the lumped disorder, is that ascertainment provides only a screening level of information that does not help to further specify etiology or management. Thus, whereas the Gubbay test and many other screening tests in the literature can well identify motor performance outcomes that deviate from developmental norms, further testing must then be carried out, if disambiguation for specific differential diagnosis and management are to be brought forward. A restricted use of term *clumsiness*, excluding adventitious movements and dyspraxia, is advocated here. This restricted use would allow the term *clumsiness* to accord with the classic "limbkinetic apraxia" that Liepmann (1908) originally described in adults with brain damage. In David et al.'s proposed classification (see Table 8.4), clumsiness is used with the more specific connotation.

The classification thus provides for separation of elemental motoric from higher order motor planning and sequencing deficits, and also distinguishes adventitious movements, in accord with the recommendations of many workers (Conrad, Cermak, & Drake, 1983; Cratty et al., 1986; Haaland, Porch, & Delaney, 1980; Manni, Martin, & Sewell, 1977).

TABLE 8.4
Proposed Criteria for Clumsiness, a Specific Type of Motor Performance Deficit

A. Motor speed of at least one (but often several, or all) groups of muscles (orofacial, hand and finger, lower extremity, shoulder or hip girdle, or trunkal) is below expectation for age and/or verbal intelligence. Effective completion of motor acts (even complex multistaged ones) is improved if prolonged performance time is allowed.
B. Discrepancy with age-expected motor speed and/or level of verbal intelligence may be used as a measure of severity.
C. Not due to another recognizable cerebral, brainstem, cerebellar, nerve, muscle, or skeletal disorder.
D. May coexist with dyspraxia and/or adventitious movements.

Note: Adapted from *Proposed Nosology of Disorders of Higher Cerebral Function in Children* (Task Force on Nosology of Disorders of Higher Cerebral Function in Children) by R. David, R. Deuel, P. Ferry, G. Gascon, G. Golden, I. Rapin, P. Rosenberger, and B. Shaywitz, 1981, Minneapolis, MN: The Child Neurology Society.

Children with clear-cut spasticity and weakness are invariably affected with elemental clumsiness but not necessarily with any sort of dyspraxia. More importantly, children with manual dyspraxia, and particularly material specific dyspraxia, may be quite dexterous in their individual finger or limb movements and so not clumsy by the proposed criteria. Simple finger-tapping and peg-moving tasks will identify clumsiness but not dyspraxia. Thus, from a clinical examination perspective, elemental clumsiness is often readily separable from dyspraxia. Clumsiness may occur without any other sign of motor system disorder and usually occurs bilaterally. If it affects only one hand, a localized lesion must be ruled out. In terms of ascertainment, clumsiness is the most common and readily testable disorder of motor performance. It may sometimes be ascertained from the history, although it is not always directly noticed by parents outside of its effects on other functional areas. Thus, it may be necessary to inquire specifically about motor dexterity in the history, and it is always important to require performance of specific tests (e.g., finger tapping) to ascertain the degree of motor performance deficit. It is important to recognize that clumsiness often affects a preschool child's entire life, particularly if there is a dexterous older sibling for comparison. Later, the child is still slow at home in self-help skills. In addition, at school, skills in tasks such as cutting, drawing, keeping materials in order, and writing are clearly deficient, and they retard academic progress, particularly in the kindergarten and early primary grades. Slowness (i.e., always last to finish), both at school and home, usually interacts with the child's social development, if only in terms of shyness and dependency, and it usually deeply and adversely affects self-esteem (Ford, 1960). These secondary consequences often color the history and chief complaint. If weakness, as manifested in clumsiness, is suspected after careful historical investigation, a full neurological examination should be performed.

Examination for Clumsiness. Examination of the hands may employ simple, well-studied tasks, such as finger tapping (Denckla, 1974) or peg moving (Rapin et al., 1960). There is ample, developmentally appropriate, normative data for finger tapping (Denckla, 1974; Goodenough, 1935; Spreen & Gaddes, 1969). It must be borne in mind that, when single, elemental movements are slow and inefficient (i.e., clumsiness is present), so are more complex ones, such as writing, drawing, buttoning, and zipping, particularly when the child is in the learning stages for these activities and when the child is being held to a speed criterion. Clumsiness may also affect gross motor activities such as throwing, kicking, and riding a bicycle in some children. Clumsiness may be separated from dyspraxia by testing for complex bimanual activities in which the clumsy child follows the correct sequence of acts but performs each constituent motion more slowly than expected. Children purely affected by clumsiness use a logical approach to completing a motor task and simply take longer to accomplish it than their peers. As clumsy children progress through school (McKinlay, 1978), they may develop a large repetoire of techniques to avoid overtaxing motor assignments, including refusal (Annell, 1949).

Differential Diagnosis of Clumsiness. Clumsiness is found in children with chronic intoxications with neuroleptics and anticonvulsants, particularly phenobarbital (Mattson & Cramer, 1982), clonazepam (Dreifuss & Sato, 1982), and Dilantin (Dam, 1982). Arthrogryposis and rheumatoid arthritis also must be considered, and weakness from neuromuscular diseases (e.g., Charcot Marie Tooth disease) should be ruled out as well. Lower extremity clumsiness (i.e., "falls alot, slow walking upstairs") should alert one to Duchenne and other progressive muscular dystrophies. Many neurological disorders that cause clumsiness are heritable conditions where genetic counseling is in order. When clumsiness is ascertained in the context of upper motor neuron signs with tone changes of spasticity, a very wide spectrum of disorders from acquired spinal cord or brain damage to degenerative white matter disorder should be entertained (Farmer, 1975).

Clumsiness in the Major Neuropsychiatric Disorders. Clumsiness with slow attainment of developmental gross and fine-motor milestones, commensurate with delay in language milestones, is certainly found in mental retardation, ADHD, and specific academic skill disorders. The mental retardation groups that have been studied the most closely are Down's syndrome and autism. Down's syndrome children are hypotonic, possibly due to deficiencies in some neurotransmitters. They have been found to have slow reaction times (Berkson, 1960), and their finger tapping is slow (Seyforth & Spreen, 1979), thus qualifying them as clumsy. However, when grip strength was measured, they developed too much force (Cole, Abbs, & Turner, 1988), suggesting an additional cerebral motor control deficit. In autism, although there recently has been interest in absence of portions of the cerebellar vermis (Bauman & Kemper, 1985; Courchesne, Yeung-Courchesne, Press, Hesselink, & Jernigan, 1988), and other structural cerebral abnormalities (Piven et al., 1989), clumsiness has not been universally found. In fact, very careful observations included in the original description of the syndrome claimed that motor learning and performance was a forté of autistic children (Kanner, 1942). Some recent more systematic evaluations of autistic children for motor skills, however, suggest that their motor development is slower than expected (DeMeyer, Hingtgen, & Jackson, 1981; Ornitz, 1985), although relatively much better than their social or communication skills. Gillberg, Stefanburg, and Jacobson (1987) found motor performance abnormalities in a group of relatively high-functioning autistics with Asberger syndrome. Although no mention is made of the details of the motor examination, the children with Asberger syndrome were said to be "more clumsy" than the other autistic children in the study, again suggesting that clumsiness is not a common problem even among lower-functioning autistic children.

A very carefully described and well-known academic skill disorder is *specific dyslexia*. The full-blown disorder is always accompanied by dysgraphia (Critchley, 1970) and at times accompanied by various other types of motor disorders. For example, Rourke (1981) and Byring and Pulliainen (1984) found finger-tapping deficits specifically in the nondominant hand of older dyslexics. The relationship of clumsiness to the other learning disability (i.e., difficulties with mathematics and spatial relations), thought perhaps to represent right hemispheric dysfunction (Denckla, 1978; Stiles-Davis, Janowsky, Engel, & Nass, 1988; Weintraub & Mesulam, 1983), is unknown. However, pure clumsiness has been cited as a cause of dysgraphia, one of the specific academic skill disorders. In such dysgraphia, the child can adequately spell orally and may benefit greatly from the use of a word processor to avoid the effect of illegible and inefficient hand writing on the teacher's understanding of the child's correctly intended spelling and grammar. In studies where academic skill disorders are less carefully categorized, most show a marked increase in the incidence of clumsiness. Prechtl and Stemmer (1962) found clumsiness in their 157 learning disabled (LD) boys but not in their group of 876 normal boys. Adams et al. (1979) found 37% of learning disabled children to have clumsiness. It is of note, however, that, within their group of normal controls, 9% had clumsiness. Peters et al. (1975) found 46% of LD children to be clumsy as against 9% of normals. Taken altogether, these various studies suggest at least triple the incidence of clumsiness within populations of children with unspecified academic skill disorder as opposed to normal control populations that are similar in IQ, age, and sex.

In at least one of the developmental language disorders, clumsiness is clearly not present. Dewey, Roy, Square-Storer, and Hayden (1988) used a specific test to ascertain developmental apraxia of speech and identified four children that fulfilled their criteria for this disorder. These four had ideomotor limb apraxia *without* clumsiness and without a deficiency in motor sequence production per se. Kornse et al.'s (1981) findings concerning lack of clumsiness in developmental apraxia of speech are replicated by the Dewey study. It is clear from these studies that children who have developmental speech apraxia are at high risk to have specific types of limb dyspraxia and oromotor apraxia but are without clumsiness. Such studies further substantiate the utility of discrete dyspraxia and clumsiness categories of motor performance deficits.

Among attention deficit hyperactivity populations, using the designation *minimal brain dysfunction* (MBD), Paine, Werry, and Quay (1968) demonstrated clumsiness in 43% of 83 affected children. Hertzig, Bortner, and Birch (1969), evaluating a group of MBD children and testing specifically for clumsiness, apraxia, and chorea, found that 57.6% of their MBD children had clumsiness, 42% had chorea, and 33.9% had apraxia. In studies that use the designation hyperactivity or *attention deficit disorder* (ADD), Lerer and Lerer (1976) reported that 77% of 40 affected children were clumsy. McMahon and Greenberg (1977) found clumsiness in 75% of 102 hyperactive children of normal IQ and synkinesis in 82%, but they obtained poor replicability of the soft signs. Denckla and Rudel (1978) also found clumsiness and synkinesis in hyperkinetic children with relatively high IQs (i.e., full scale $\geq$ 90). Thus, in the MBD, ADD, or ADHD populations, disorders of motor performance are found in a much higher percentage than in populations of normal controls, and when clumsiness has been separately ascertained, it is of higher incidence than dyspraxia or adventitious movements.

In the light of this fact, it seems worthwhile to consider the hypothesis that the motor performance deficit came first and that behavioral components of the ADDH, MBD, or hyperactivity were secondary to the motor problem, at least in some instances. Chronic anxiety over motor performance failure could lead to distractibility, for instance. On a more conscious level, the child Annell (1949) described presents an example of clumsiness leading to sociopathy: The boy went truant rather than embarrass himself by manipulating his clothing ineffectively while under scrutiny of his more dexterous classmates.

Adventitious Movements

The second category of motor performance deficit in the David typology is adventitious movements. These are unwilled movements that occur in addition to whatever purposeful motor activity is ongoing, and

TABLE 8.5
Proposed Criteria for Adventitious Movements, a Specific Type of Motor Performance Deficit

A. At least one variety of adventitious movement is more frequent than expected for the chronological age of the subject. Adventitious movements may coexist with clumsiness and/or dyspraxia.
  1a. Synkinetic movement disorder
      Excess homologous (mirror) or heterologous synkinesis for age and/or verbal IQ.
  b. Not due to another recognizable cerebral, brainstem, cerebellar, nerve, or muscle disorder.
  2a. Chorea
      Excessive sudden rapid muscular movements of limbs or face for age and irrespective of developmental level over 20 years.
  b. Must be differentiated from athetosis (slow writhing muscular movements of limbs excessive for age).
  c. Not due to another recognizable cerebral, brainstem, cerebellar, nerve, or muscle disorder.
  3a. Tremor
      Rhythmic rapid oscillations of a body part.
  b. Not due to another recognizable cerebral, brainstem, cerebellar or metabolic cause.
  4a. Tic
      Excessive rapid repetitive involuntary movements of one or more muscle groups irrespective of age or verbal developmental level.
  b. Not due to another recognizable cerebral, cerebellar, neural, muscular, or skeletal disorder.
  c. Specific types of tic disorder include Gilles de la Tourette syndrome and chronic motor tics.

Note: Adapted from *Proposed Nosology of Disorders of Higher Cerebral Function in Children* (Task Force on Nosology of Disorders of Higher Cerebral Function in Children) by R. David, R. Deuel, P. Ferry, G. Gascon, G. Golden, I. Rapin, P. Rosenberger, and B. Shaywitz, 1981, Minneapolis, MN: The Child Neurology Society.

sometimes they even supplant purposeful motor activity. Table 8.5 outlines the four types of adventitious movements considered: synkineses (i.e., overflow movements), chorea, tremor, and tic.

Synkinesis

Synkineses are unwilled activities of voluntary musculature that occur during the course of a voluntary action but that occur in a different set of muscles than those in play for the willed (voluntary) movement. There are different kinds of synkinesis. "Mirror" or homologous synkineses occur in muscle groups homologous to those in voluntary play. For example, when the left hand is voluntarily tapping, if the right hand involuntarily performs the same activity or parts of it, then that is a mirror synkinesis. Mirror synkinesis is distinguished from "heterologous" synkinesis, in which muscle groups unrelated to those in voluntary play are active. For example, the child voluntarily opens his mouth as wide as possible, and his eyes involuntarily open wide too. Synkinesis is a normal phenomenon and has been shown to occur in almost any individual who is making a maximal effort (Todor & Lazarus, 1986; Waterland & Hellebrandt, 1964). A few individuals of normal intelligence exhibit synkinesis with overlearned, common, everyday activities (Bobele, Bodensteiner, Marks, & Hamza, 1988, Somers, Levin, & Hamay, 1976), and these individuals are deemed highly abnormal. Several studies have been devoted to defining the developmental trajectory of synkinesis in normal and developmentally disabled populations (Cohen, Taft, Mahadeviah, & Birch, 1967; Connolly & Stratton, 1968; Fog & Fog, 1963; Lazarus & Todor, 1987; Wolff et al., 1983).

Examination for Synkinesis. Observation and direct examination are required, as synkinesis is almost never a component of the history, even when the most observant of guardians are directly questioned. Thus, historical information is difficult to obtain in a reliable manner. Fortunately, observation and testing are both informative.

For practical clinical examination without special equipment, progressive finger-on-thumb tapping and wrist turning are useful (Denckla, 1974, 1978; Wolff et al., 1983). The Fog feet-hands test, or Fog "gait," elicits heterologous adventitious movements of the hands when the child is told to walk with the feet inverted (i.e., on the medial aspect of the sole) or everted (i.e., on the lateral aspect of the sole). The original data showed marked suppination of the hand in normals before the age of 8, with a rapid drop-off to age 11, when pronation became more prominent. In mentally retarded and brain damaged children, these synkineses persisted much later (Fog & Fog, 1963).

To more formally assess for synkinesis, the tests Fog and Fog (1963) described, and Wolff et al. (1983), Todor and Lazarus (1986), and Taylor et al. (1988) adapted, are often used. Their hand-to-hand test requires pressure of thumb against forefinger with one hand and observation of the hand not in voluntary play by the examiner. In the course of development, normal children show a decreasing number of homologous synkineses movements, particularly between 7 and 10 years, but more heterologous synkineses between 11 and 13 years (Fog & Fog, 1983). Recent investigations with the Fog hand-to-hand test (Todor & Lazarus, 1986; Lazarus & Todor, 1987) reveal that the degree of homologous or heterologous movements developed in the nonsqueezing hand (i.e., the hand not in voluntary play) depends on the degree of effort expended by the squeezing hand (Todor & Lazarus, 1986; Lazarus & Todor, 1987). These studies also confirm that the number and kind of synkineses rapidly decreases in normal children between 6½ and 8½ years (Fog & Fog, 1963; Lazarus & Todor, 1987) and that heterologous associated motions (e.g., extension of the fingers and all other motions of the nonsqueezing hand except squeezing the thumb and forefinger together) become prominent at about 16 years of age.

Differential Diagnosis of Synkinesis. Homologous and heterologus synkineses occur more frequently in agenesis of the corpus callosum, a disorder that may result in mild mental retardation, seizure disorder, some learning and nonspecific behavioral abnormalities, and synkineses (Dennis, 1976). Children with this syndrome are in fact often referred to the physician on account of learning and behavioral difficulties. The most appropriate neurodiagnostic procedure in suspected absence of the corpus callosum is an MRI scan because of its high sensitivity to white matter. Homologous synkineses may also be seen in the Klippel–Feil syndrome that often results in a short neck and cervical vertebral anomalies (Gunderson & Solitaire, 1958). Homologous synkineses also have been reported with hypogonadism and anosmia and midline facial deficits (Conrad, Kriebel, & Hetzel, 1978). In addition, there appears to be a very rare familial form of synkinesis that is not accompanied by any dysraphism or midline defect (Somers et al., 1976). Bobele et al. (1988) presented a mother and two children who all had such severe manual synkineses that handwriting was very difficult. Even subjects with the worst variety of mirror synkineses do not necessarily have slow and inefficient movements of the fingers in all activities (Conrad et al., 1978); that is, they are not necessarily clumsy. Nonetheless, both synkinesis and clumsiness often do appear together in the same person. Acquired lesions that disrupt the corpus callosum or the supplementary motor area could also potentially cause synkinesis as well as a lesion at any level of the nervous system that makes motor activity more effortful.

Synkinesis in Neuropsychiatric Disorders. Using a modification of the Fog feet-hands test, Satzmari and Taylor (1984) tested 138 primary school boys, and their degree of adventitious movements was correlated with behaviors in the classroom. The degree of age-inappropriate synkinetic movements correlated with classroom hyperactivity and antisocial behavior. In a subsequent study, these researchers (Taylor et al., 1988) evaluated primary school girls using a modification of the Fog test and more formalized behavioral assessments. Again, the deviant motor performers had more behavior problems at school. These studies are similar to those of Wolfe et al., (1983), who also used the Fog test, plus additional tests for upper extremity overflow movements. They detected several types of overflow in 5- to 7-year-olds.

The conceptual bias of many studies of overflow movement or synkinesis (as well as clumsiness) is that some degree of brain damage or dysfunction leads to a decreased efficiency of those neural systems that promote bimanual coordination by inhibition of inappropriate motor excitation. Once abnormal motor activity is found, it is used as proof of abnormal brain structure or function, and any aberrant behavioral activities are also attributed to a brain abnormality (Barlow, 1974; Gomez, 1967). However, one alternative possibility, that the perfectly appropriate frustration and anguish suffered by the child because of his inappropriate involuntary motor behaviors may be the medium through which his emotional well-being is disrupted, should probably also be considered.

Chorea

A completely different category of adventitious involuntary movement is one that appears whether or not the subject is attempting a voluntary movement and may in fact interrupt a voluntary movement by moving the limb in a differently directed fashion than is desired by the subject. Chorea is the most common of this sort of adventitious movement, but tics and some types of tremor are others.

Examination for Choreiform and Tic-Type Adventitious Movements. These types of movement generally do not require specific testing for identification and often form a complaint in the presenting history. To detect chorea clinically on examination, one need only observe the patient. Sudden, very rapid displacements (jerks) of face or limb segments will appear in the course of routine voluntary activities.

These movements often disturb the ongoing behavior intended by the child or interrupt a posture being held. The movements usually can be voluntarily inhibited for brief periods of time (seconds). Hypotonia is often noted to accompany chorea. A common test for choreiform movements is to have the child stand, eyes closed, extend the arms forward in space (with pronated and extended wrists and extended fingers) for 30 seconds (Wolff & Hurwitz, 1966). There is a normal amount of choreiform activity that reaches a peak in 6- to 7-year old boys and consists of twitching finger movements. Movements of proximal muscles and entire limbs, however, are not expected even in this age group. Very young children (2 to 4 yrs.) normally have frequent athetotic movements but not choreiform ones (Prechtl, 1987), and the two must be distinguished.

Differential Diagnosis of Chorea. Sydenham's chorea, an acquired disorder of movement that accompanies rheumatic disease, is manifest generally after the age of 6. The onset of this acquired movement disorder may be directly accompanied by a change in personality and behavior. If Syndenham's chorea is suspected, typically presenting after the age of 6 years, an antistreptolysin titer (ASL) and an electrocardiogram (EKG) should be conducted and clinical evidence for rheumatic disease sought (Lockman, 1982). If a rheumatic basis is found, the patient should be placed on prophylactic penicillin to prevent further involvement with streptococcus and possible repeated bouts of carditis (Markowitz, 1987). Hypothroidism and benign familial chorea are other entities to consider and, of course, stroke with sudden onset of unilateral movements. Chorea may also be seen in juvenile lupus erythematosis. A variety of low incidence, degenerative disorders of childhood may also result in chorea, chorea-athetosis, dystonia, and personality changes. These include Wilson's disease, Hallervorden–Spatz disease, Bassen Kornzweig disease, homocystinuria, and others (Eldridge & Fahn, 1976). It is less likely that chorea will be found in Huntington's chorea of the juvenile type, but rather another movement disorder, tremor, is more often found. Chorea, athetosis, and dystonia may result from drug intoxications including haloperidol and pimozide (Shapiro, Shapiro, & Fulop, 1987). These medications also lead to personality and learning changes. Thus, there is a broad differential diagnosis of choreiform movements when encountered in children who are referred because of behavioral and emotional difficulties. Each member of this differential diagnostic list may require specific testing. Treatment of significant chorea with neuroleptics is well known and should be managed through neurological consultation.

Tremor

Oscillations of a body part, another distinct type of adventitious movement, may also occur regardless of the patient's voluntary movements. Some variations of tremor are elicited by voluntary movements and are identified in the standard neurological exam (i.e., by the finger-to-nose test) as "cerebellar" signs. For the finger-to-nose test, proper execution demands full extension of the elbow when touching the examiner's finger and reaching at a moderate pace (rather than very fast ballistic poking) to at least three different positions in front of the patient. Clear appendicular cerebellar or intention tremor in a child, despite good evidence for emotional, cognitive, and behavioral difficulties, should be investigated with a full neurological examination and possibly an imaging study, as tumors of the posterior fossa are of fairly high incidence of childhood. Children also suffer from essential benign or familial tremor, determined by exam and history. Dilantin toxicity can produce nystagmus and tremor. Although hyperthyroidism is of low incidence in childhood, it also can lead to tremor and referral for emotional difficulties.

Tics

Tics are a form of adventitious movement that, like chorea, appear regardless of voluntary activity. Tics are usually readily observed in the patient *en passant* and often form part of the history. Even more than chorea, tics can be voluntarily suppressed by the subject so that fairly long periods of close observation are in order. Tics can be differentiated from chorea by their relatively stable form. Tics usually involve the same set of muscles each time. They also can be distinguished by their relatively complex form, as tics often involve functionally related muscles and result in a motion that could be voluntary (e.g., eye blinking or shoulder shrugging). Both tics and choreiform involuntary movements are performed more rapidly than similar voluntary movements.

Simple tics and chronic motor tics may be differentiated from tics in Gilles de la Tourette syndrome (APA, 1987). Respiratory tics (e.g., sniffing, snorting, throat clearing, vocalizing, including coprolalilia and barking) are a symptom that clearly differentiates simple, chronic motor tics from full-blown Tourette syndrome (Ehrenberg, Cruse, & Rothner, 1986). Often, during an observation period, the patient will exhibit some tics, but not the characteristic respiratory ones. Thus, collecting an accurate history is important in the endeavor of separating chronic motor tic from Gilles de la Tourette syndrome.

TABLE 8.6
Proposed Diagnostic Criteria for Dyspraxia, a Specific Type of Motor Performance Deficit

A. Dyspraxia is recognized when there is inability to learn or perform serial voluntary movements to complete skilled acts at an expected level for age and/or verbal intelligence. Deficits in volition, strength, coordination, motor speed, and sensation must not be sufficient to explain the poor performance, and the poor performance should persist irrespective of amount of time allowed for completion. Dyspraxia may be of the ideomotor, or ideational variety, or both.
B. Discrepancy with age-expected motor performance of serial voluntary movements, and/or verbal intelligence may be used as a measure of severity.
C. May coexist with clumsiness (in which case motor speed will also be impaired) and/or adventitious movements.
D. Not due to another recognizable cerebral, brainstem, cerebellar, nerve, or muscle disorder.

Note: Adapted from *Proposed Nosology of Disorders of Higher Cerebral Function in Children* (Task Force on Nosology of Disorders of Higher Cerebral Function in Children) by R. David, R. Deuel, P. Ferry, G. Gascon, G. Golden, I. Rapin, P. Rosenberger, and B. Shaywitz, 1981, Minneapolis, MN: The Child Neurology Society.

Dyspraxia

The third general category of motor performance disturbance suggested in the David categorization is dyspraxia, or the inability to learn and perform age-appropriate sequences of voluntary movements in the face of preserved coordination, strength, and sensation (see Table 8.6). Although it occurs together with clumsiness at times, the dyspraxic abnormality may be observed in isolation in the individual who cannot learn to perform age-appropriate motor sequences but yet has age-appropriate efficiency of fine-motor coordination. Dyspraxia may affect facial and oral musculature primarily, or hand use particularly, or involve the trunk and lower extremities. In fact, in some cases, all effector muscle groups may be involved (David et al., 1981). Several varieties of dyspraxia, including ideomotor and ideational dyspraxia, may be delineated in children as well as adults (Deuel et al., 1984; Liepmann 1900, 1908; Manni et al., 1977). A child with pure dyspraxia will perform each component movement of a complex action rapidly and dexterously but often in the wrong order, so that the outcome is incorrect. For example, in reassembling a lunch box, the dyspraxic child rapidly tries to put the cork in the thermos upside down, turns the screw cap the wrong way, places the wire clip in the up position, and then cannot close the lid but nonetheless tries to close the strap, which does not, because the lid will not close, and finally gives up in frustration. A child with pure clumsiness and no dyspraxia will slowly and logically perform each component of the complex action, taking a long time but ending up with a lunch box that can be carried away from the table. The clumsy plus dyspraxic child will be inhibited by both difficulties, with perhaps double the resultant frustration. In the dyspraxic child, motor *performance* is only part of the problem: Motor *learning* is also severely compromised (Cermak, 1985). An example of differentiating clumsiness from apraxia is presented in the study of Dewey et al. (1988). Speech disabled subjects received a finger-tapping test in which they were normal, a test for pantomime on command, and imitation of gestures (where their performance was less than that normals), and a test for their use of actual objects (where again their performance compared to normals). Thus, these findings demonstrate ideomotor (deficits in pantomime and imitation) but not ideational (deficits in use of actual objects) manual apraxia in the study subjects and, in addition, show a lack of clumsiness.

Examination for Dyspraxia. When dyspraxia occurs in isolation, it may be a silent symptom going unsuspected except by the most sophisticated evaluators. Thus, specific historical evidence may not be volunteered or even rendered in response to directed questions. When motor problems are a complaint, the generic clumsiness is usually cited (Gubbay, 1975). Nonetheless, apraxic difficulties can be elicited by a careful historian, if one can dissociate complex serial movements from acts that require no novel, motor strategies. Generally, dyspraxia cannot be determined without actually observing motor performance for praxis, especially in sequential motor tasks.

To test for apraxia for most clinical purposes, having the child imitate hand postures and then pantomime manual activities (e.g., "show me how you would pour the milk into the glass and drink it"), and then use actual objects (e.g., "show me how you would fold the paper so it fits like a letter into this envelope") constitute sufficient tasks (Deuel et al., 1984). Some developmental normative values for the pantomime and use-of-objects activities are also available but are still sparse (Kools & Tweedie, 1975). A formal test of imitation of gestures is part of the Lincoln–Oseretsky test and has developmental normative values (Sloan, 1948). To demonstrate that pure dyspraxia is present apart from clumsiness, age-appropriate speed and dexterity in simple finger tapping and peg placement (as discussed in the section on clumsiness) must be shown in the face of deficiency when different sequences of fingers are to be tapped or when the pegs are to be moved in a specified order with specified hands. The major fallacy in examination of motor performance is neglecting tests specifically for dyspraxia and only testing fine motor speed (that ascertains only clumsiness).

The clinical importance of testing for and recognizing dyspraxia, apart from its differential diagnostic value, seems to lie in the potential for improving motivation and conduct disorders in a manner similar to what occurs in the diagnosis and treatment of clumsiness (Gubbay, 1978). The differences are, however, that the simple expedient of allowing more time to complete a given act often does not improve the dyspraxic child's performance (Gubbay, 1978), and further, although practice may result in more efficiency in a given task (e.g., tying shoes), when a related task is attempted (e.g., tying an apron behind ones back), the dyspraxic deficit is all too apparent again. These simple motor learning tasks are sometimes useful in ascertaining apraxia as well as indicating management strategies that may include teaching the child to avoid engaging in new complex motor tasks in public, because, even after a given motor sequence has been practiced, and its performance appears dexterous and normal, new sequences likely will remain an insurmountable and embarrassing problem (Cermak, 1985).

Oral and facial dyspraxias often accompany verbal dyspraxia (Ferry, Hall, & Hicks, 1975), one of the developmental language disorders, and may also be simply tested by pantomime and use of actual objects (e.g., "Show me how you would blow out the candles on your birthday cake;" "Show me how you would drink from the straw").

Differential Diagnosis of Dyspraxia. This motor performance deficit may appear early in dementing disorders such as subacute sclerosing pancencephalatis (SSPE), acquired immune deficiency syndrome (AIDS) encephalopathy, Rett's syndrome, and other degenerative disorders of the nervous system either due to inborn or acquired metabolic errors. In children as in adults, it may follow stroke, head injury, or encephalitis. Some children with cerebral palsy, but not all, have manual, oral-facial, and/or other apraxia (Molnar & Taft, 1977).

Dyspraxia in Neuropsychiatric Disorders. Children with mental retardation do demonstrate true dyspraxia (Henderson & Hall, 1982) in addition to clumsiness. For example, the study of Newell (1985) shows children with Down's syndrome to be truly dyspraxic. Some investigators considered the oddities of motor skill learning and performance of autistic children to be dyspraxic (Jones & Prior, 1985; Ohta, 1987). Given that there is little evidence for clumsiness in autism, and the reported higher incidence of dyspraxia, it seems likely that autistic groups differ from ADHD groups on these motor parameters. Clearly related to the diagnosis of autism is Developmental Language Disorder. Many of the various language disorders seen in autism appear to be more expressive than receptive. Some of them have peculiarities of intonation and tempo that might be regarded as motor disorders, (e.g., the robot-like speech of some autistics). As for children with specific academic skill disorders, much evidence suggests that manual dyspraxia is highly represented. Deuel et al. (1984) tested for dyspraxia in 20 learning disabled children who had been selected by the school systems because they had "motor concerns" in conjunction with learning disabilities, and they compared results with those of age-, grade-, and sex-matched controls. The LD group demonstrated ideational and ideomotor apraxia. Among learning disabled children that were not selected for "motor concerns" but with an index problem of Reading Disability, 25% of children had some motor soft sign (Deuel, 1981). Klipcera, Wolff, and Drake (1981) tested 30 reading disabled boys on a complex bimanual rhythm task and found marked differences in motor sequencing compared to normal (non-reading-disabled) controls. Peters et al. (1975) noted 29% of learning disabled (versus 14% of controls) to be dyspraxic. The differing percentages in these studies reflect the basic problem of comparing studies with different criteria for dyspraxia. All of the studies use different means of ascertainment of learning disabilities as well as of dyspraxia and clumsiness, making generalization even more difficult. Nonetheless, the incidence of motor performance deficit is high, and in studies with specific ascertainment of both clumsiness and manual dyspraxia, dyspraxia seems to be of a lower incidence than clumsiness in specific learning disabilities.

Material-Specific Dyspraxia

This is properly a subcategory of dyspraxia, but, because it is so specialized and has been infrequently demonstrated and recognized, it is treated as a separate entity. The term *material-specific dyspraxia* implies that motor execution becomes inefficient only when certain cognitive materials are utilized to direct the motor output (see Table 8.7).

The most well known type of material-specific dyspraxia is *specific dysgraphia*. This is characterized by handwritten letters that are rudimentary, poorly formed, with wavering of the hand during their execution (reflected in irregular lines and curves) and letters that are usually placed improperly on the page, with the outcome that the child's handwriting is very difficult to decipher (Critchley, 1970; Critchley & Critchley, 1978). However, the same child may use the same dominant hand dexterously, adroitly, and with excellent perspective to draw a picture in which lines and curves are straight and smooth. Thus, the motor output in the latter exercise, when visuospatial material is driving the motor output, is perfectly normal, whereas in the former instance,

TABLE 8.7
Proposed Criteria for Material Specific Dyspraxia, a Specific Type of Motor Performance Deficit

A. Motor execution below expected for age and/or verbal IQ but only in performance of activities utilizing particular cognitive material such as language or spatial representation, while motor execution remains at expected level in performance of similar activities not utilizing that specific cognitive material.
B. Discrepancy with age, verbal IQ, or observed motor performance utilizing other cognitive material may be used as a measure of severity.
C. Not due to lack of appropriate instruction, or an identifiable focal cerebral abnormality.

Note: Adapted from *Proposed Nosology of Disorders of Higher Cerebral Function in Children* (Task Force on Nosology of Disorders of Higher Cerebral Function in Children) by R. David, R. Deuel, P. Ferry, G. Gascon, G. Golden, I. Rapin, P. Rosenberger, and B. Shaywitz, 1981, Minneapolis, MN: The Child Neurology Society.

when verbal material is directing the actual hand movements, motor performance is much less efficient. The reasons for this are unclear. A child with dyslexic dysgraphia does not spell correctly via the oral modality either. One possible reason for this difficulty with manual motor execution during writing words is that writing demands the simultaneous application of neural processing to two tasks (both the task of specifying which letter to execute and the task of selecting the motor output that is appropriate—Schneider & Fisk, 1983; Schneider & Shiffrin, 1977), whereas, for speech, the motor output is more automated (McKay, 1981, 1982, 1983).

Abnormality of drawing (and sometimes copying drawings) is a second type of material-specific dyspraxia. Here, manual spelling and writing are done relatively efficiently, but a great deal of difficulty in drawing and copying pictorial material as well as constructing an object out of its components occurs. This may be called *constructional dyspraxia* (Benton, 1984). Both constructional dyspraxia and general dyspraxia may occur with an apparent poor intuitive understanding of space as the basis of a "spatial cognition deficit' (Weintraub & Mesulam, 1983). A test of matchstick (Yule & Taylor, 1987) or block construction (Baum, Edwards, Leavitt, Grant, & Deuel, 1988) may help to delineate this form of dyspraxia. Care must be taken when using the specialized drawing tests that Goodenough initiated and first tabulated (1926) so as not to mistake constructional dyspraxia for overall decreased cognitive power, just as the cognitive power of a language disordered child is incorrectly judged if only verbal tests are used.

A third type of material-specific dyspraxia is the entity known as *verbal dyspraxia*. It has been carefully investigated for motor performance deficits (Ferry et al., 1975; Yoss & Darley, 1974). Taking into account that the diagnostic criteria for verbal apraxia differ among investigators (Dewey et al., 1988; Ferry et al., 1975; Rapin & Allen, 1988), it is interesting to see that Aram and Horwitz (1983) found that most verbal dyspraxics did not have limb apraxia, although some had oral apraxia of varying degrees of severity. A well-known developmental language disorder, this type of "cluttered or dilapidated" speech (Edwards, 1973) is often associated with oral motor and even limb dyspraxias, but it sometimes occurs in the absence of other related dyspraxias. In pure form, therefore, it qualifies as a material- (in this case oral language-) specific dyspraxia (Aram & Horwitz, 1983; Ferry et al., 1975; Kornse et al., 1981; Milner, 1971).

Examination for Material-Specific Dyspraxia. It should be clear from the foregoing discussion that ascertainment and testing for material-specific dyspraxias is a matter of determination of what the child *can* do (e.g., write but not draw, spell orally but not in written form) once a more general evaluation has confirmed what they *cannot* do. Again, age-appropriate normative values would be most helpful here, and, although they do exist in some tests for some activities (Kelsey, 1980), more are needed. In the meantime, without detailed normative values, only material-specific dyspraxias in which there are gross discrepancies between abilities may be confidently ascertained. Material-specific dyspraxias serve to remind us that motor execution deficits may be highly specialized and imply an interaction of neural motor programming processes at the level of higher brain function. This further justifies the separation of the less specific limb and orofacial dyspraxias from more elemental clumsiness.

Differential Diagnosis of Material-Specific Dyspraxias. In childhood, dysgraphia and oral motor dyspraxia are less likely to be directly related to a specific disease or stroke syndrome than in adulthood and in fact have been reported very infrequently despite a modest prevalence of stroke and hemiplegic migraine in childhood (Capildeo, 1979). If there is a sudden onset of such a disorder in childhood, however, and a stroke has occurred, sickle-cell disease, Von Recklinghausen's neurofibromatosis, congenital heart disease, and homocystinuria are some relatively common etiologies that are associated as well with high incidence of behavior disorders and learning disabilities (Capildeo, 1979). In regard to poor drawing as a material-specific dyspraxia, children with Williams' syndrome (Jones & Smith, 1975) have been found to have a disproportionately severe specific deficit in drawing and block construction (Bellugi, Sabo, & Vald, 1988). Verbal dyspraxia, the material- (speech-) specific dyspraxia just described, has been reported by Hayes et al. (1988) in patients who have survived the first year of life with galactosemia, and evidence for verbal dyspraxia has been found in Fragile-X mental retardation (Wolf-Schein et al., 1987; Borghgraef, Fryns, Dielkens, Pyck, & Van Der Berghe, 1987). Thus, among the other utilities of determining material-specific disorders of motor performance in children with neuropsychiatric syndromes of childhood, indications of differential diagnostic possibilities must be taken into account.

Manual Dominance Variations

This domain of motor performance is clearly not an elemental motoric one but involves the choice of the child in the leading hand for skilled fine-motor performance. Handedness in the normally developing human is largely genetically determined in a fashion that is not yet fully understood (Annett, 1986) and is highly conditioned by cognitive factors including learning. Current evidence suggests that, around the age of 12 months, a child begins to prefer one hand over the other for activities that require unimanual manipulation and for the leading role in bimanual activities. Prior to 1 year of age, the hand that is closest to the manipulandum is often employed (Corballis, 1983; Deuel & Moran, 1980), but, between 12 and 24 months, hand preference is normally declared (Ramsey et al., 1979; Rice et al., 1984).

There is a long history of disagreements over the pathological significance of left-hand preference (sinistrality) (Geschwind & Behan, 1984; Hicks & Dusek, 1980; Zurif & Bryden, 1969). It has become clear lately that there are two types of left-handedness, and confusion of the two probably fueled the old argument. One type is familial and may be associated with full, right cerebral dominance for language, speech, and hand preference functions. More usually it is associated with left hemisphere dominance for speech and language or varying degrees thereof (Zurif & Bryden, 1969). Such left-handedness occurs in about 10% of all human populations studied to date and appears to be both normal and heritable (Annett, 1973, 1986). In terms of cerebral control of function, the most common, normal variation seems to be left cerebral dominance for language with right cerebral dominance for hand use in the normal left hander, but the hemispheric lateralization of selected cognitive processes is not always predictable from handedness (Milner, 1971). Left-handedness has been tested in students by Hicks and Dusek (1980), who showed that there is a higher proportion of left handers among gifted than ordinary students.

The other type of left-handedness is pathological and acquired. It occurs in the setting of early injury to the left hemisphere with subsequent right-hand paralysis (however subtle), leaving the right hand impossible or at least difficult to use for manipulative activities. Pathological or acquired right-handedness has not been reported. If there is a right hemiparesis and no genetic disposition among primary relatives for left-handedness, pathological left-handedness can be assumed. Incidence of pathological left-handedness of a subtle variety not necessarily accompanied by clear signs of right paresis is probably more common in populations of children with neuropsychiatric syndromes (Geschwind & Behan, 1984), especially developmental language disorder (Aram, 1988). Certainly, left-handedness or "incomplete dominance" was found to be of higher incidence in children with other motor performance deficits in the perinatal project (Nichols, 1987) and has been demonstrated in autistic populations (Tsai, 1982), where it appears to be of the nongenetic variety.

Obviously, it is difficult to discriminate genetic from minor degrees of pathological acquired left-handedness in populations, and when it comes to individuals, in certain individual patients it may prove impossible to distinguish acquired from heritable hand preference in the absence of obvious brain damage and sufficient genetic information. In sum, although "atypical dominance" may indicate brain damage or other factors adverse to normal cognitive and emotional development, the individual instances in which this is the case are often difficult to ascertain.

In adults, testing for hand preference is very readily accomplished by asking the subject which hand they use for a variety of activities (Oldfield, 1971). In children, testing for handedness more often reveals discrepancies between stated preference and performance (Deuel & Moran, 1980). The Harris laterality test, in which unimanual and bimanual acts are pantomimed on verbal command, is often used (Harris, 1958). Indeed, in the normally intelligent school-aged child, observing hand preference in pantomime activities is quite sufficient ("Show me how you eat with a fork;" "Show me how you swing a baseball bat;" etc.). Once the hand preference is documented, however, as stated earlier, the conclusion that it is atypical for this particular individual may be tenuous.

In terms of differential diagnosis of atypical lateral dominance, anomalies of hand preference development are seen in infantile stroke where hand preference may be well developed before a year of age and may betoken a subtle hemiparesis of the other side. An infantile or early childhood stroke with hemiparesis as a result carries a similar differential diagnosis as a stroke resulting in material specific dyspraxia. Atypical dominance has been noted in groups of children with a variety of neuropsychiatric disorders (Aram, 1988; Bellugi et al., 1988;

Belmont & Birch, 1965; Gillberg & Rassmussen, 1982; Tsai, 1982). Unlike an individual, a population may be readily determined to have more or less right-handedness than the normal population, so the studies cited previously do have the degree of reliability that their design permits.

SUMMARY AND CONCLUSION

In summary, since Orton first mentioned motor incoordination in children with developmental neuropsychiatric disorders, there have been major changes in opinion concerning the significance of motor abnormalities. Before the 1950s, motor abnormalities were largely ignored. Particularly in the late 1950s through the 1970s, motor performance deficits, soft neurological signs, were considered difficult to elicit but generally indicative of some sort of brain damage or permanent brain dysfunction. Recent literature of the late 1970s and 1980s points to the absence of a specific profile of motor performance deficits in most specific neuropsychiatric disorders of childhood but continues to support a high incidence of deficits when active motor performance rather than fixed motor signs are tested in children with neuropsychiatric disorders.

Presently, the DSM-III-R provides a single, separate motor skill disorder category. From a clinical point of view, to enhance detection and discussions of motor performance deficits, a more detailed and comprehensive categorization of motor performance deficits over that provided by DSM-III-R is suggested. The motor skill disorders can be provisionally divided into three general categories of performance deficit: clumsiness, adventitious movements, and dyspraxia. These categories provide comprehensive coverage of the motor execution deficits to be encountered in children with neuropsychiatric disorders. Once comprehensive coverage of motor performance disorders is achieved by the three main categories of clumsiness, dyspraxia, and adventitious movements, further analysis into specific subtypes that may be quantitated as to severity allows for differential diagnosis of specific disease entities. Such analysis allows management of each child to be based on a weighting of the influence of behavioral, cognitive, and motor performance deficits in regard to the child's total burden of dysfunction. In turn, allocation of remediative resources can become more effective. Finally, once broad specific and quantitative clinical analysis is available, it should provide objective, functionally defined entities for research endeavors.

Theoretically, on the basis of speculations concerning the significance of motor performance deficits for neural system interactions, it is suggested that motor performance deficits often result from interactions among variously competent neural networks when an affected individual is faced with certain performance demands rather than from simple, fixed brain damage or invariant neural dysfunction. It is speculated that more refined assessment techniques may provide precise identification of the interactions that result in motor and other dysfunctions. With further understanding of the interactions of neural networks under various learning and performance contingencies, some of the conduct and emotional deficits seen in the neuropsychiatric disorders may also become more understandable.

In conclusion, the study of the neural substrate of motor performance deficits that are found in neuropsychiatric syndromes of childhood has really just begun. The task ahead includes comprehensive and detailed study of normal developmental trajectories and the clinical aspects of motor performance deficits in individuals with the neuropsychiatric disorders, with the subsequent unraveling of the interactions between the different conditions and the observed motor performance deficits. This effort may in turn help to discern the interactions of the neural mechanisms that define both the motor performance deficits that are of such high incidence in these disorders and possibly the neuropsychiatric syndromes themselves.

REFERENCES

Adams, R., Kocsis, J., & Estes, R. (1979). Soft neurological signs in learning disabled children and controls. *American Journal of Diseases of Children, 128*, 614–618.

Almli, C.R., & Lawter, A. (1984). A description of movement in premature human infants. *International Society of Developmental Psychobiology Abstracts, 17*, 2.

Almli, C. R., Miller, J., Orup, I., & Galiano, R. (1988). Movement patterns and cyclicity of newborn rats. *Society for Neuroscience Abstracts, 14*, Abstract No. 401.2.

American Psychiatric Association. (1987). *Diagnostic and statistical manual of mental disorders* (3rd ed.–rev.). Washington, DC: Author.

Annell, A. (1949). School problems in children of average or superior intelligence: A preliminary report. *Journal of Mental Sciences, 95*, 901–909.

Annett, M. (1973). Handedness in families. *Annals of Human Genetics, 37*, 93–105.

Annett, M. (1986). Left, right, hand and brain: The right shift theory. *Developmental Medicine and Child Neurology, 28*, 550–553.

Aram, D. M., & Horwitz, S. J. (1983). Sequential and non-speech praxic abilities in developmental verbal apraxia. *Developmental Medicine and Child Neurology, 25*, 197–206.

Aram, D. (1988). Language sequelae of unilateral brain lesions in children. *Research Publications: Association for Research in Nervous and Mental Disease, 66*, 171–197.

Bakwin, H. (1968). Symposium on developmental disorders of motility and language. *Pediatric Clinics of North America, 15*, 565–567.

Barlow, C. (1974). "Soft signs" in children with learning disorders. *American Journal of Diseases of Childhood, 128*, 605–606.

Bates, E., & Marchman, V. (1988). What is and what is not universal in language acquisition? *Research Publications: Association for Research in Nervous and Mental Disease, 66*, 19–36.

Baum, C., Edwards, D., Leavitt, K., Grant, E., & Deuel, R. K. (1988). Performance components in senile dementia of the Alzheimer's type: Motor planning, language and memory. *Occupational Therapy Journal of Research, 8*, 356–359.

Bauman, M., & Kemper, T. (1985). Histoanatomic observations of the brain in early infantile autism. *Neurology, 35*, 866–874.

Bellugi, U., Klima, E., & Poizner, H. (1988). Sign language and the brain. In F. Plum (Ed.), *Language, communication, and the brain* (pp. 39–56). New York: Raven.

Bellugi, U., Sabo, H., & Vald, J. (1988). Spatial defect in children with Williams' syndrome. In J. Stiles-Davis, M. Kritchevsky, & U. Bellugi (Eds.), *Spatial cognition, brain bases and development* (pp. 321–349). Hillsdale, NJ: Lawrence Erlbaum Associates.

Belmont, L., & Birch, H. (1965). Lateral dominance, lateral awareness, and reading disabilities. *Child Development, 36*, 57–71.

Bender, L. (1970). *Psychopathology of children with organic brain disorders*. Springfield, IL: Thomas.

Benton, A. (1984). Constructional apraxia: Some unanswered questions. In A. Benton (Ed.), *Contributions to clinical neuropsychology* (pp. 129–141). Chicago: Aldine.

Berkson, G. (1960). An analysis of reaction time in normal and mentally deficient young men: Variation of stimulus and response complexity. *Journal of Mental Deficiency Research, 4*, 69–77.

Bigler, E. (1988). The role of neuropsychological assessment in relation to other types of assessments with children. In M. G. Tramontana & S. R. Hooper (Eds.), *Assessment issues in child neuropsychology* (p. 75). New York: Plenum.

Bobath, B. (1971). Motor development, its effects on general development, and applications to the treatment of cerebral palsy. *Physiotherapy, 57*, 526–532.

Bobele, G. B., Bodensteiner, J. B., Marks, W. A., & Hamza, M. (1988). Familial congential mirror movements. *Annals of Neurology, 24*, Abstract No. 343.

Bord, L., & Gascon, G. (1988). Rett syndrome: Review and discussion of current diagnostic criteria. *Journal of Child Neurology, 3*, 263–268.

Borghgraef, M., Fryns, J., Dielkens, A., Pyck, K., & Van Der Berghe, H. (1987). Fragile–X syndrome: A study of the psychological profile in 23 prepubertal patients. *Clinical Genetics, 32*, 179–186.

Byring, R., & Pulliainen, V. (1984). Neurological and neuropsychological deficiencies in a group of older adolescents with dyslexia. *Developmental Medicine and Child Neurology, 26*, 765–773.

Capildeo, R. (1979). Cerebrovascular Diseases. In C. Rose (Ed.), *Pediatric neurology* (pp. 509–551). Oxford: Raven.

Cermak, S. (1985). Developmental dyspraxia. *Advances in Psychology, 23*, 225–243.

Cermak, S., Coster, W., & Drake, C. (1983). Differentiation of praxis among children. *Advances in Psychology, 37*, 466–473.

Clements, S. (1966). *Minimal brain dysfunction in children* (USPHS Publication No. 141). Washington, DC: United States Government Printing Office.

Clinton, L., & Boyce, K. (1975). Acquisition of simple motor imitative behavior in mentally retarded and nonretarded children. *American Journal of Mental Deficiency, 79*, 695–700.

Cogan, D. G. (1966). *Neurology of the visual system*. Springfield, IL: Thomas.

Cohen, H. J., Taft, L. T., Mahadeviah, M. S., & Birch, H. G. (1967). Developmental changes in overflow in normal and aberrantly functioning children. *Journal of Pediatrics, 71*, 39–47.

Cole, K. J., Abbs, J. H., & Turner, G. S. (1988). Deficits in the production of grip forces in Down syndrome. *Developmental Medicine and Child Neurology, 30*, 752–758.

Connolly, K., & Stratton, P. (1968). Developmental changes in associated movements. *Developmental Medicine and Child Neurology, 10*, 49–56.

Conrad, B., Kriebel, J., & Hetzel, W. (1978). Hereditary bimanual synkinesis combined with hypogonadotropic, hypogonadism and anosmia in four brothers. *Journal of Neurology, 318*, 263–274.

Conrad, K. E., Cermak, S. A., & Drake, C. (1983). Differentiation of praxis among children: Apraxia, learning disabilities, motor planning. *American Journal of Occupational Therapy, 37*, 466–473.

Corballis, M. (1983). *Human laterality*. New York: Academic Press.

Courchesne, E., Yeung-Courchesne, R., Press, G. A., Hesselink, J. R., & Jernigan, T. L. (1988). Hypoplasia of cerebellar vermal lobules VI and VII in autism. *New England Journal of Medicine, 318*, 1349–1354.

Cratty, B. (1986). *Perceptual and motor development in infants and children* (3rd ed., pp. 124–137). Englewood Cliffs, NJ: Prentice-Hall.

Cratty, B. J., Cratty, I. J., & Cornell, S. (1986). Motor planning abilities in deaf and hearing children. *American Annals of the Deaf, 131*, 281–284.

Critchley, M. (1970). *The dyslexic child*. Springfield, IL: Thomas.

Critchley, M., & Critcheley, L. (1978). *Dyslexia defined*. Springfield, IL: Thomas.

Dam, M. (1982). Phenytoin toxicity. In D. M. Woodbury, J. K. Penny, & C. E. Pippenger (Eds.), *Antiepileptic drugs* (pp. 247–257). New York: Raven.

Dare, M., & Gordon, N. (1970). Clumsy children: A disorder of perception and motor organization. *Developmental Medicine and Child Neurology, 12*, 178–185.

David, R., Deuel, R., Ferry, P., Gascon, G., Golden, G., Rapin, I., Rosenberger, P., & Shaywitz, B. (1981). *Proposed nosology of disorders of higher cerebral function in children* (Task Force on Nosology of Disorders of Higher Cerebral Function in Children). Minneapolis, MN: Child Neurology Society.

Deeke, L., Kornhuber, H., Lang, W., Lang, M., & Schreiber, H. (1985). Timing function of the frontal cortex in sequential motor and learning tasks. *Human Neurobiology, 4*, 143–154.

DeMeyer, M., Hingtgen, J., & Jackson, R. (1981). Infantile autism reviewed. A decade of research. *Schizophrenia Bulletin, 7*, 88–451.

Denckla, M. (1974). Development of motor coordination in normal children. *Developmental Medicine and Child Neurology, 16*, 729–741.

Denckla, M. (1978). Minimal brain dysfunction. In J. Chall, A. Mirsky, & K. Rekage (Eds.), *Education and the brain* (pp. 73–91). Chicago: University of Chicago Press.

Denckla, M., LeMay, M., & Chapman, C. (1985). Few CT scan abnormalities found even in neurologically impaired learning disabled children. *Journal of Learning Disabilities, 18*, 132–135.

Denckla, M., & Rudel, R. (1978). Anomalies of motor development in hyperactive boys. *Annals of Neurology, 3*, 231–233.

Dennis, M. (1976). Impaired sensory and motor differentiation with corpus callosum agenesis: A lack of callosal inhibition during ontogeny. *Neuropsychologia, 14*, 455–469.

Deuel, R. (1981). Minimal brain dysfunction, hyperkinesis, learning disabilities, attention deficit disorder. *Journal of Pediatrics, 98*, 912–915.

Deuel, R., & Moran, C. (1980). Cerebral dominance and cerebral asymmetries on computed tomogram in childhood. *Neurology, 30*, 934–938.

Deuel, R. K. (1988). Treatment of attention problems with stimulant medication. *Journal of Pediatrics, 113*, 68–71.

Deuel, R. K., & Collins, R. C. (1984). The functional anatomy of frontal lobe neglect in the monkey: Behavioral and quantitative 2–Deoxyglucose studies. *Annals of Neurology, 15*, 521–529.

Deuel, R. K., Feeley, C., & Bonskowski, C. (1984). Manual apraxia in learning disabled children. *Annals of Neurology, 16*, Abstract No. 388.

Deuel, R. K., & Robinson, D. (1987). Developmental motor signs. In D. Tupper (Ed.), *Soft neurological signs* (pp. 95–129). New York: Grune & Stratton.

DeVries, J., Visser, G., & Prechtl, H. (1985). The emergence of fetal behaviors. *Early Human Development, 7*, 301–322.

Dewey, D., Roy, E., Square-Storer, P., & Hayden, D. (1988). Limb and oral apraxic abilities of children with verbal sequencing deficits. *Developmental Medicine and Child Neurology, 30*, 743–751.

Dreifuss, F., & Sato, S. (1982). Clonazepam. In D. M. Woodbury, J. K. Penny, & C. E. Pippenger (Eds.), *Antiepileptic drugs* (pp. 737–753). New York: Raven.

Edwards, M. (1973). Developmental verbal dyspraxia. *British Journal of Disorders of Communication, 8*, 64–70.

Ehrenberg, G., Cruse, R., & Rothner, A. (1986). Tourette syndrome. An analysis of 200 pediatric and adolescent cases. *Cleveland Clinic Quarterly, 53*, 127–131.

Eldridge, R., & Fahn, S. (1976). Dystonia. *Advances in Neurology: Volume 14*. New York: Raven.

Fantz, R., & Fagan, J. (1975). Visual attention to size and number of pattern details by term and pre-term infants during the first six months. *Child Development, 46*, 3–18.

Farmer, T. (1975). *Pediatric neurology* New York: Harper & Row.

Ferry, P., Hall, S., & Hicks, J. (1975). Dilapidated speech: Developmental verbal dyspraxia. *Developmental Medicine and Child Neurology, 17*, 749–756.

Fletcher, J., & Taylor, H. (1982). Neuropsychological approaches to children. *Journal of Clinical Neuropsychology, 6*, 39–56.

Fog, E., & Fog, M. (1963). Cerebral inhibition examination by associated movements. In M. C. O. Bax (Ed.), *Little club clinics in developmental medicine: Volume 10. Minimal cerebral dysfunction* (pp. 52–57). London: Heinemann.

Ford, F. R. (1960). *Diseases of the nervous system in infancy, childhood, and adolescence* (4th ed.). Springfield, IL: Thomas.

Forssberg, H., & Nashner, L. (1982). Ontogenetic development of postural control in man. *Journal of Neuroscience, 2*, 545–552.

Frei, H. (1986). The clumsy child: Differential diagnosis and therapy indications. *Schweizerische Medizinische Wochenschrift, 116*, 294–299.

Geschwind, N., & Behan, P. (1984). Hormone, handedness, and immunity. *Immunology Today, 5*, 190–191.

Getting, P. (1989). Emerging principles governing the operations of neural networks. *Annual Review of Neuroscience, 12*, 185–204.

Gevins, A. S., Bressler, N., Morgan, N. H., Cutillo, B. A., White, R. M., Greer, D. S., & Illes, J. (1989). Event related covariances during a bimanual visuomotor task. *Electroencephalography and Clinical Neurophysiology, 74*, 58–75.

Gillberg, C., Stefanburg, S., & Jacobson, G. (1987). Neurological findings in 20 relatively gifted children with Kanner-type autism or Asperger syndrome. *Developmental Medicine and Child Neurology, 29*, 641–649.

Gillberg, C., & Rasmussen, P. (1982). Perceptual, motor and attentional deficits in seven-year-old children: Background factors. *Developmental Medicine and Child Neurology, 24*, 752–770.

Godfrey, B., & Kephart, N. (1969). *Movement patterns and motor education.* New York: Appleton-Century-Croft.

Gomez, M. (1967). Minimal cerebral dysfunction (maximal neurologic confusion). *Clinical Pediatrics, 6,* 589–591.

Goodenough, F. L. (1926). *The measurement of intelligence by drawings.* Chicago: World Book Co.

Goodenough, F. L. (1935). A further study of speed of tapping in early childhood. *Journal of Applied Psychology, 19,* 309–315.

Grimm, R. (1983). Program disorders of movement. In J. Desmedt (Ed.), *Motor control mechanisms of health and diseases* (pp. 1–12). New York: Raven.

Gubbay, S. (1973). A standardized test battery for the assessment of clumsy children. *Proceedings of the Australian Association of Neurologists, 10,* 19–25.

Gubbay, S. (1975). *The clumsy child: A study of developmental apraxic and agnostic ataxia.* London: Saunders.

Gubbay, S. (1978). The management of developmental apraxia. *Developmental Medicine and Child Neurology, 20,* 643–646.

Gubbay, S., Ellis, E., Walton, J., & Court, S. (1965). Clumsy children. A study of apraxic and agnostic defects in children. *Brain, 88,* 295–312.

Gunderson, C., & Soltaire, G. (1968) Mirror movements in patients with the Klippel–Feil syndrome. *Archives of Neurology, 18,* 675–679.

Gur, R. C., Gur, R. E., Rosen, A. D., Warach, S., Alavi, A., Greenberg, J., & Reivich, M. (1983). A cognitive-motor network demonstrated by positron emission tomography. *Neuropsychologia, 21,* 601–606.

Haaland, K. Y., Porch, B. E., & Delaney, H. D. (1980). Limb apraxia and motor performance. *Brain and Language, 9,* 315–323.

Hamburger, V. (1971). Development of embryonic motility. In E. Tobach (Ed.), *The biopsychology of development.* New York: Academic Press.

Harris, A. J. (1958). *Harris tests of lateral dominance* (3rd ed.). New York: Psychological Corporation.

Hart, Z., Rennick, P., & Klinge, U. (1974). A pediatric neurologist's contribution to evaluations of school underachievers. *American Journal of Diseases of Children, 128,* 319–323.

Hayes, A., Bowling, T., Fraser, D., Krimmer, H., Marrionan, A., & Claque, A. (1988). Neonatal screening and an intensive management program for galactosemia: Early evidence of benefits. *Medical Journal of Australia, 149,* 21–25.

Henderson, E., & Hall, D. (1982). Concomitants of clumsiness in young school children. *Developmental Medicine and Child Neurology, 24,* 448–460.

Hertzig, M. (1981). Neurological soft signs in low-birth weight children. *Developmental Medicine and Child Neurology, 23,* 778–791.

Hertzig, M., & Shapiro, T. (1987). The assessment of nonfocal neurological signs in school aged children. In D. Tupper (Ed.), *Neurological soft signs* (pp. 71–94). New York: Grune & Stratton.

Hertzig, M., Bortner, M., & Birch, H. (1969). Neurological findings in children educationally designated as "brain-damaged." *American Journal of Orthopsychiatry, 39,* 437–446.

Hicks, R., & Dusek, C. (1980). The handedness distributions of gifted and nongifted children. *Cortex, 16,* 479–481.

Hulme, C., Biggerstaff, A., & Morgan, G. (1982). Visual, kinaesthetic and cross-modal judgements of length by normal and clumsy children. *Developmental Medicine and Child Neurology, 24,* 461–471.

Jones, K., & Smith, D. (1975). The Williams' elfin facies syndrome. *Journal of Pediatrics, 86,* 718–723.

Jones, V., & Prior, M. (1985). Motor imitation abilities and neurological signs in autistic children. *Journal of Autism and Developmental Disorders, 15*, 37–46.

Kanner, L. (1942). Autistic disturbances of affective contact. *Nervous Child, 2*, 217–250.

Kelsey, R. W. (1980). *Screening Test for Developmental Apraxia of Speech*. Tigard, OR: C. C. Publishers.

Kelso, J. A. S., & Tuller, B. (1981). Toward a theory of apractic syndromes. *Brain and Language, 12*, 224–245.

Kennard, M. (1960). Value of equivocal signs in neurologic diagnosis. *Neurology, 10*, 753–764.

Keogh, B. (1986). Future of the LD field: Research and practice. *Journal of Learning Disabilities, 19*, 455–460.

Kinsbourne, M., & Cook, J. (1971). Generalized and lateralized effects of concurrent verbalization on a unimanual skill. *Journal of Experimental Psychology, 23*, 341–345.

Klipcera, C., Wolff, P., & Drake, C. (1981). Bimanual coordination in adolescent boys with reading retardation. *Developmental Medicine and Child Neurology, 23*, 617–625.

Knuckey, N., & Gubbay, S. (1983). Clumsy children: A prognostic study. *Australian Paediatric Journal, 19*, 9–13.

Kools, J. A., & Tweedie, E. (1975). Development of praxis in children. *Perceptual and Motor Skills, 40*, 11–19.

Kornse, D., Manni, J., Rubenstein, H., & Gratziani, L. (1981). Developmental apraxia of speech and manual dexterity. *Journal of Communication Disorders, 14*, 321–330.

Lazarus, C., & Todor, J. I. (1987). Age differences in the magnitude of associated movement. *Developmental Medicine and Child Neurology, 29*, 726–733.

Lerer, R., & Lerer, M. (1976). The effects of methylphenidate on the soft neurological signs of hyperactive children. *Pediatrics, 57*, 521–525.

Levine, M. (1987). *Developmental variation and learning disorders*. Cambridge, MA: Educators Publishing Service.

Liepmann, H. (1900). Das Krankheitsbild der Apraxie [The clinical profile of apraxia]. *Monatsschrift Für Psychiatrie und Neurologie, 8*, 15–44, 102–132, 182–192.

Liepmann, H. (1908). *Drie aufsätze aus dem apraxiegebriet* [Three essays on apraxia]. Berlin: Karger.

Lockman, L. (1982). Movement disorders. In Swaiman & Wright (Eds.), *The practice of pediatric neurology* (pp. 287–308). St. Louis, MO: Mosby.

Manni, J., Martin, R., & Sewell, T. (1977). Imitation of gestures technique: A preliminary report on a preschool test of visual-motor integration. *Perceptual and Motor Skills, 44*, 1067–1072.

Markowitz, M. (1987). Rheumatic Fever. In R. Behrman & V. Vaughan (Eds.), *Nelson textbook of pediatrics* (15th ed.). Philadelphia: Saunders.

Mattson, R. H., & Cramer, J. P. A. (1982). Phenobarbital toxicity. In D. M. Woodbury, J. K. Penry, & C. E. Pippenger (Eds.), *Antiepileptic drugs* (2nd ed.) (pp. 351–363). New York: Raven Press.

McKay, D. (1981). The problem of rehearsal or mental practice. *Journal of Motor Behavior, 13*, 274–285.

McKay, D. (1982). The problem of flexibility, fluency and speed accuracy trade-off in skilled behavior. *Psychology Review, 89*, 483–506.

McKay, D. (1983). A theory of representation and enactment of intentions. *Advances in Psychology, 12*, 217–230.

McKinlay, I. (1978). Strategies for clumsy children. *Developmental Medicine and Child Neurology, 20*, 494–501.

McMahon, S., & Greenberg, L. (1977). Serial neurological examination of hyperactive children. *Pediatrics, 59,* 584–587.

Mesulam, M. M. (1981). A cortical network for directed attention and unilateral neglect. *Annals of Neurology, 10,* 309–325.

Milner, B. (1971). Interhemispheric differences in the localization of psychological processes in man. *British Medical Bulletin, 19,* 421–446.

Molnar, G., & Taft, P. (1977). Cerebral palsey. *Current Problems in Pediatrics, 7,* 6–26.

Newell, K. (1985). Motor skill acquisition and mental retardation. *Motor Development, 1,* 183–192.

Nichols, P., & Chen, T. (1981). *Minimal brain dysfunction: A prospective study.* Hillsdale, NJ: Lawrence Erlbaum Associates.

Nichols, P. (1987). Minimal brain dysfunction and soft signs: The collaborative perinatal project. In D. Tupper (Ed.), *Soft neurological signs* (pp. 179–201). New York: Grune & Stratton.

Ohta, M. (1987). Cognitive disorders of infantile autism: A study employing the WISC, spatial relationship conceptualization, and gesture imitations. *Journal of Autism and Developmental Disorders, 17,* 45–62.

Oldfield, R. C. (1971). The assessment and analysis of handedness: The Edinburgh Inventory. *Neuropsychologia, 9,* 97–113.

Ornitz, E. (1985). Neurophysiology of infantile autism. *Journal of the American Academy of Child Psychiatry, 24,* 251–262.

Orton, S. (1925). "Word-blindness" in school children. *Archives of Neurology, 14,* 582–615.

Orton, S. (1937). *Reading, writing, and speech problems in children.* New York: Norton.

Paine, R., Werry, J., & Quay, H. (1968). A study of minimal cerebral dysfunction. *Developmental Medicine and Child Neurology, 10,* 505–520.

Parmelee, A. H., & Sigmon, M. (1976). Development of visual behavior and neurological organization in pre-term, full-term infants. In A. D. Pich (Ed.), *Proceedings of the Minnesota Symposium in Child Psychology* (Vol. 10, pp. 165–179). Minneapolis: University of Minnesota Press.

Peiper, A. (1963). *Cerebral Function in Infancy and Childhood* (trans., pp. 248–250). New York: Consultants Bureau.

Peters, J. (1987). A special neurological examination for school aged children. In D. Tupper (Ed.), *Soft neurological signs* (pp. 370–371). New York: Grune & Stratton.

Peters, J., Romine, J., & Dykman, R. (1975). A special neurological examination of children with learning disabilities. *Developmental Medicine and Child Neurology, 17,* 63–78.

Piven, J., Berthier, M., Starkstein, S., Nehme, E., Perlson, G., & Folstein, S. (1989). Magnetic resonance in autism: Evidence for defect of cerebral cortical development. *Neuroscience Abstracts, 15,* 1334, Abstract No. 526.4.

Prechtl, H. (1987). Choreiform movements. In D. Tupper (Ed.), *Soft neurological signs.* New York: Grune & Stratton.

Prechtl, H., & Stemmer, C. (1962). The choreiform syndrome in children. *Developmental Medicine and Child Neurology, 4,* 119–127.

Ramsey, D. C., Campos, J. J., & Fenson, L. (1979). Onset of bimanual handedness in infants. *Infant Behavior and Development, 2,* 69–77.

Rapin, I., & Allen, D. (1988). Syndromes in development dysphasia and adult aphasia. *Research Publications: Association for Research in Nervous and Mental Disease, 66,* 57–75.

Rapin, J., Tourke, K., & Costa, L. (1960). Evaluation of the Purdue Peg Board as a screening test for brain damage. *Developmental Medicine and Child Neurology, 8,* 45–54.

Rice, T., Plomin, R., & DeFries, J. (1984). Development of hand preference in the Colorado Adoption Project. *Perceptual and Motor Skills, 58*, 686–689.
Roach, E., & Kephart, N. (1966). *The Purdue Perceptual Motor Survey*. Columbus, OH: Merrill.
Rourke, B. (1981). Neuropsychological assessment of children with learning disabilities. In S. B. Filskov & T. J. Boll (Eds.), *Handbook of clinical neuropsychology* (pp. 89–102). New York: Wiley.
Rude ', R., Healey, J., & Denckla, M. (1984). Development of motor coordination by normal left handed children. *Developmental Medicine and Child Neurology, 26*, 104–111.
Rutter, M. (1981). Psychological sequelae of brain damage in children. *American Journal of Psychiatry, 138*, 1533–1544.
Rutter, M. (1982). Syndromes attributed to "minimal brain dysfunction" in childhood. *American Journal of Psychiatry, 139*, 21–33.
Rutter, M. (1983). Behavioral studies: Questions and findings on the concept of distinctive syndrome: In M. Rutter (Ed.), *Developmental neuropsychiatry* (pp. 259–279). New York: Guilford.
Satzmari, P., & Taylor, D.C. (1984). Overflow movements and behavior problems. *Developmental Medicine and Child Neurology, 26*, 297–310.
Schneider, W., & Shiffrin, R. (1977). Controlled and automatic information processing. I. Detection, search and attention. *Psychological Review, 84*, 1–66.
Schneider, S., & Fisk, A. (1983). Attention theory and mechanisms for skilled performance. *Advances in Psychology, 12*, 119–143.
Seyforth, B., & Spreen, O. (1979). Two-plated tapping performance by Down syndrome and non-Down syndrome retardates. *Journal of Child Psychology and Psychiatry, 20*, 351–355.
Shafer, S. Q., Shaffer, D., O'Connor, P., & Stokman, C. (1983). Hard thoughts on neurological "soft signs." In M. Rutter (Ed.) *Developmental neuropsychiatry* (pp. 133–143). New York: Guilford.
Shaffer, D., Schonfeld, I., O'Connor, P. A., Stokman, C., Trautman, P., Shafer, S., & Ng, S. (1985). Neurological soft signs: Their relationship to psychiatric disorder and intelligence in childhood and adolescence. *Archives of General Psychiatry, 42*, 342–352.
Shapiro, A., Shapiro, E., & Fulop, G. (1987). Pimozide treatment of tic and Tourette syndrome. *Pediatrics, 79*, 1032–1039.
Sloan, W. (1948). *The Lincoln adaptation of the Oseretsky tests: A measure of motor proficiency*. Lincoln: IL: Lincoln State School & Colony.
Somers, A., Levin, H., & Hamay, H. (1976). Neuropsychological study of a family with hereditary mirror movements. *Developmental Medicine and Child Neurology, 8*, 791–798.
Spreen, O., & Gaddes, W. (1969). Developmental norms for 15 neuropsychological tests age 6–15. *Cortex, 5*, 170–181.
Stiles-Davis, J., Janowsky, J., Engel, M., & Nass, R. (1988). Drawing ability in four young children with congenital unilateral brain lesions. *Neuropsychologia, 26*, 359–371.
Strauss, A., & Lehtinen, N. C. (1947). *Psychopathology and education of the brain injured child*. New York: Grune & Stratton.
Taft, L., & Barowsky, E. (1989). Clumsy child. *Pediatric Research, 10*, 247–252.
Taylor, D., Powell, R., Cherland, E., & Vaughan, C. (1988). Overflow movements and cognitive, motor and behavioral disturbances. *Developmental Medicine and Child Neurology, 30*, 759–768.
Thielen, E. (1979). Rhythmic stereotypes in normal human infants. *Animal Behavior, 27*, 99–715.

Thompson, R., & O'Quinn, A. (1979). *Developmental disabilities* (pp. 200–201). New York: Oxford University Press.

Todor, J., & Lazarus, C. (1986). Exertion level and the intensity of associted movements. *Developmental Medicine and Child Neurology, 28,* 278–281.

Tsai, L. Y. (1982). Brief report: Handedness in autistic children and their families. *Journal of Autism and Developmental Disorders, 12,* 421–423.

Uzqiris, I. (1967). Ordinality in the development of schemas for relating to objects. In J. Hellmuth (Ed.), *The exceptional infant Volume 1. The normal infant* (pp. 44–61). Seattle: Special Child Publications.

Voeller, K. (1981). A proposed extended behavioral, cognitive, and sensorimotor pediatric neurological examination. In R. Ochroch (Ed.), *Diagnosis and treatment of minimal brain dysfunction* (pp. 65–90). New York: Human Sciences Press.

Walton, J. (1961). Clumsy children. *Spastics Quarterly, 10,* 9–21.

Walton, J. N., Ellis, E., & Court, S. D. (1962). Clumsy children: Developmental apraxia and agnosia. *Brain, 85,* 603–612.

Waterland, J. C., & Hellerbrandt, F. A. (1964). Involuntary patterning associated with willed movement performed against progressively increasing resistance. *American Journal of Physical Medicine, 43,* 13–30.

Weintraub, S., & Mesulam, M. M. (1983). Developmental learning disabilities of the right hemisphere: Emotional, interpersonal and cognitive components. *Archives of Neurology, 40,* 463–468.

Wise, S., & Desimone, R. (1988). Behavioral neurophysiology: Insights into seeing and grasping. *Science, 242,* 736–741.

Wolff, P., Gunnoe, C., & Cohen, C. (1983). Associated movements as a measure of developmental age. *Developmental Medicine and Child Neurology, 25,* 417–429.

Wolff, P., & Hurwitz, I. (1966). The choreiform syndrome. *Developmental Medicine and Child Neurology, 4,* 160–165.

Wolf-Schein, E., Sudhartha, V., Cohen, I., Fisch, G., Hanson, D., Pfadt, A., Hagerman, R., Jenkins, E., & Brown, T. (1987). Speech language and the fragile X syndrome: Initial findings. *Journal of the American Speech Hearing Association, 29,* 35–38.

World Health Organization. (1989). *Mental disorders: Glossary and guide to their classification in accordance with the ninth revision of the International Classification of Diseases* (3rd ed.). Geneva: Author.

Yoss, D., & Darley, F. L. (1974). Developmental apraxia of speech in children with defective articulation. *Journal of Speech and Hearing Research, 17,* 399–416.

Yule, W., & Taylor, D. (1987). Classification of soft signs. In D. Tupper (Ed.), *Soft neurological signs* (pp. 19–44). New York: Grune & Stratton.

Zurif, E., & Bryden, M. (1969). Familial handedness and left–right differences in auditory and visual perception. *Neuropsychologia, 7,* 179–187.

Epilogue: Developmental Disorders

Stephen R. Hooper
University of North Carolina School of Medicine

Relatively recently, Kuhn (1970) observed that the field of psychopathology had no single paradigm that was dominant, and consequently the study of psychopathology was described as a "preparadigmatic" science. Although the field has continued to evolve in a preparadigmatic manner, the DSM-III and DSM-III-R represent significant movement toward a more integrated perspective with respect to the classification and diagnosis of child and adolescent psychopathology. In particular, these gains were evident in the evolution of the developmental disorders and subsequent expansion in the sheer number of these disorders described in the DSM-III and DSM-III-R. This volume provides expert views with respect to many of the major developmental disorders described in the current version of the DSM.

One of the major tasks for each of the contributors to this volume was to evaluate the diagnostic criteria for a specific diagnosis from a critical perspective and then to offer suggested changes in diagnostic criteria, if necessary, based on their review of the current literature. It is clear from the preceding chapters that all of the contributors made substantive suggestions that were based on clinical as well as scientific parameters. These suggestions include changes and refinements in the specific criteria for specific diagnoses, concerns with respect to tightening up operational definitions for many of the developmental disorders, and improved linkages between diagnostic criteria and assessment methods. Issues of heterogeneity, comorbidity, and etiology also are mentioned by many of the contributors.

Hooper (chapter 1) has described many of the issues inherent in the classification of developmental disorders. Although many different models for classification research have been proposed (e.g., Cantwell, 1975; Quay, 1986; Skinner, 1986), it is clear that the field of psychopathology has become more interested in the scientific merits of diagnoses over the past 20 years. Issues of reliability and validity now pervade the field, as illustrated by discussion of these issues by many of the authors (e.g., Mesibov & Van Bourgondien, chapter 3; Stanovich, chapter 6), and diagnosis by pure clinical judgment appears to be in the twilight of its existence. Some investigators even have proposed specific "rules of thumb" for considering particular diagnoses for inclusion in the upcoming DSM-IV (e.g., Blashfield, Sprock, & Fuller, 1990; Quay, 1986). In particular, Skinner's model for validating psychiatric disorders is robust with respect to addressing the preceding concerns, and given the dynamic interconnection of its components it has potential applications to nearly any diagnosis, psychiatric or developmental, in any taxonomy.

To date, there have been numerous classification attempts for a variety of the developmental disorders (e.g., reading, arithmetic, language, motor skills). Although few of these efforts were conducted in conjunction with the DSM-III or DSM-III-R criteria for these disorders, there has been an evolving literature devoted to classification of many of the developmental disorders (e.g., Hooper & Willis, 1989). Unfortunately, the current version of the DSM has not taken advantage of much, if any, of this literature, and this information would help in refining criteria for the developmental disorders to be included in the DSM-IV.

In addition to this obvious neglect of a wealth of information pertaining to many of the developmental disorders, particularly the specific developmental disorders, some investigators have questioned the rationale for including developmental disorders in a psychiatric nosology (e.g., Garfield, 1986). Although the DSM-III-R indirectly addresses this issue conceptually by shifting all developmental disorders to Axis II, including mental retardation (see Reschly, chapter 2) and autistic disorder (see Mesibov & Van Bourgondien, chapter 3), the justification for including these diagnoses within a classification framework pertaining to psychiatric disorders remains suspect. There does appear to be a firm commitment to include these diagnoses in a system on "mental disorders," but the rationale for their inclusion must become stronger if they are to remain within such a classification framework. Perhaps the DSM definition of mental disorder should be reassessed so as to include these disorders more appropriately.

Despite these concerns, there has been clear progress in the evolution of many of the developmental disorders. In addition to the increase in the number of these diagnoses, the criteria for many of the developmental disorders have been improved in the DSM-III-R. For example, the IQ criteria for mental retardation have shifted with each successive version of the DSM, with the current version allowing for more liberal ranges of IQ scores to be considered diagnostically. Further, Mesibov and Van Bourgondien discussed the subtle changes noted in the DSM-III-R for autistic disorder (e.g., elimination of age of onset, removal of residual state category, emphasis on lack of reciprocity rather than on absence of social behavior, inclusion of difficulties in imaginative abilities), one of the most understood childhood diagnoses in the DSM system, and they asserted that these changes should improve the reliability of the diagnosis, reduce the diagnostic confusion in distinguishing the significance of behaviors in younger children, increase the likelihood that higher functioning clients will be included, and lessen the implication that autism can be cured. Mesibov and Van Bourgondien also called for greater clarification of the pervasive developmental disorder category in general. This evolution also has been observed in the specific developmental disorders, although the emphasis of the DSM has been more on increasing the number of these disorders (e.g., increasing from one in DSM-II to six in DSM-III, to nine in DSM-III-R) as opposed to refining associated criteria.

Although major steps were taken by the DSM-III and DSM-III-R to improve the comprehensiveness of the classification system by inclusion of a variety of developmental disorders and refinement of selected diagnostic domains, the criteria for many of the developmental disorders were woefully inadequate. This was particularly obvious for the specific developmental disorders, although Reschly (chapter 2) and Mesibov and Van Bourgondien (chapter 3) challenged specific aspects of the current diagnostic criteria for mental retardation and autistic disorder, respectively, as well. Consequently, it is no surprise that new diagnostic criteria have been offered for nearly all of the specific developmental disorders. Several of these offerings include more clearly defined operational criteria, such as in developmental arithmetic disorder, language and speech disorders, and motor skills disorder, where variations of the "ability–achievement discrepancy" are utilized. Despite the many problems inherent within the ability–achievement conceptualization (Stanovich, chapter 6), these efforts do attempt to assist the clinician and the researcher in defining criteria more specifically.

Several others approached the operational aspects of selected criteria differently by challenging the concept of IQ as the major benchmark of ability. For example, despite the fact that mental retardation has been one of the least controversial diagnostic categories, Reschly argued that the use of traditional IQ tests with lower functioning individuals added little to the prescriptive planning for these individuals. Similarly, in a more creative vein, Stanovich (chapter 6) called for a serious re-examination of IQ as the only measure of abilities and suggested that other discrepancy conceptualizations should be considered that are more directly related to the construct(s) in question and less likely to be influenced by the actual reading process itself (i.e., "Matthew Effects"—the rich get richer and the poor get poorer). Stanovich illustrated this by suggesting that the ability–achievement discrepancy for developmental reading disorder should be a discrepancy between a measure of listening comprehension (i.e., ability) and reading comprehension (i.e., achievement). Such creative solutions also will tend to give more power to the behavior in question (i.e., developmental reading disorder) as opposed to the predictor.

Two of the most recent additions to the DSM system were developmental expressive writing disorder and motor skills disorder. Although their inclusion would seem to improve the diagnostic coverage of the DSM system, these diagnoses were included with poor diagnostic criteria. Gregg (chapter 5) and Deuel (chapter 8) provided extensive discussions pertaining to these domains. Not only did they highlight the multidimensionality of these domains, but both contributors called for more improved operational criteria for each of these diagnostic categories. The detailed chapter by Gregg is particularly noteworthy, given the relatively recent emergence of work pertaining to children and adolescents with written language disorders, as it represents one of the most comprehensive reviews of this domain to date. If the DSM-IV would incorporate only a fraction of the information contained within these two chapters, the improvement in these domains likely would be significant.

More broadly, Reschly (chapter 2) boldly argued that the entire diagnostic concept of *mild* mental retardation should be modified significantly to another term, such as "general educational handicap," so as to represent more accurately the disabilities characteristic of individuals so diagnosed. This would possibly strengthen educational programming efforts, reduce the chances that the characteristics of more-impaired children would be attributed to the mildly affected children, and increase expectations for this group of individuals. Relatedly, Paul (chapter 7) expressed a similar argument for the language and speech disorders, contesting that the DSM-III-R diagnosis of developmental receptive language disorder is more indicative of an adult disorder (i.e., Wernicke's Aphasia) and may not actually exist in children. She suggested that this diagnostic term be changed to *developmental language disorder–global type* so as to be more in line with current thinking with respect to language development in children. Paul (chapter 7) even called for a renaming and reorganization of an entire category, with specific changes being noted for several of the diagnoses within that category. Specifically, Paul suggested a new category that would organize the speech and language disorders from a stronger conceptual perspective ("Disorders Affecting Communication"), noting the unexplained inclusion of developmental articulation disorder under the specific developmental disorders.

As with many of the chapters in this text's companion volume (*Child Psychopathology*), issues of heterogeneity, comorbidity, and etiology are mentioned for nearly all of the developmental disorders. From the chapters in this volume, it is clear that the current literature is conceptualizing many of these disorders as heterogeneous in nature. This is especially true for the specific learning disorders (e.g., reading, arithmetic), where investigators have been discussing their multidimensionality for more than 30 years. The DSM-IV must devote more attention to this literature if these diagnoses are to prove clinically relevant. Comorbidity also is discussed in many of the chapters, with several chapters devoting time to the issue of the impact of psychiatric disorders on developmental disorders (e.g., Semrud-Clikeman & Hynd, chapter 4), several chapters discussing the co-occurrence of developmental disorders (e.g., Paul, chapter 7), and others discussing the prevalence of specific developmental disorders in more psychiatrically defined samples (e.g., Deuel, chapter 8). Finally, etiological concerns continue to plague many of the developmental disorders; however, it is interesting to note the increased interest in the suspected neurological basis for many of these disorders (e.g., developmental arithmetic disorder; language and speech disorder; motor skills disorder). Although this latter speculation will continue to require investigation, strong evidence is presented in many of these chapters in support of this contention.

Given the preceding points of discussion, it should come as no surprise that nearly all of the chapters call for collecting diagnostic data in a more systematic and standardized fashion. As noted by Hooper (chapter 1), the sheer number of assessment devices available to professionals performing diagnostic work with children and adolescents with developmental disabilities far exceeds the number available to those working with psychiatric disorders. Further, given the more objective nature of many of these devices (e.g., IQ tests), test reliability and validity actually may be higher than that of the psychiatric assessment tools (e.g., structured interviews). What is clearly lacking for the study of developmental disorders, however, is a direct linkage of these assessment devices to diagnostic constructs. Most of the chapters in this volume directly address this concern by specifically listing possible criterion-assessment linkages, such as in autistic disorder (Mesibov &

Van Bourgondien) and developmental expressive writing disorder (Gregg). In one instance (developmental expressive writing disorder), the DSM-III-R calls for the use of standardized assessment tools to diagnose the condition, but it is clear that there are few such measures available to clinicians and researchers as yet. In all of the chapters, there is a unanimous outcry to: (a) improve assessment strategies in an effort to improve the operational definitions for each of the disorders, (b) refine diagnostic criteria more clearly, (c) address the developmental components of each of these disorders, and (d) help with the differential diagnosis in many of these disorders. More systematic assessment techniques will serve to increase the likelihood of increasing the reliability and validity of these diagnostic categories so as to justify their inclusion within a classification framework for psychopathology from a scientific perspective. Further, many of the chapters note that careful selection of these assessment strategies will serve to improve the linkages between diagnosis and prescription, one of the basic tenets of any classification system for psychopathology. These latter issues were poignantly discussed by Reschly in his strong suggestion that adaptive behavior measures be given more importance in the diagnostic assessment of individuals suspected of having mental retardation, and by Stanovich, who called for a closer examination of the constructs that are attempted to be measured in his discussion of developmental reading disorder.

Although the evolution of all of the psychopathological categories of childhood and adolescence undoubtedly will continue, giant strides are necessary for many of the developmental disorders. There has been a proliferation of research for many of these disorders, particularly the specific developmental disorders, but it is clear that the DSM did not take advantage of this wealth of information in its most recent revision. Conversely, much of the literature on developmental disorders has not attempted to relate research findings to formal classification efforts such as the DSM, and this has hindered the development of a productive interrelationship. A greater integration of these two bodies of literature is clearly in order. The inclusion of this information into a psychiatric nosology, however, should be considered carefully. Not only should the basic premises for including developmental disorders within a psychiatric nosology be re-examined critically (i.e., should they be considered "mental disorders"?), but the nosology must be a balance between scientific integrity and parsimony so as to preserve its clinical utility. Further, should the developmental disorders continue to be included within a psychiatric nosology—and it is expected that they will be for DSM-IV and ICD-10—then a greater emphasis must be placed on the complex interplay between social, emotional, and developmental disabilities.

Finally, all of the chapters within this volume address significant issues related to the developmental disorders. It is hoped that this information will contribute to the ongoing evolution of the developmental disorders, particularly if they continue to be conceptualized within a psychiatric nosology. As these dynamic developments evolve in the field of childhood and adolescent psychopathology, it is hoped that the field will move from a "preparadigmatic science," where conflicts abound, toward "normal science," where there are increased efforts to work in a collaborative fashion (Kuhn, 1970). With greater clarity beginning to be realized for the newer developmental disorders, and increased refinement continuing to occur for the diagnoses of longer standing, it is suspected that the assessment, diagnosis, and treatment for children and adolescents with developmental disorders will continue to improve, with perhaps many of these improvements having a stronger scientific foundation.

REFERENCES

Cantwell, D. P. (1975). A model for the investigation of psychiatric disorders of childhood: Its application in genetic studies of the hyperkinetic syndrome. In E. J. Anthony (Ed.), *Explorations in child psychiatry* (pp. 57–79). New York: Plenum.

Blashfield, R. K., Sprock, J., & Fuller, A. K. (1990). Suggested guidelines for including or excluding categories in the DSM-IV. *Comprehensive Psychiatry, 31*, 15–19.

Garfield, S. L. (1986). Problems in diagnostic classification. In T. Millon & G. L. Klerman (Eds.), *Contemporary directions in psychopathology. Towards the DSM-IV* (pp. 99–114). New York: Guilford.

Hooper, S. R., & Willis, W. G. (1989). *Learning disability subtyping. Neuropsychological foundations, conceptual models, and issues in clinical differentiation.* New York: Springer-Verlag

Kuhn, T. (1970). *The structure of scientific revolutions* (2nd ed., Vol. 2). Chicago: University of Chicago Press.

Quay, H. C. (1986). A critical analysis of DSM-III as a taxonomy of psychopathology in childhood and adolescence. In T. Millon & G. L. Klerman (Eds.), *Contemporary directions in psychopathology. Towards the DSM-IV* (pp. 151–165). New York: Guilford.

Skinner, H. A. (1986). Construct validation approach to psychiatric classification. In T. Millon & G. L. Klerman (Eds.), *Contemporary directions in psychopathology. Towards the DSM-IV* (pp. 307–330). New York: Guilford.

Author Index

Dunn, L., 50, 64, 230, 233, 235

G

H

I

J

M

N

R

S

Y

Z

Subject Index